Green Energy and Technology

Climate change, environmental impact and the limited natural resources urge scientific research and novel technical solutions. The monograph series Green Energy and Technology serves as a publishing platform for scientific and technological approaches to "green"—i.e. environmentally friendly and sustainable—technologies. While a focus lies on energy and power supply, it also covers "green" solutions in industrial engineering and engineering design. Green Energy and Technology addresses researchers, advanced students, technical consultants as well as decision makers in industries and politics. Hence, the level of presentation spans from instructional to highly technical.

Indexed in Scopus.

Indexed in Ei Compendex.

Mohammadreza Kamali · Tejraj M. Aminabhavi ·
Maria Elisabete V. Costa · Shahid Ul Islam ·
Lise Appels · Raf Dewil

Advanced Wastewater Treatment Technologies for the Removal of Pharmaceutically Active Compounds

Springer

Mohammadreza Kamali
Process and Environmental Technology
Lab, Department of Chemical Engineering
KU Leuven
Sint-Katelijne-Waver, Belgium

Maria Elisabete V. Costa
Department of Materials and Ceramics
Engineering
University of Aveiro
Aveiro, Portugal

Lise Appels
Process and Environmental Technology
Lab, Department of Chemical Engineering
KU Leuven
Sint-Katelijne-Waver, Belgium

Tejraj M. Aminabhavi
School of Advanced Sciences
KLE Technological University
Hubballi, Karnataka, India

Shahid Ul Islam
Department of Biological and Agricultural
Engineering
University of California, Davis
Davis, CA, USA

Raf Dewil
Process and Environmental Technology
Lab, Department of Chemical Engineering
KU Leuven
Sint-Katelijne-Waver, Belgium

ISSN 1865-3529 ISSN 1865-3537 (electronic)
Green Energy and Technology
ISBN 978-3-031-20805-8 ISBN 978-3-031-20806-5 (eBook)
https://doi.org/10.1007/978-3-031-20806-5

This Springer imprint is published by the registered company Springer Nature Switzerland AG
The registered company address is: Gewerbestrasse 11, 6330 Cham, Switzerland

Darkness cannot drive out darkness, only light can do that. Hate cannot drive out hate, only love can do that.

Martin Luther King

Preface

Water pollution is one of the most serious environmental threats of the twenty-first century, creating much disturbance to the benign nature of the environment. The toxic effects of this phenomenon on aquatic life and its deleterious impacts on maintaining the balance of the ecosystem have been widely investigated in recent years, as reported by scientists around the world. The scarcity of clean water resources is therefore an outcome of this global issue, leading to severe health, economic, and social concerns. The detection and remediation of contaminants of emerging concern (CECs) in water bodies in particular have added further challenges to the scientific community worldwide. These issues have created innumerable risks to humans and the environment; such aspects have not yet been deeply investigated and fully understood. To solve these issues, enormous efforts have been initiated by the scientific community to explore and develop efficient and economic methods to remove such compounds from polluted waters.

The present book covers an overview of the fundamental aspects related to the detection, quantification, and removal of pharmaceutically active compounds (PhACs) as an important class of contaminants of emerging concern. Critical discussions are provided regarding the fate of PhACs using a variety of treatment systems and technologies as well as the mechanisms involved in their removal using a wide range of biological and physico-chemical methods. The book is aimed at discussing the sustainability aspects of various methods developed and used in the elimination of PhACs in efforts to help decision-makers select the best available technique among the existing alternatives.

The fundamentals presented in various chapters of this book will aid readers and researchers in designing innovative future studies to address the remaining gaps in the literature for further developing sustainable wastewater treatment technologies to deal with toxic PhACs. To achieve these goals, the latest achievements of the scientific community are carefully retrieved, analyzed, and critically discussed from the most reputable platform of ever-increasing science, Web of Science (WoS; previously known as Web of Knowledge), for critical analysis and discussion. Furthermore, many complementary references are included in each chapter of the book to help readers and researchers search for more detailed information regarding the fundamentals and applicability of the technologies discussed in this book. We sincerely hope that this book will benefit a wide range of academicians, researchers, industrialists, and policy-makers, seeking further development and implementation of sustainable wastewater treatment technologies to remove pharmaceutically active compounds as well as other types of contaminants of emerging concern.

Sint-Katelijne-Waver, Belgium Mohammadreza Kamali

About This Book

This book provides an overview of the most important biological and physico-chemical (waste)water treatment technologies developed from time to time in the literature in efforts to remove pharmaceutically active compounds (PhACs). Chapter 1 of the book summarizes and discusses the available literature on the occurrence, environmental concentrations, fate, possible effects of the typical PhACs after these are introduced into the receiving environments. Chapter 2 introduces the advanced techniques for the detection of various PhACs, their quantification, and methods employed to identify the mechanisms involved in removing the PhACs using various physico-chemical and biological treatment approaches. Chapter 3 covers a discussion on the scientometric analysis for the identification, retrieval, and analysis of the scientific documents published from the Web of Science (WoS) on the application of various biological and physico-chemical treatments to deal with the PhACs. Chapters 4–7 of the book address the critical discussion of the applicability of the most popular biological wastewater treatment technologies, including activated sludge, anaerobic digestion, microbial fuel cells, and constructed wetlands, to remove various types of PhACs from water streams. The mechanisms involved in the removal of PhACs using these technologies and possible interactions between such compounds and the microbial communities are elegantly discussed. The mechanisms involved in the application of membrane separation and adsorption technologies and their applications for the removal of PhACs are critically evaluated with the relevant examples in Chaps. 8 and 9 of the book. The last two chapters (i.e., 10 and 11) are aimed at discussing the potential of homogeneous (Chap. 10) and heterogeneous (Chap. 11) advanced oxidation processes (AOPs) used in the elimination of PhACs. These two chapters deeply discuss the mechanisms involved in the removal of various types of PhACs along with the pros and cons involved in the application of both energy-free and energy-intensive AOPs. Overall, the entire book outlines the existing research gaps involved in the development of sustainable technologies for the removal of pharmaceutically active compounds and provides valuable recommendations for further future studies.

Contents

Abbreviations

3DPT	Three-dimensional printing technology
AC	Activated carbon
ACI	Average citation per item
ACMFCs	Air cathode microbial fuel cells
AD	Anaerobic digestion
AERs	Anion-exchange resins
AnMBRs	Anaerobic membrane bioreactors
AOPs	Advanced oxidation processes
AOXs	Halogenated organic compounds
ARBs	Antibiotic-resistant bacteria
ARGs	Antibiotic resistance genes
AS	Activated sludge
Ass	Active species
ATP	Adenosine triphosphate
BC	Biochar
BDD	Boron-doped diamond
BET	Brunauer–Emmett–Teller (theory)
BOD	Biological oxygen demand
CB	Conduction band
CD	Corona discharge
cDNA	Complementary DNA
CECs	Contaminants of emerging concern
CFs	Carbon fibers
CMC	Critical micelle concentration
CMs	Conductive materials
CNTs	Carbon nanotubes
COD	Chemical oxygen demand
CW	Constructed wetlands
DBD	Dielectric barrier discharge
DHA	Dehydrogenase activity
DHEA	Dehydroepiandrosterone

DIET	Direct electron transfer
DO	Dissolved oxygen
EC	Electrical conductivity
EET	Extracellular electron transfer
EMEA	European Medicine Agency
EO-AOPs	Electrochemical advanced oxidation processes
EPR	Electron paramagnetic resonance
EPSs	Extracellular polymeric substances
ESI	Electrospray ionization
ESR	Electron spin resonance
FO	Forward osmosis
FTIR	Fourier transform infrared spectroscopy
FWS-CWs	Water surface flow constructed wetlands
GA	Gamma irradiation
GAC	Granular activated carbon
GADP	Gliding arc discharge
GC	Gas chromatography
GC–MS	Gas chromatography with mass spectrometry
GC–MS/MS	Gas chromatography with tandem mass spectrometry
GDP	Glow discharge plasma
GO	Graphene oxide
HDL	High-density lipoprotein
HE-AOPs	Heterogeneous advanced oxidation processes
HLR	Hydraulic loading rate
HO-AOPs	Homogeneous advanced oxidation processes
HPLC	High-performance liquid chromatography
HRs	Hydroxyl radicals
HRT	Hydraulic retention time
HSF-CWs	Horizontal subsurface flow constructed wetlands
IF	Infrared
LC	Liquid chromatography
LC–MS	Liquid chromatography with mass spectrometry
LC–MS/MS	Liquid chromatography with tandem mass spectrometry
LECA	Light expanded clay aggregates
LOEC	Lowest observed effect concentration
LTQ	Linear trap quadrupole
MAs	Metal-based adsorbents
MBBRs	Moving-bed biofilm reactors
MBRs	Membrane bioreactors
MEUF	Micellar-enhanced ultrafiltration
MFC	Microbial fuel cells
MGEs	Mobile genetic elements
MOFs	Metal-organic frameworks
MOx	Metal oxides
MS	Mass spectrometry

MUVP	Microwave-UV plasma
NAC	NH_4Cl-triggered activation
NF	Nanofiltration
NGS	Next-generation sequencing
OC	Oseltamivir carboxylate
OLR	Organic loading rate
ORR	Oxygen reduction reaction
ORs	Oxidative radicals
OUR	Oxygen uptake rate
PC	Photocatalysis
PCOz	Photocatalytic ozonation
PCR	Polymerase chain reaction
PEC	Photoelectrocatalytic
PEM	Proton-exchange membrane
PhACs	Pharmaceutically active compounds
PI	Periodate
PL	Photolysis
PMS	Peroxymonosulfate
PPCPs	Pharmaceutical and personal care products
PS	Persulfate
QIA	Quantitative image analysis
qPCR	Quantitative PCR
Q-TOF-MS	Quadrupole time-of-flight mass spectrometry
rGO	Reduced graphene oxide
RO	Reverse osmosis
ROS	Reactive oxygen species
SDGs	Sustainable Development Goals
SEM	Scanning electron microscopy
SPE	Solid-phase extraction
SRT	Solid retention time
SSA	Specific surface area
STAs	Spin-trapping agents
TEM	Transmission electron microscopy
TFC	Turbulent flow chromatography
TFCMs	Thin-film composite membranes
TMCs	Transition metal carbides
TOC	Total organic carbon
TSS	Total suspended solids
VB	Valence band
VFAs	Volatile fatty acids
VSF-CWs	Vertical subsurface flow constructed wetlands
WoS	Web of Science
WWTPs	Wastewater treatment plants
XPS	X-ray photoelectron spectroscopy

List of Figures

List of Tables

Chapter 1
Pharmaceutically Active Compounds in Water Bodies—Occurrence, Fate, and Toxicity

1.1 Introduction

According to the classic definitions, pollution is defined as the introduction of undesirable amounts of any element in the forms of chemicals (such as organics and inorganics) and energy (such as light and noise) into the natural environment in concentrations that can cause adverse effects to the ecosystem and living organisms [1]. There are classifications for environmental pollution based on the source, type, and possible effects on the biota and abiota. EP can originate from natural events (such as natural forest fires, volcanic activities, or natural release of greenhouse gases from soils) or can be introduced by human activities such as the release of industrial pollutants.

Environmental pollution is among the most challenging issues of the twenty-first century. This is even more dramatic in developing countries where raw materials are being increasingly consumed to produce industrial products as the growth engine of the economy [2]. Pollution of air, light, noise, soil, and water can be considered as the consequence of such activities, which have negatively influenced the quality of the environment as well as the living standards of the affected communities. Among them, water pollution is of high significance because water is an essential element for human and ecosystem life. WP can be classified into three main categories: (a) organic pollutants, (b) heavy metals, and (c) nutrients [3].

The adverse effects of WP have been observed and reported in various countries all over the world, especially in rapidly developing countries, where huge amounts of raw materials and water are being consumed, especially in industrial processes. For instance, rapid environmental contamination has been experienced in China since the 1970s as a consequence of rapid industrial and economic development [4]. According to Han et al. [5], "China's war on water pollution has just begun, and it will be a fight that will take decades". The problem is more dramatic when polluted waters originating from industrial and nonindustrial activities are being released into water bodies without any appropriate treatment and decontamination [6].

M. Kamali et al., *Advanced Wastewater Treatment Technologies for the Removal of Pharmaceutically Active Compounds*, Green Energy and Technology,
https://doi.org/10.1007/978-3-031-20806-5_1

In addition to pollutants that have been well known for their effects on the environment and living organisms (such as halogenated compounds, including adsorbable halogenated organic compounds (AOXs) [2]), some contaminants have caused concerns in recent years due to their increasing concentrations in the environment [7, 8], as well as their unknown risks to humans and the environment, known as contaminants of emerging concern (CECs) [7, 9, 10]. CECs can be divided into biological agents (such as pathogens) and natural and artificial chemicals and their by-products, including pharmaceutical and personal care products (PPCPs), pesticides, flame retardants, nanoparticles, artificial sweeteners, and microplastics. Figure 1.1 demonstrates the main sources and routes for the release of CECs into the environment. In addition, antibiotic-resistant bacteria (ARB), antibiotic resistance genes (ARG), and, more recently, the SARS-CoV-2 virus have been categorized among the CECs (see Sect. 1.3).

Pharmaceutically active compounds (PhACs) are among the CECs that are released in huge amounts to water bodies. This issue has been highlighted, especially under the current COVID-19 pandemic, and this is expected to have an increase in the environmental concentrations of PhACs.

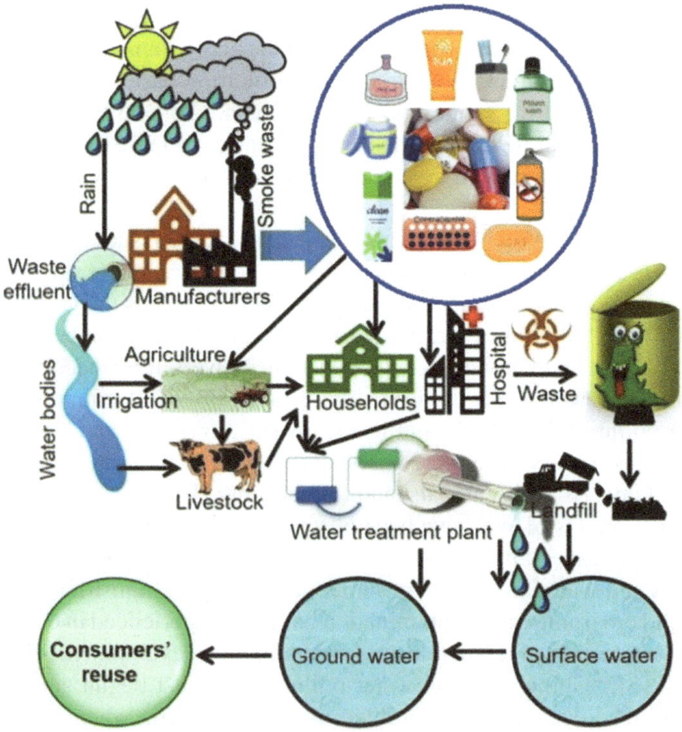

Fig. 1.1 Various origins of CECs in water bodies, adapted from Rasheed et al. [11]

1.2 Occurrence and Environmental Concentrations

PHACs cover a wide range of compounds that differ in structure and function. They are used in both humans and animals for curing infections and diseases and to mitigate symptoms. Because most PhACs are not completely metabolized in the human and animal bodies, there is a risk of the release of such compounds to water bodies through the sewage system. The main concern in this regard is that many pharmaceuticals can pass through conventional sewage wastewater treatment systems (such as activated sludge, as the most common wastewater treatment technology [12, 13]), and hence, they can be easily transferred into the environment through effluents [14, 15]. Due to their potential benefits for humans and animals, as well as their economic importance, it is expected that the consumption and release of such compounds will increase, especially with the increasing average age of the population and the need for more pharmaceuticals in daily life.

Pharmaceuticals in the environment were first detected in the 1970s [16]. Since then, more than 200 human and animal pharmaceuticals have been identified in aquatic environments. Figure 1.2 represents the most important routes for the release and occurrence of PhACs into the environment.

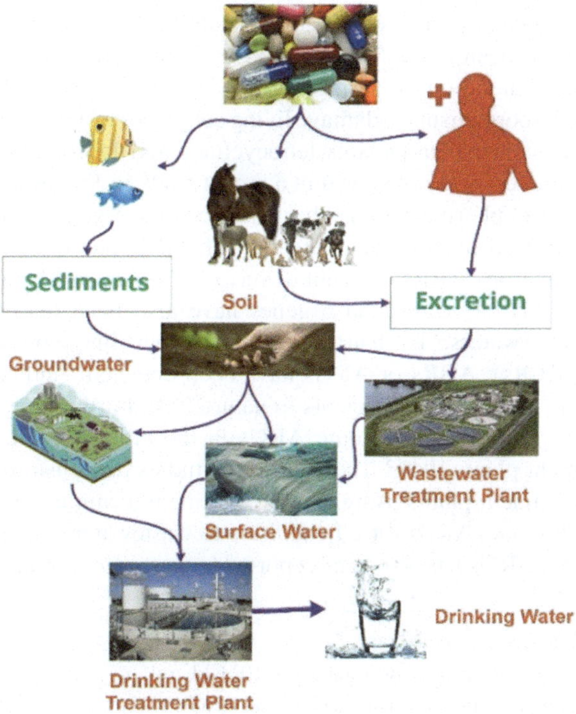

Fig. 1.2 Routes and fate of PhACs into the environment

There have been efforts to control the concentrations of PhACs in the environment to minimize their possible risks to humans and living organisms. For instance, Decision 2015/495/EU issued by the European Union [17] has established a watch list for compounds with possible environmental impacts, such as PhACs. Furthermore, the European Medicine Agency (EMEA) has issued a guideline to identify and estimate the possible environmental risks of PhACs according to a tiered approach [18]. However, there has been a need for more information on the environmental concentrations of such compounds, their acute and chronic toxic effects, and their fate and transformation products in real environmental conditions where the interactions between various PhACs need to be considered. Table 1.1 presents the properties of the typical pharmaceuticals, their properties, and their environmental concentrations in surface and groundwater bodies.

1.3 Fate and Toxicity

As stated before, pharmaceutical compounds are being released in relatively high quantities into the environment from industrial and nonindustrial points, such as the pharmaceutical, hospital, household, and veterinary industries. The fate of PhACs has been the subject of various studies all over the world, especially in recent years. It has been noted that the release of antibiotics in municipal wastewaters results in the generation of antibiotic resistance and virulence (as the ability of pathogens or microorganisms to damage living cells). For instance, noticeable amounts of bacteria[1] resistant to beta-lactams, tetracyclines, and fluoroquinolones in municipal wastewater samples[2] were observed in a recent study in Poland [89]. Resistance of *Escherichia coli* to pharmaceuticals has also been observed for cephalothin, streptomycin, and amoxicillin [90]. The detection of ABRs has been reported since the 1930s, and numerous ABRs resistant to various PhACs have been observed thus far (Fig. 1.3).

Antibiotic resistance genes have also been detected in effluents of municipal wastewaters.[3] Even after treatment with technologies such as membrane bioreactors (MBRs), ABRs or ARGs have been detected,[4] indicating a high degree of risk of the release of such agents to nature [93]. Furthermore, some studies have indicated the enrichment of some ARGs in the effluents of conventional wastewater treatment plants [94].[5] There are also studies demonstrating that biological wastewater treatment plants using aerobic treatment technologies can remove ARGs and fail to eliminate ARBs [96]. In addition to wastewater streams, the presence of both ABRs and ARGs has also been reported in wastewater treatment sludge [97]. Hence, it can

---

[1] *Escherichia coli* strains.

[2] 6.4×10^4, 4.2×10^4, and 3.1×10^3 CFU/mL, respectively.

[3] For example, the gene *intI*1 and all ARGs, except $bla_{CTX\text{-}M}$ in influent samples in the municipal wastewater treatment plants utilizing activated sludge process [92].

[4] Especially sulfonamide resistance.

[5] Such as blaOXA-48 [95].

Table 1.1 Typical pharmaceuticals, their properties, and concentrations in surface and groundwater bodies

Pharmaceutical	Chemical structure	Properties	Concentration in water bodies
Acetaminophen ($C_8H_9NO_2$)	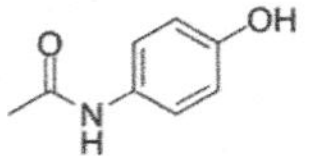	– An antipyretic, analgesic drug which represents moderate anti-inflammatory characteristics [19] – Reported as a water pollutant in 29 countries all over the world [20]	461 μg/L (in effluents) and 45–868 μg/L (in influent) of municipal wastewater treatment plants (WWTPs), Canada [21, 22]
Propyphenazone ($C_{14}H_{18}N_2O$)	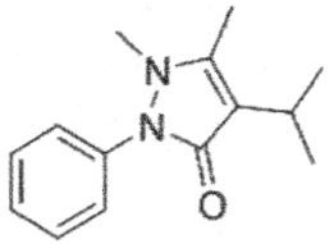	A type of pyrazolone compounds which are used as an analgesic, antipyretic, and anti-inflammatory drugs [23]	0.01 to 1.2 μg/L in sewage effluents (Greece) [24–27]
Clofibric acid ($C_{10}H_{11}O_3Cl$)	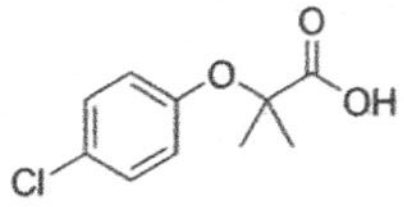	A lipid-lowering drug biologically active metabolite [28]	Approximately 40 μg/L at municipal wastewater treatment plants in many countries [29]
Gemfibrozil ($C_{15}H_{22}O_3$)	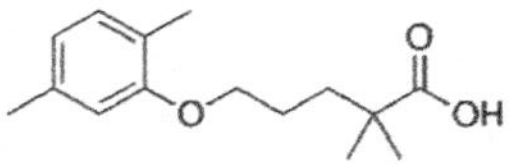	An acidic lipid regulator drug that is used to reduce plasma triglycerides as well as total cholesterol. It is also used to increase the levels of high-density lipoprotein (HDL) cholesterol in humans, preventing coronary heart disease [30]	Ranging from 0.08–19.4 μg/L in treated wastewater, 0.009~0.51 μg/L in surface waters, and 0.07 μg/L in drinking waters [31–33]

(continued)

Table 1.1 (continued)

Pharmaceutical	Chemical structure	Properties	Concentration in water bodies
Famotidine ($C_8H_{15}N_7O_2S_3$)	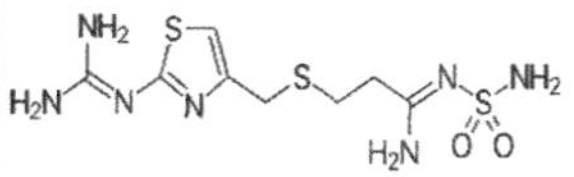	An H_2 blocker drug widely used for gastroesophageal reflux disease [34]	240 μg/L in river water samples (Poland) [35]
Bezafibrate ($C_{19}H_{20}ClNO_4$)	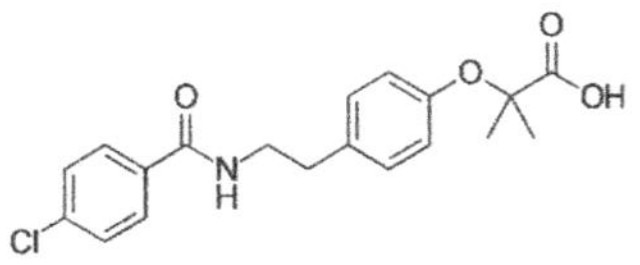	A widely used antilipemic drug, especially in developed countries [36]	Up to 4.6 μg/L with the median value of 2.2 μg/L in WWTP (Germany) [37]
Pravastatin ($C_{23}H_{36}O_7$)	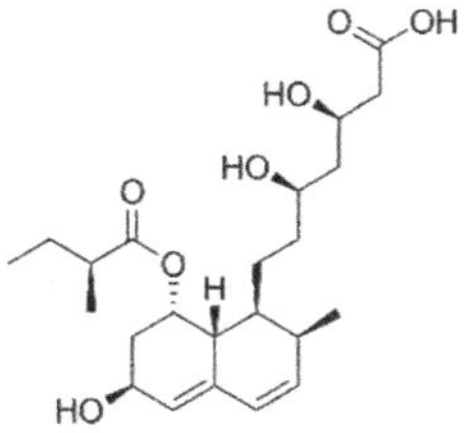	A pharmaceutical used for dyslipidemia treatment and to prevent cardiovascular disease [38]. It can also display high anti-inflammatory effects compared with other lipophilic statins [39]	4–49 ng/L and 1–59 ng/L in untreated and treated sewage samples, respectively [40]
Mevastatin ($C_{25}H_{38}O_5$)	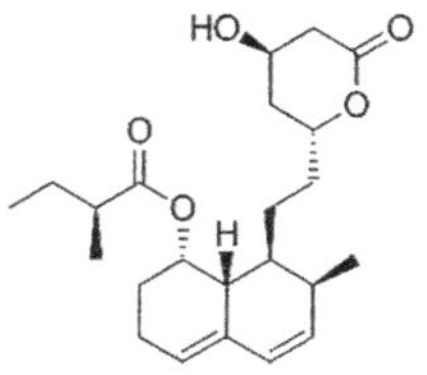	A hypolipidemic statin drug that inhibits the hydroxymethylglutaryl~SCoA (HMG-CoA) reductase [41] There is also evidence that mevastatin can inhibit the growth of methanogenic microorganisms (e.g., Methanobrevibacter) [42]	Below detection range [43]

(continued)

Table 1.1 (continued)

Pharmaceutical	Chemical structure	Properties	Concentration in water bodies
Ketoprofen ($C_{16}H_{14}O_3$)	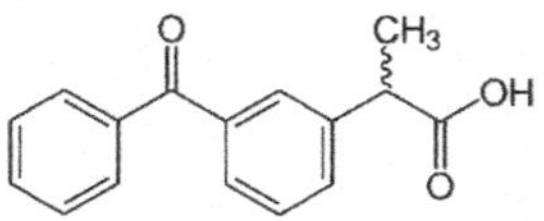	A nonsteroidal anti-inflammatory drug that is typically used to treat pain, fever, and inflammation [44]	Ranging from ng/L up to μg/L in sewage effluents and surface waters [45]
Naproxen ($C_{14}H_{14}O_3$)	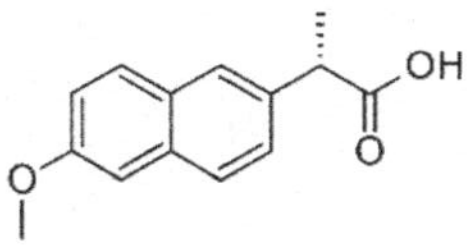	A nonsteroidal anti-inflammatory drug that represents analgesic, antipyretic, and anti-inflammatory actions. This drug is also widely used to mitigate fever, pain, and inflammation [46]	Naproxen is commonly found in the effluents of WWTPs as well as the surface waters ng/L-μg/L [47]
Ibuprofen ($C_{13}H_{18}O_2$)	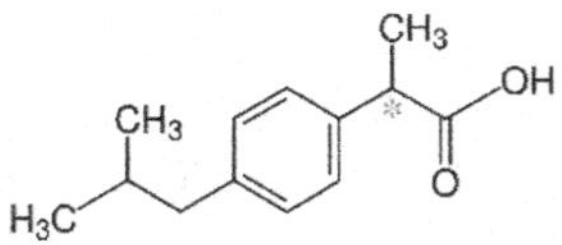	An anti-inflammatory, analgesic, and antipyretic drug which is widely used to mitigate fever and pain [48]	This pharmaceutical can be often detected in various water sources in concentrations. In the UK and the USA surface waters, it can be found in concentrations ranging from 2 ng/L and 8.77 μg/L [49, 50]
Indomethacin ($C_{19}H_{16}ClNO_2$)	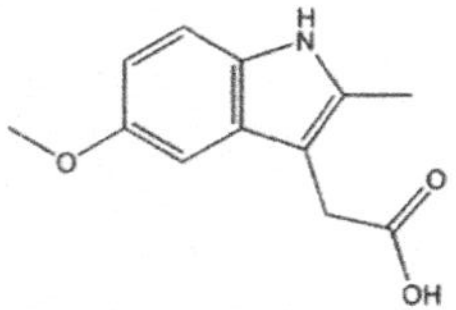	A common stable nonsteroid anti-inflammatory drug is used during the neonatal period to prevent the persistent patent ductus arteriosus [51]	It has been frequently detected in WWTPs and drinking water with concentrations ranging from 19 to 200 ng/L [52]

(continued)

Table 1.1 (continued)

Pharmaceutical	Chemical structure	Properties	Concentration in water bodies
Sotalol ($C_{12}H_{20}N_2O_3S$)	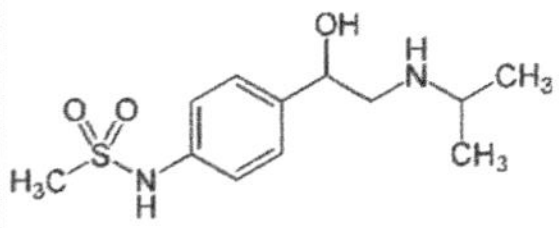	A beta-blocker drug which is used to cure cardiac arrhythmias of the heart [53]. Highly persistent in water bodies	Up to 150 ng/L at river waters [54]
Metoprolol ($C_{15}H_{25}NO_3$)	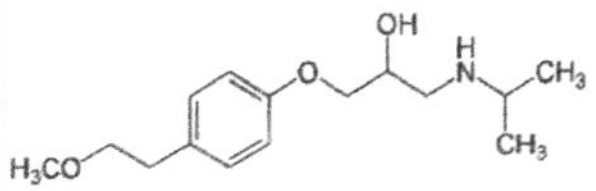	A highly prescribed beta-blocker which is used to treat hypertension, tachycardia, and heart diseases [55]	0.6–1.4 μg/L in WWTP influents
Diclofenac ($C_{14}H_{10}Cl_2NO_2$)	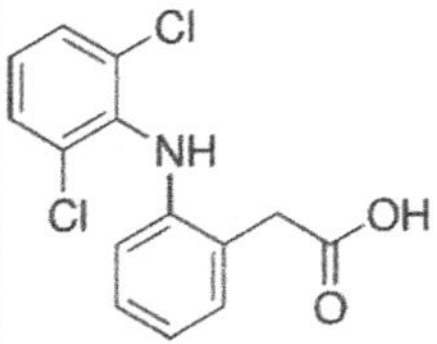	A nonsteroidal anti-inflammatory drug used widely to mitigate pain as well as inflammatory diseases [56]	It is frequently detected in WWTPs and drinking waters (4.7 μg/L and 1.2 μg/L, respectively) [57]
Meclofenamic acid $C_{15}H_{15}NO_2$	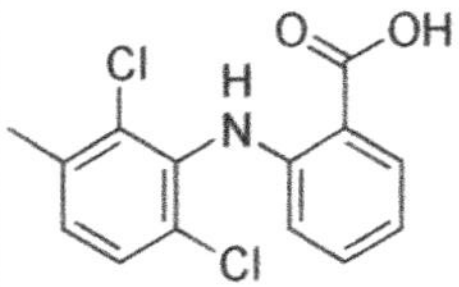	A nonsteroidal anti-inflammatory drug that is commonly used in the treatment of musculoskeletal diseases such as osteoarthritis and rheumatoid arthritis [58]	In concentrations from ng/L to μg/L [59]

(continued)

Table 1.1 (continued)

Pharmaceutical	Chemical structure	Properties	Concentration in water bodies
Carbamazepine ($C_{15}H_{12}NO$)	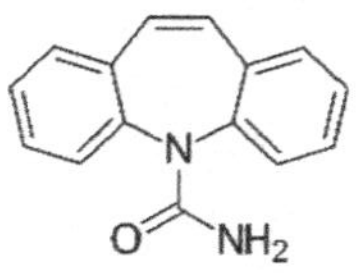	An antiepileptic and mood-stabilizing drug which is used to treat bipolar disorder, trigeminal neuralgia, and epilepsy [60]. Among the most frequently detected drugs in the environment which is persistent and of very low biodegradability	It has been detected in concentrations of 610 ng/ L and up to 18 ng L^{-1} in groundwater and drinking, respectively [61]
Fluoxetine ($C_{17}H_{18}F_3NO$)	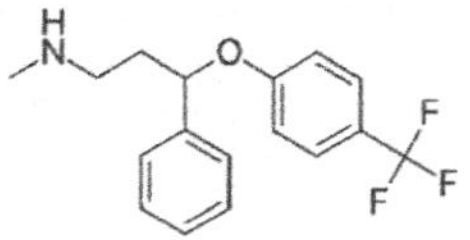	A selective serotonin reuptake inhibitor which is used to cure depression [62]	Ranging from 34.8 to 105 ng/L in hospital effluents in Portugal [63] 21 ng/L in a psychiatric hospital in China [64]
Paroxetine ($C_{19}H_{20}FNO_3$)	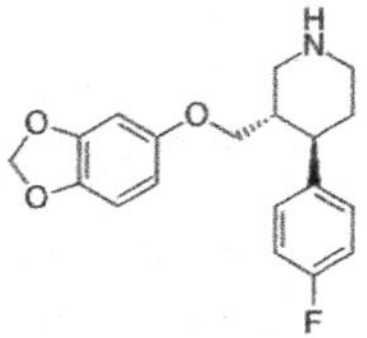	A drug widely used in curing humor disorders which acts as an allosteric serotonin selective reuptake inhibitor [65]	Approximately 25 ng/L in surface waters (Portugal) [66]
Loratadine ($C_{22}H_{23}ClN_2O_2$)	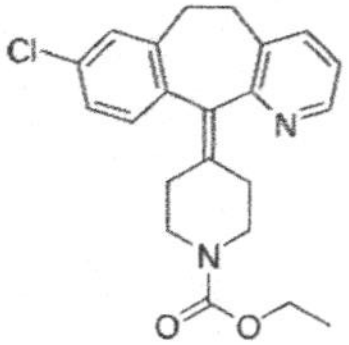	A drug used to treat allergies including allergic rhinitis and hives [67]	0.713–10.2 ng/L in urban wastewater treatment facilities [63]

(continued)

Table 1.1 (continued)

Pharmaceutical	Chemical structure	Properties	Concentration in water bodies
Ranitidine ($C_{13}H_{22}N_4O_3S$)	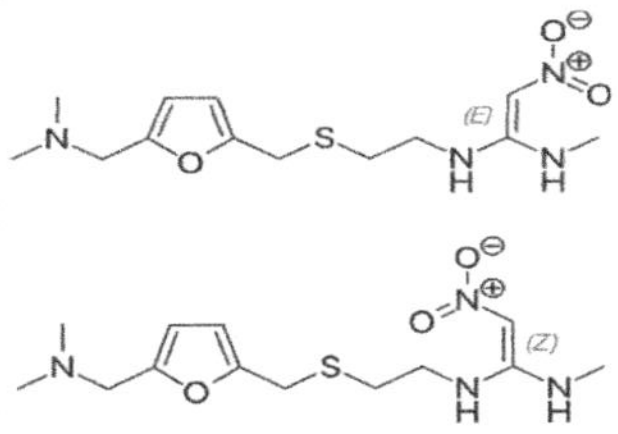	A common histamine-2 blocker is an antiulcer drug which is used to reduce acid production in the stomach. It represents a high water solubility (79.5 mg/L) [68]	It is commonly found in water bodies in various countries all over the world 70–540 ng/L in WWTPs Approximately 10 ng/L in surface waters [68–70]
Atenolol ($C_{14}H_{22}N_2O_3$)	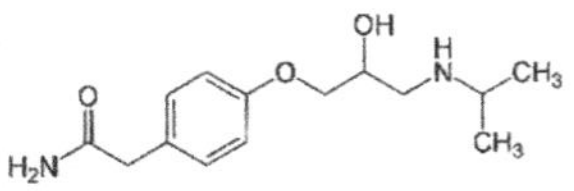	A β-adrenolytic, cardio-selective drug. It has been widely applied for the treatment of diseases such as cardiovascular [71]	From ng/L to μg/L in most WWTPs [72]
Erythromycin ($C_{37}H_{67}NO_{13}$)	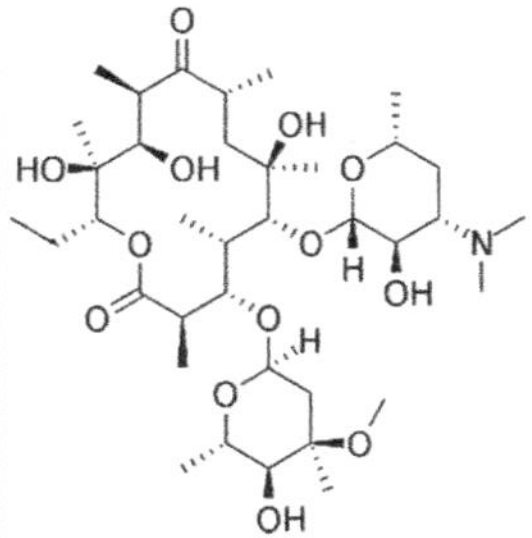	A macrolide antibiotic which is used for human and veterinary medicines as an antibiotic. It is also applied in aquaculture and livestock production [73]	200–6000 ng/L in WWTPs 10–1700 ng/L in surface waters Approximately 50 ng/L in groundwater [74]

(continued)

Table 1.1 (continued)

Pharmaceutical	Chemical structure	Properties	Concentration in water bodies
Azithromycin ($C_{38}H_{72}N_2O_{12}$)	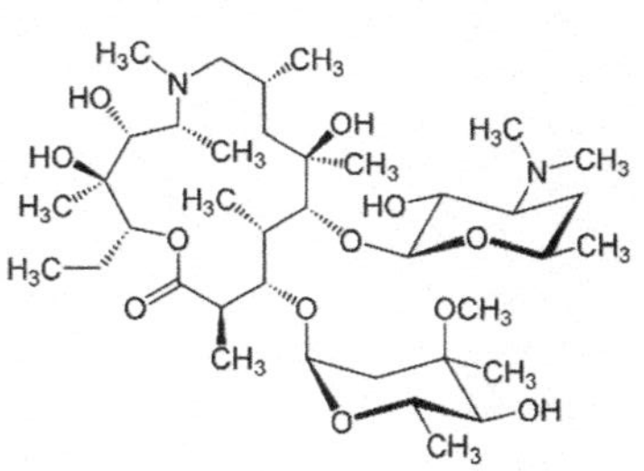	A semisynthetic derivative of erythromycin. It is a macrolide antibiotic having a wide range of activities against Gram-positive and Gram-negative bacteria. It has been widely used for the clinical treatment of sexually transmitted diseases as well as respiratory infections [75]	Up to 70 ng/L in WWTPs [74]
Sulfamethoxazole ($C_{10}H_{11}N_3O_3S$)	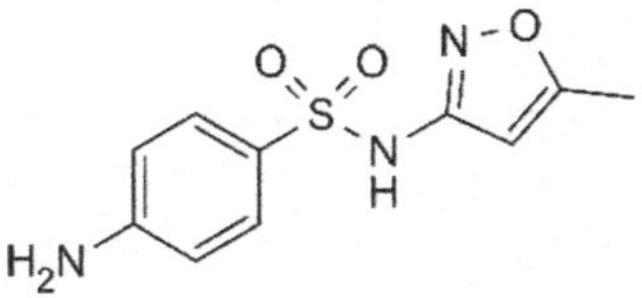	The most frequently prescribed antibiotic in the US. It interferes with dihydrofolate production which is normally prescribed along with trimethoprim (TMP), an antibiotic that acts by binding to bacterial dihydrofolate reductase [76]	– Up to 50 μg/L in countries such as Mozambique and Kenya [77]
Trimethoprim ($C_{14}H_{18}N_4O_3$)	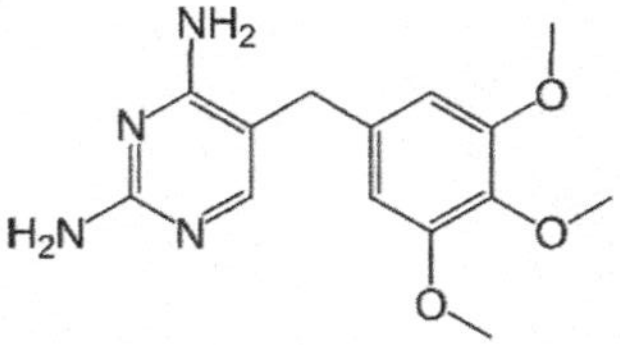	A representative antibiotic which is commonly used for the treatment of infections in the respiratory tract and urinary tract, etc. [78]	– 2 mg/L in WWTPs and up to 0.48 mg/L in surface waters [79–81]

(continued)

Table 1.1 (continued)

Pharmaceutical	Chemical structure	Properties	Concentration in water bodies
Ofloxacin ($C_{18}H_{20}FN_3O_4$)	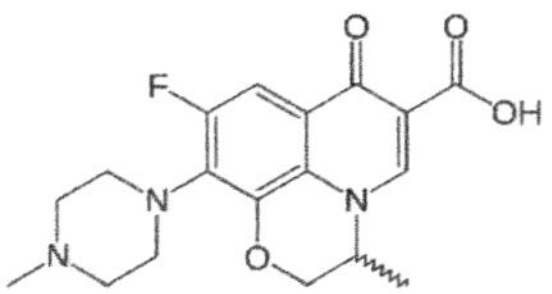	A fluorinated quinolone-type antibiotic that has been widely used for curing serious bacterial infections. It has demonstrated high activities against both Gram-positive and Gram-negative bacteria [82]	– 0.2–0.3 μg/L in the WWTPs in the USA [83] – 0.09–31.7 μg/L in the influents of a Spanish wastewater treatment plant [84]
Propranolol ($C_{16}H_{21}NO_2$)	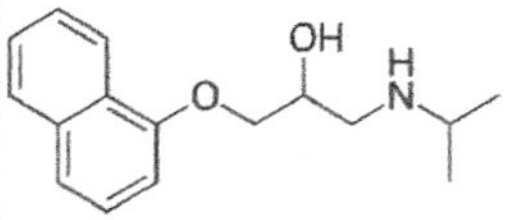	A nonselective β-blocker has been widely used in various countries for curing cardiac malfunctions such as hypertension or angina pectoris [85]	Up to 0.5 μg/L in WWTPs [86]
Thioridazine ($C_{21}H_{26}N_2S_2$)	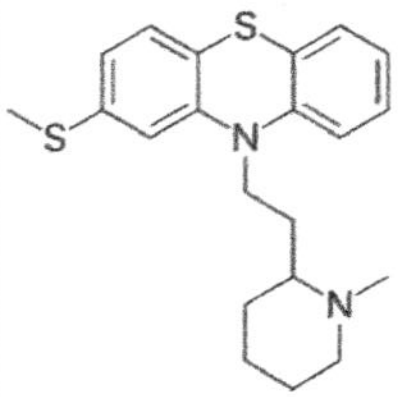	An antipsychotic (neuroleptic) drug that has been extensively prescribed for the treatment of psychiatric disorders. It has been reported with anticancer properties [87]	6–22 ng/L in river Medway, UK [88]

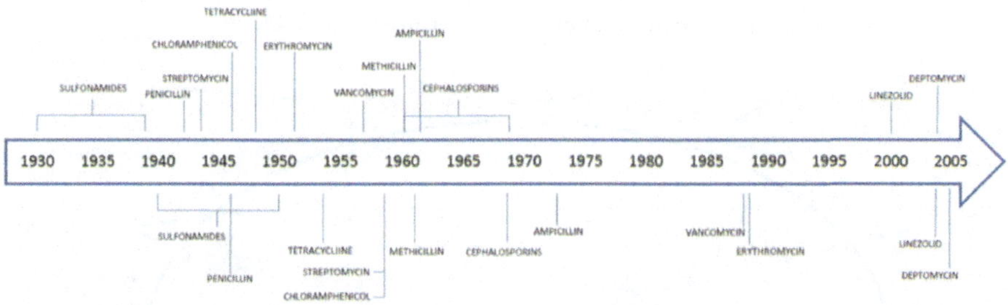

Fig. 1.3 First observations of antibiotic-resistant bacteria, adapted from Pazda et al. [91]

be expected to release such agents into the environment with the potential to create severe environmental and health issues (Fig. 1.4).

According to Fig. 1.5, four mechanisms can be involved in the resistance of the microbial communities, including (a) removal of the antibiotics actively from the bacteria cell utilizing an efflux pump, (b) establishing alternative metabolic pathways, (c) alteration in the target of the antibiotic, and (d) enzymatic inactivation of the PhAC [91].

The environmental concentration and the possible impacts of the PhACs depend on the type, pattern of use, the mechanisms involved in their action, and the way they are released into the environment. For instance, ofloxacin or its metabolism products are excreted via urine and feces and discharged into aquatic environments [100]. Hence, this PhAC can be frequently found in surface waters [101]. A low degree of

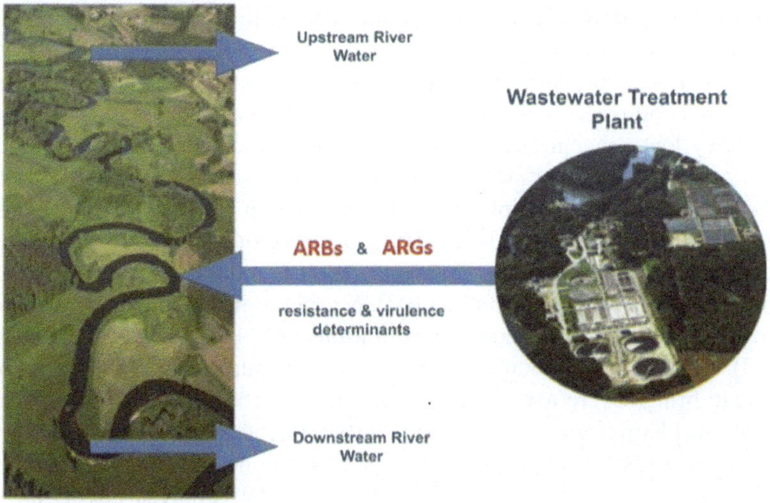

Fig. 1.4 Release of ABRs and ARGs from municipal wastewater treatment plants is among the most important routes of the presence of such agents in the environment with possible severe health and environmental impacts, adapted from Osińska et al. [98]

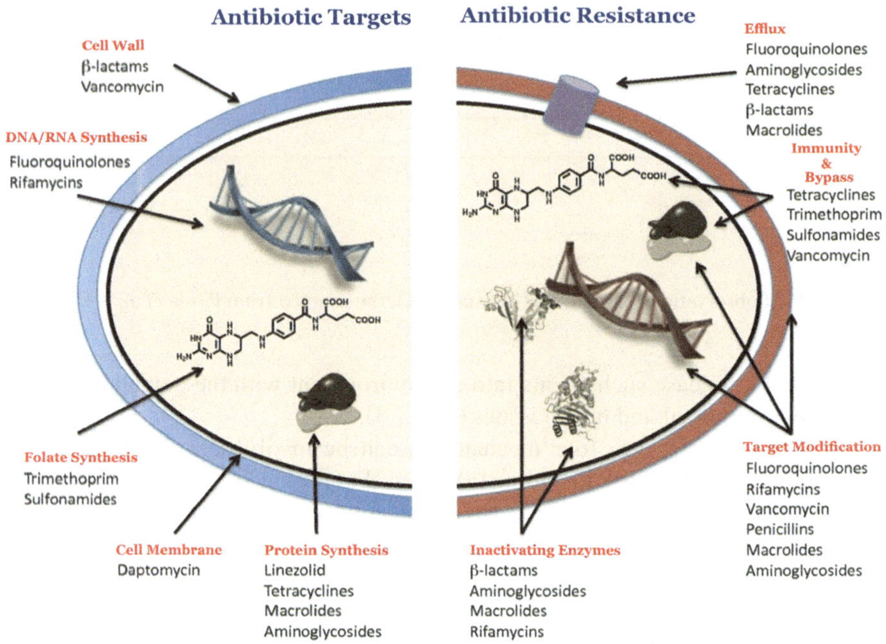

Fig. 1.5 Various mechanisms for the resistance of microbial communities to antibiotics. As can be observed in this figure, the type of mechanism involved in this process is highly dependent on the type of PhACs, adapted from Pazda et al. [91], and Wright [99]

metabolism for compounds such as atenolol has led to its frequent detection in its original form in water bodies [102].

Various studies have indicated the possible toxic effects of PhACs on various environmental compartments, such as aquatic microorganisms [103, 104] (Fig. 1.6).

Even very low concentrations of such compounds can create toxic effects. For instance, atenolol and propranolol resulted in an increase in caspase-3 activity and a decrease in the adenosine triphosphate (ATP) level in cardiomyocytes in rats after exposure [106]. Concentrations below 100 μg/L of ketoprofen in surface waters have caused negative effects on living organisms, such as oxidative stress and endocrine disruption in zebrafish (*Danio rerio*) [107]. Morphological changes in the zebrafish embryos were also detected under concentrations of 1, 10, and 100 μg/L ketoprofen (Fig. 1.7). For fluoxetine, there are reports that exposure to even very low concentrations can bring severe toxic effects to aquatic microorganisms. For instance, 1–5 μg/L of this pharmaceutical can significantly influence egg fertilization in medaka fish during 4 weeks of exposure [108]. A lower concentration of this drug can even increase lethargy in mosquitofish [109]. Oxidative stress effects have also been observed in *Oncorhynchus mykiss* (rainbow trout) induced by erythromycin [73]. Such effects have also been observed in *M. flosaquae* (0.1–40 μg/L) [110]

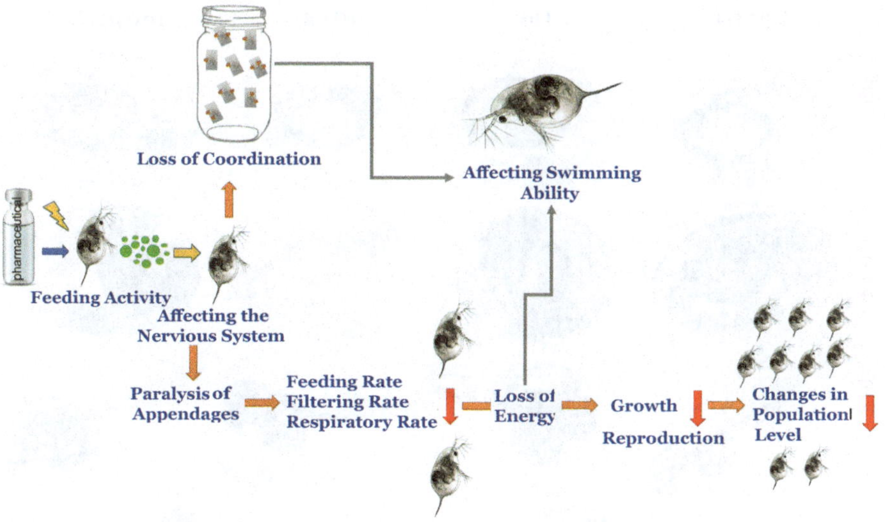

Fig. 1.6 *Daphnia magna* has been used as a model in the toxicity assessment of PhACs and indicated effects such as immobilization, lethality, and reproductive, behavioral, physiological, and biochemical changes when exposed to PhACs, reprinted with permission from Tkaczyk et al. [105]

and *Pseudokirchneriella subcapitata* (0.06–0.3 mg/L) [111]. In addition to oxidative stress, inhibition of physiological processes has been observed in *Selenastrum capricornutum* (0.06–0.3 mg/L) [112].

In this regard, concern is rising, especially for widely used pharmaceuticals, due to their increasing concentrations in aquatic media. Diclofenac is among the mentioned drugs, which can be expected in surface waters with increasing concentrations. Such a concentration can threaten rainbow trout, with the lowest observed effect concentration (LOEC) of 5 μg/L [109].

There are also drugs that are commonly consumed in European countries. Ranitidine is an example of such a compound that has been listed among the most sold prescribed drugs in Europe. There is evidence for the toxicity of these drugs. For instance, growth inhibition in rotifers and crustaceans is a result of chronic exposure to this drug [69].

In addition to the mentioned parameters, global health issues such as pandemic conditions can considerably influence the occurrence and probable impacts of PhACs. For instance, COVID-19 has caused a dramatic change in the pattern of the use of specific pharmaceutical compounds. In particular, the consumption of drugs such as antibiotics, antivirals (as therapeutic agents), antiprotozoals, and antiparasitics, which are being used to treat this illness, has increased considerably, leading to the release of large amounts of such pharmaceuticals in water bodies. For instance, it has been reported that the concentration of azithromycin in surface waters has been elevated from approximately 4 ng/L to above 900 ng/L [114].

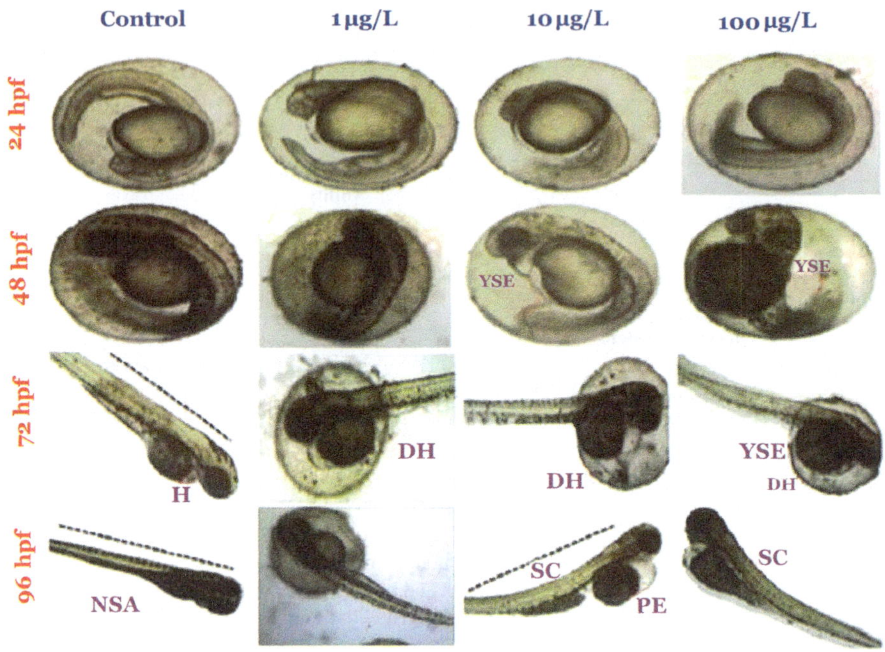

Fig. 1.7 Morphological changes in zebrafish embryos as a result of exposure to ketoprofen (1, 10, and 100 µg/ml at 24, 48, 72, and 96 h). H, PE, YES, SC, DH, and NSA represent heart, pericardial edema, yolk sac edema, scoliosis, delayed hatching, and normal spine axis, respectively. Reprinted with permission from Rangasamy et al. [113]

Among antivirals, favipiravir, ribavirin, lopinavir, and remdesivir have been reported to have elevated concentrations after the COVID-19 pandemic. Favipiravir is generally used to cure influenza [115]. Higher concentrations of this pharmaceutical have been observed in water bodies in the influenza season (February and March) in many countries (e.g., Japan [116]). Clinical studies have revealed that favipiravir can result in a 30% reduction in the mortality caused by COVID-19 [117]. Environmental concentrations of over 60 ng/L have been reported for this drug during the current pandemic [118]. Ribavirin has also been reported for the treatment of COVID-19 [119] and to have a positive effect on the reduction in mortality caused by COVID-19, especially when used together with lopinavir and ritonavir. The concentration of this pharmaceutical has also increased considerably during the current pandemic, reaching over 50 ng/L in June 2020 in surface waters [114].

Table 1.2 Further reading suggestions for more detailed coverage of the literature on the origin, presence, and possible effects of PhACs on humans, the environment, and living organisms

Reference	Item	Subject
Starling et al. [101]	Table 1	Occurrence of PhACs in the environment
Kasonga et al. [120]	Section 4	Presence of endocrine-disruptive chemicals in wastewater treatment plants
[89]	Table 2	Expected concentrations of PhACs in wastewater treatment plants
Bilal et al. [9]	Section 9	Enzyme-assisted biodegradation of PhACs
Joseph et al. [121]	Section 4	Metal–organic frameworks for the removal of PhACs
Kim et al. [7]	Table 3	A summary of the results of the removal of PhACs by membrane-based technologies
Pazda et al. [91]	Table 1	Beta-lactam ARGs[a]
	Table 2	Macrolide ARGs[a]
	Table 3	Quinolone ARGs[a]
	Table 4	Sulfonamide and trimethoprim ARGs[a]
	Table 5	Tetracycline ARGs[a]
Kaur [122]	Supplementary information	Chronic toxicity valued of selected PhACs
Tkaczyk et al. [105]	Table 1	Acute immobilization tests of pharmaceuticals on *D. magna*
	Table 2	Effects of the toxicity of PhACs on swimming behavioral and physiological parameters of *D. magna*

[a]Detected in wastewater treatment plants using conventional wastewater treatment methods

1.4 Further Reading

Table 1.2 contains items from the recent literature that the reader can consult for more detailed information regarding the presence of the PhACs in the environment and their possible effects on humans, the environment, and living organisms.

1.5 Summary

The presence of pharmaceutically active compounds (PhACs) in the environment has been recently considered an issue of global concern. This is mainly due to some reasons, including the vast consumption pattern of most PhACs and their probable environmental and health impacts. It has been indicated that these compounds can lead to the generation of antibiotic-resistant bacteria (ARBs) and antibiotic resistance genes (ARGs). Such compounds can also lead to toxic effects on aquatic organisms even under current environmental considerations (from ng/L to μg/L). The adverse environmental and health impacts of such compounds are directly related to the type, pattern of use, molecular structure, metabolism pathway, and by-products. Considering the current increasing patterns of the use of PhACs, especially under current COVID-19 pandemic conditions, this is highly expected to have an increase in their environmental concentrations. Hence, there is a need to implement efficient and cost-effective (waste)water techniques at real scales to reduce and control the release of such compounds into the environment.

References

1. Manisalidis I et al (2020) Environmental and health impacts of air pollution: a review. Front Public Health 8:1–13. https://doi.org/10.3389/fpubh.2020.00014
2. Kamali M, Khodaparast Z (2015) Review on recent developments on pulp and paper mill wastewater treatment. Ecotoxicol Environ Safety 114:326–342. https://doi.org/10.1016/j.ecoenv.2014.05.005
3. Kamali M et al (2021) Biochar in water and wastewater treatment—a sustainability assessment. Chem Eng J 420:129946. https://doi.org/10.1016/j.cej.2021.129946
4. Wang Q, Yang Z (2016) Industrial water pollution, water environment treatment, and health risks in China. Environ Poll 218:358–365. https://doi.org/10.1016/j.envpol.2016.07.011
5. Han D, Currell MJ, Cao G (2016) Deep challenges for China's war on water pollution. Environ Poll 218:1222–1233. https://doi.org/10.1016/j.envpol.2016.08.078
6. Azizullah A et al (2011) Water pollution in Pakistan and its impact on public health—a review. Environ Int 37:479–497. https://doi.org/10.1016/j.envint.2010.10.007
7. Kim S et al (2018) Removal of contaminants of emerging concern by membranes in water and wastewater: a review. Chem Eng J 335(Nov 2017):896–914. https://doi.org/10.1016/j.cej.2017.11.044
8. Quesada HB et al (2019) Surface water pollution by pharmaceuticals and an alternative of removal by low-cost adsorbents: a review. Chemosphere 222:766–780. https://doi.org/10.1016/j.chemosphere.2019.02.009
9. Bilal M et al (2019) Emerging contaminants of high concern and their enzyme-assisted biodegradation—a review. Environ Int 124:336–353. https://doi.org/10.1016/j.envint.2019.01.011
10. Lee BCY et al (2021) Emerging contaminants: an overview of recent trends for their treatment and management using light-driven processes. Water (Switzerland) 13:2340. https://doi.org/10.3390/w13172340
11. Rasheed T et al (2019) Environmentally-related contaminants of high concern: potential sources and analytical modalities for detection, quantification, and treatment. Environ Int 122:52–66. https://doi.org/10.1016/j.envint.2018.11.038
12. Kamali M et al (2019) Enhanced biodegradation of phenolic wastewaters with acclimatized activated sludge—a kinetic study. Chem Eng J 378. https://doi.org/10.1016/j.cej.2019.122186
13. Xin X et al (2019) An integrated approach for waste activated sludge management towards electric energy production/resource reuse. Bioresour Technol 274:225–231. https://doi.org/10.1016/j.biortech.2018.11.092
14. Müller B et al (2012) Pharmaceuticals as indictors of sewage-influenced groundwater. Hydrogeol J 20:1117–1129. https://doi.org/10.1007/s10040-012-0852-4

15. Zhang Z, Grover DP, Zhou JL (2009) Monitoring of pharmaceutical residues in sewage effluents. In: Handbook of water purity and quality. Elsevier Inc. https://doi.org/10.1016/B978-0-12-374192-9.00014-5
16. Daughton CG (2016) Pharmaceuticals and the Environment (PiE): evolution and impact of the published literature revealed by bibliometric analysis. Sci Tot Environ 562:391–426. https://doi.org/10.1016/j.scitotenv.2016.03.109
17. European Commission (EC) (2015) DECISION (EU) 2015/495. Official J Eur Union L78/40(C(2015) 1756):20–30
18. European Medicines Agency (2006) Guideline No. 4447/00 on the environmental risk assessment of medicinal products for human use. Committee for Medicinal Products for Human Use (June), pp 1–12. Available at: http://www.emea.eu.int
19. Patel M et al (2021) Ciprofloxacin and acetaminophen sorption onto banana peel biochars: environmental and process parameter influences. Environ Res 201:111218. https://doi.org/10.1016/j.envres.2021.111218
20. aus der Beek T et al (2016) Pharmaceuticals in the environment: global occurrence and potential cooperative action under the Strategic Approach to International Chemicals Management (SAICM), German Environment Agency. http://ovidsp.ovid.com/ovidweb.cgi?T=JS&PAGE=reference&D=emed18a&NEWS=N&AN=72198002
21. Balakrishna K et al (2017) A review of the occurrence of pharmaceuticals and personal care products in Indian water bodies. Ecotoxicol Environ Safety 137(Oct 2016):113–120. https://doi.org/10.1016/j.ecoenv.2016.11.014
22. Kleywegt S et al (2019) Environmental loadings of active pharmaceutical ingredients from manufacturing facilities in Canada. Sci Tot Environ 646:257–264. https://doi.org/10.1016/j.scitotenv.2018.07.240
23. Costa D et al (2006) Inhibition of human neutrophil oxidative burst by pyrazolone derivatives. Free Radical Biol Med 40:632–640. https://doi.org/10.1016/j.freeradbiomed.2005.09.017
24. Koutsouba V et al (2003) Determination of polar pharmaceuticals in sewage water of Greece by gas chromatography-mass spectrometry. Chemosphere 51:69–75. https://doi.org/10.1016/S0045-6535(02)00819-6
25. Reddersen K, Heberer T, Dünnbier U (2002) Identification and significance of phenazone drugs and their metabolites in ground- and drinking water. Chemosphere 49:539–544. https://doi.org/10.1016/S0045-6535(02)00387-9
26. Zuehlke S, Duennbier U, Heberer T (2007) Investigation of the behavior and metabolism of pharmaceutical residues during purification of contaminated ground water used for drinking water supply. Chemosphere 69:1673–1680. https://doi.org/10.1016/j.chemosphere.2007.06.020
27. Zühlke S, Dünnbier U, Heberer T (2004) Detection and identification of phenazone-type drugs and their microbial metabolites in ground and drinking water applying solid-phase extraction and gas chromatography with mass spectrometric detection. J Chromatogr A 1050(2):201–209. https://doi.org/10.1016/j.chroma.2004.08.051
28. Tauxe-Wuersch A et al (2005) Occurrence of several acidic drugs in sewage treatment plants in Switzerland and risk assessment. Water Res 39:1761–1772. https://doi.org/10.1016/j.watres.2005.03.003
29. Salgado R et al (2011) Assessing the diurnal variability of pharmaceutical and personal care products in a full-scale activated sludge plant. Environ Poll 159:2359–2367. https://doi.org/10.1016/j.envpol.2011.07.004
30. Ma J et al (2016) Photodegradation of gemfibrozil in aqueous solution under UV irradiation: kinetics, mechanism, toxicity, and degradation pathways. Environ Sci Pollut Res 23:14294–14306. https://doi.org/10.1007/s11356-016-6451-5
31. Fang Y et al (2012) Occurrence, fate, and persistence of gemfibrozil in water and soil. Environ Toxicol Chem 31:550–555. https://doi.org/10.1002/etc.1725
32. Heberer T (2002) Tracking persistent pharmaceutical residues from municipal sewage to drinking water. J Hydrol 266:175–189. https://doi.org/10.1016/S0022-1694(02)00165-8

33. Zhao JL et al (2010) 'Occurrence and a screening-level risk assessment of human pharmaceuticals in the pearl river system, South China. Environ Toxicol Chem 29:1377–1384. https://doi.org/10.1002/etc.161
34. Rad TS et al (2018) Synthesis of pumice-TiO_2 nanoflakes for sonocatalytic degradation of famotidine. J Clean Prod 202:853–862. https://doi.org/10.1016/j.jclepro.2018.08.165
35. Kiszkiel-Taudul I, Starczewska B (2019) Dispersive liquid-liquid microextraction of famotidine and nizatidine from water samples. J Chromatogr Sci 57:93–100. https://doi.org/10.1093/chromsci/bmy087
36. Koopal C et al (2017) Effect of adding bezafibrate to standard lipid-lowering therapy on post-fat load lipid levels in patients with familial dysbetalipoproteinemia. A randomized placebo-controlled crossover trial. J Lipid Res 58:2180–2187. Â© 2017 ASBMB. Currently published by Elsevier Inc; originally published by American Society for Biochemistry and Molecular Biology. https://doi.org/10.1194/jlr.M076901
37. Shi XT et al (2018) Kinetics and pathways of Bezafibrate degradation in UV/chlorine process. Environ Sci Pollut Res 25:672–682. https://doi.org/10.1007/s11356-017-0461-9
38. Ledeţi I et al (2015) Kinetic analysis of solid-state degradation of pure pravastatin versus pharmaceutical formulation. J Therm Anal Calorim 121:1103–1110. https://doi.org/10.1007/s10973-015-4842-3
39. Mao Z et al (2018) Pravastatin alleviates interleukin 1β-induced cartilage degradation by restoring impaired autophagy associated with MAPK pathway inhibition. Int Immunopharmacol 64:308–318. https://doi.org/10.1016/j.intimp.2018.09.018
40. Razavi B et al (2011) Treatment of statin compounds by advanced oxidation processes: kinetic considerations and destruction mechanisms. Radiat Phys Chem 80:453–461. https://doi.org/10.1016/j.radphyschem.2010.10.004
41. Akul NB et al (2021) Effects of mevastatin on electricity generation in microbial fuel cells. Pol J Environ Stud 30:5407–5412. https://doi.org/10.15244/pjoes/133402
42. Gottlieb K et al (2016) Inhibition of methanogenic archaea by statins as a targeted management strategy for constipation and related disorders. Aliment Pharmacol Ther 43:197–212. https://doi.org/10.1111/apt.13469
43. Hapeshi E et al (2015) Licit and illicit drugs in urban wastewater in Cyprus. Clean—Soil, Air, Water 43:1272–1278. https://doi.org/10.1002/clen.201400483
44. Azusano IPI, Caparanga AR, Chen BH (2020) Degradation of ketoprofen using iron-supported ZSM-5 catalyst via heterogeneous Fenton oxidation. IOP Conf Ser: Earth Environ Sci 612:012048. https://doi.org/10.1088/1755-1315/612/1/012048
45. Feng Y et al (2017) Degradation of ketoprofen by sulfate radical-based advanced oxidation processes: kinetics, mechanisms, and effects of natural water matrices. Chemosphere 189:643–651. https://doi.org/10.1016/j.chemosphere.2017.09.109
46. Chin CJM et al (2014) Effective anodic oxidation of naproxen by platinum nanoparticles coated FTO glass. J Hazar Mater 277:110–119. https://doi.org/10.1016/j.jhazmat.2014.02.034
47. Sétifi N et al (2019) Heterogeneous Fenton-like oxidation of naproxen using synthesized goethite-montmorillonite nanocomposite. J Photochem Photobiol, A 370:67–74. https://doi.org/10.1016/j.jphotochem.2018.10.033
48. Tran N et al (2015) Optimization of sono-electrochemical oxidation of ibuprofen in wastewater. J Environ Chem Eng 3:2637–2646. https://doi.org/10.1016/j.jece.2015.05.001
49. Richardson SD, Kimura SY (2020) Water analysis: emerging contaminants and current issues. Anal Chem 92(1):473–505. https://doi.org/10.1021/acs.analchem.9b05269
50. Roberts PH, Thomas KV (2006) The occurrence of selected pharmaceuticals in wastewater effluent and surface waters of the lower Tyne catchment. Sci Total Environ 356(1–3):143–153. https://doi.org/10.1016/j.scitotenv.2005.04.031
51. Perron N et al (2013) Deleterious effects of indomethacin in the mid-gestation human intestine. Genomics 101:171–177. https://doi.org/10.1016/j.ygeno.2012.12.003
52. Chen H et al (2020) Significant role of high-valent iron-oxo species in the degradation and detoxification of indomethacine. Chemosphere 251:126451. https://doi.org/10.1016/j.chemosphere.2020.126451

53. Heli H et al (2009) Copper nanoparticles-carbon microparticles nanocomposite for electrooxidation and sensitive detection of sotalol. Sens Actuators, B Chem 140:245–251. https://doi.org/10.1016/j.snb.2009.04.021
54. González-Mariño I et al (2015) Transformation of methadone and its main human metabolite, 2-ethylidene-1,5-dimethyl-3,3-diphenylpyrrolidine (EDDP), during water chlorination. Water Res 68:759–770. https://doi.org/10.1016/j.watres.2014.10.058
55. Benner J, Ternes TA (2009) Ozonation of metoprolol: elucidation of oxidation pathways and major oxidation products. Environ Sci Technol 43:5472–5480. https://doi.org/10.1021/es900280e
56. Kronacher C, Hogreve F (1936) Röntgenologische Skelettstudien an Dahlemer Binder-Drillingen und -Zwillingen. Zeitschrift für Züchtung Reihe B, Tierzüchtung und Züchtungsbiologie einschließlich Tierernährung 36(3):281–294. https://doi.org/10.1111/j.1439-0388.1936.tb00094.x
57. Bae S, Kim D, Lee W (2013) Degradation of diclofenac by pyrite catalyzed Fenton oxidation. Appl Catalysis B, Environ 134–135:93–102. https://doi.org/10.1016/j.apcatb.2012.12.031
58. Li J et al (2017) The degradation efficiency and mechanism of meclofenamic acid in aqueous solution by UV irradiation. J Adv Oxidation Technol 20:1–8. https://doi.org/10.1515/jaots-2016-0188
59. Maskaoui K, Zhou JL (2010) Colloids as a sink for certain pharmaceuticals in the aquatic environment. Environ Sci Pollut Res 17:898–907. https://doi.org/10.1007/s11356-009-0279-1
60. French J (2006) Introduction. Epilepsy Res 68:21–22. https://doi.org/10.1016/j.eplepsyres.2005.09.011
61. Ding Y et al (2017) Chemical and photocatalytic oxidative degradation of carbamazepine by using metastable Bi^{3+} self-doped $NaBiO_3$ nanosheets as a bifunctional material. Appl Catalysis B: Environ 202:528–538. https://doi.org/10.1016/j.apcatb.2016.09.054
62. de Lima Perini JA et al (2017) Photo-Fenton degradation of the pharmaceuticals ciprofloxacin and fluoxetine after anaerobic pre-treatment of hospital effluent. Environ Sci Poll Res 24:6233–6240. https://doi.org/10.1007/s11356-016-7416-4
63. Santos LHMLM et al (2013) Contribution of hospital effluents to the load of pharmaceuticals in urban wastewaters: Identification of ecologically relevant pharmaceuticals. Sci Tot Environ 461–462:302–316. https://doi.org/10.1016/j.scitotenv.2013.04.077
64. Yuan S et al (2013) Detection, occurrence and fate of 22 psychiatric pharmaceuticals in psychiatric hospital and municipal wastewater treatment plants in Beijing, China. Chemosphere 90:2520–2525. https://doi.org/10.1016/j.chemosphere.2012.10.089
65. Ferreira APG, Pinto BV, Cavalheiro TG (2018) Thermal decomposition investigation of paroxetine and sertraline. J Anal Appl Pyrolysis 136:232–241. https://doi.org/10.1016/j.jaap.2018.09.022
66. Paíga P et al (2016) Presence of pharmaceuticals in the Lis river (Portugal): sources, fate and seasonal variation. Sci Tot Environ 573:164–177. https://doi.org/10.1016/j.scitotenv.2016.08.089
67. Hunto ST et al (2020) Loratadine, an antihistamine drug, exhibits anti-inflammatory activity through suppression of the NF-kB pathway. Biochem Pharmacol 177:113949. https://doi.org/10.1016/j.bcp.2020.113949
68. Patel M et al (2019) Pharmaceuticals of emerging concern in aquatic systems: chemistry, occurrence, effects, and removal methods. Chem Rev 119:3510–3673. https://doi.org/10.1021/acs.chemrev.8b00299
69. Isidori M et al (2009) Effects of ranitidine and its photoderivatives in the aquatic environment. Environ Int 35:821–825. https://doi.org/10.1016/j.envint.2008.12.002
70. Zuccato E et al (2006) Pharmaceuticals in the environment in Italy: causes, occurrence, effects and control. Environ Sci Pollut Res 13:15–21. https://doi.org/10.1065/espr2006.01.004
71. Patil RH, Hegde RN, Nandibewoor ST (2009) Voltammetric oxidation and determination of atenolol using a carbon paste electrode. Ind Eng Chem Res 48:10206–10210. https://doi.org/10.1021/ie901163k

72. Ji Y et al (2013) Photocatalytic degradation of atenolol in aqueous titanium dioxide suspensions: Kinetics, intermediates and degradation pathways. J Photochem Photobiol A: Chem 254:35–44. https://doi.org/10.1016/j.jphotochem.2013.01.003
73. Rodrigues S et al (2019) Toxicity of erythromycin to Oncorhynchus mykiss at different biochemical levels: detoxification metabolism, energetic balance, and neurological impairment. Environ Sci Pollut Res 26:227–239. https://doi.org/10.1007/s11356-018-3494-9
74. Voigt M, Jaeger M (2017) On the photodegradation of azithromycin, erythromycin and tylosin and their transformation products—a kinetic study. Sustain Chem Pharm 5:131–140. https://doi.org/10.1016/j.scp.2016.12.001
75. Parnham MJ et al (2014) Azithromycin: mechanisms of action and their relevance for clinical applications. Pharmacol Therapeutics 143:225–245. https://doi.org/10.1016/j.pharmthera.2014.03.003
76. Kielhofner G (2005) Rethinking disability and what to do about it: disability studies and its implications for occupational therapy. Am J Occup Ther 59:487–496. https://doi.org/10.5014/ajot.59.5.487
77. Avramiotis E et al (2021) Oxidation of sulfamethoxazole by rice husk biochar-activated persulfate. Catalysts 11:850. https://doi.org/10.3390/catal11070850
78. Wang Q et al (2019) Study of the degradation of trimethoprim using photo-Fenton oxidation technology. Water (Switzerland) 11:207. https://doi.org/10.3390/w11020207
79. Brown KD et al (2006) Occurrence of antibiotics in hospital, residential, and dairy effluent, municipal wastewater, and the Rio Grande in New Mexico. Sci Total Environ 366:772–783. https://doi.org/10.1016/j.scitotenv.2005.10.007
80. Kovalakova P et al (2020) Occurrence and toxicity of antibiotics in the aquatic environment: a review. Chemosphere 251:126351. https://doi.org/10.1016/j.chemosphere.2020.126351
81. Yilmaz G et al (2017) Characterization and toxicity of hospital wastewaters in Turkey. Environ Monit Assess 189:55. https://doi.org/10.1007/s10661-016-5732-2
82. Weng X, Owens G, Chen Z (2020) Synergetic adsorption and Fenton-like oxidation for simultaneous removal of ofloxacin and enrofloxacin using green synthesized Fe NPs. Chem Eng J 382:122871. https://doi.org/10.1016/j.cej.2019.122871
83. Renew JE, Huang CH (2004) Simultaneous determination of fluoroquinolone, sulfonamide, and trimethoprim antibiotics in wastewater using tandem solid phase extraction and liquid chromatography-electrospray mass spectrometry. J Chromatogr A 1042:113–121. https://doi.org/10.1016/j.chroma.2004.05.056
84. Pi Y et al (2014) Oxidation of ofloxacin by Oxone/Co^{2+}: identification of reaction products and pathways. Environ Sci Pollut Res 21:3031–3040. https://doi.org/10.1007/s11356-013-2220-x
85. Baydum VPA et al (2012) Pre-treatment of propranolol effluent by advanced oxidation processes Valderice. AfinidAd LXiX 559:211–216
86. Anquandah GAK et al (2013) Ferrate(VI) oxidation of propranolol: kinetics and products. Chemosphere 91:105–109. https://doi.org/10.1016/j.chemosphere.2012.12.001
87. Chu CW et al (2019) Thioridazine enhances p62-mediated autophagy and apoptosis through Wnt/β-catenin signaling pathway in glioma cells. Int J Mol Sci 20:473. https://doi.org/10.3390/ijms20030473
88. Ebele AJ, Abou-Elwafa Abdallah M, Harrad S (2017) Pharmaceuticals and personal care products (PPCPs) in the freshwater aquatic environment. Emerg Contaminants 3:1–16. https://doi.org/10.1016/j.emcon.2016.12.004
89. Krzeminski P et al (2019) Performance of secondary wastewater treatment methods for the removal of contaminants of emerging concern implicated in crop uptake and antibiotic resistance spread: a review. Sci Tot Environ 648:1052–1081. https://doi.org/10.1016/j.scitotenv.2018.08.130
90. Lopes et al (2016) Antibiotic resistance in E. coli isolated in effluent from a wastewater treatment plant and sediments in receiver body. Int J River Basin Manag 14:441–445. https://doi.org/10.1080/15715124.2016.1201094

91. Pazda M et al (2019) Antibiotic resistance genes identified in wastewater treatment plant systems—a review. Sci Tot Environ 697:134023. https://doi.org/10.1016/j.scitotenv.2019.134023
92. Rafraf ID et al (2016) Abundance of antibiotic resistance genes in five municipal wastewater treatment plants in the Monastir Governorate, Tunisia. Environ Poll 219:353–358. https://doi.org/10.1016/j.envpol.2016.10.062
93. Ben W et al (2017) Distribution of antibiotic resistance in the effluents of ten municipal wastewater treatment plants in China and the effect of treatment processes. Chemosphere 172:392–398. https://doi.org/10.1016/j.chemosphere.2017.01.041
94. Karkman A et al (2016) High-throughput quantification of antibiotic resistance genes from an urban wastewater treatment plant. FEMS Microbiol Ecol 92:1–7. https://doi.org/10.1093/femsec/fiw014
95. Zanotto C et al (2016) Identification of antibiotic-resistant *Escherichia coli* isolated from a municipal wastewater treatment plant. Chemosphere 164:627–633. https://doi.org/10.1016/j.chemosphere.2016.08.040
96. Li J et al (2016) Occurrence and removal of antibiotics and the corresponding resistance genes in wastewater treatment plants: effluents' influence to downstream water environment. Environ Sci Pollut Res 23:6826–6835. https://doi.org/10.1007/s11356-015-5916-2
97. Pallares-Vega R et al (2019) Determinants of presence and removal of antibiotic resistance genes during WWTP treatment: a cross-sectional study. Water Res 161:319–328. https://doi.org/10.1016/j.watres.2019.05.100
98. Osińska A et al (2017) The prevalence and characterization of antibiotic-resistant and virulent *Escherichia coli* strains in the municipal wastewater system and their environmental fate. Sci Total Environ 577:367–375. https://doi.org/10.1016/j.scitotenv.2016.10.203
99. Wright GD (2010) Q&A: Antibiotic resistance: where does it come from and what can we do about it? BMC Biol 8:123. https://doi.org/10.1186/1741-7007-8-123
100. Ballesteros O, Vílchez JL, Navalón A (2002) Determination of the antibacterial ofloxacin in human urine and serum samples by solid-phase spectrofluorimetry. J Pharm Biomed Anal 30:1103–1110. https://doi.org/10.1016/S0731-7085(02)00466-1
101. Starling MCVM, Amorim CC, Leão MMD (2019) Occurrence, control and fate of contaminants of emerging concern in environmental compartments in Brazil. J Hazar Mater 372:17–36. https://doi.org/10.1016/j.jhazmat.2018.04.043
102. Head A (1997) Exercise metabolism in healthy volunteers taking celiprolol, atenolol, and placebo. Br J Sports Med 31:120–125. https://doi.org/10.1136/bjsm.31.2.120
103. Prata JC et al (2018) Influence of microplastics on the toxicity of the pharmaceuticals procainamide and doxycycline on the marine microalgae *Tetraselmis chuii*. Aquatic Toxicol 197:143–152. https://doi.org/10.1016/j.aquatox.2018.02.015
104. Santos A, Veiga F, Figueiras A (2020) Dendrimers as pharmaceutical excipients: synthesis, properties, toxicity and biomedical applications. Materials. https://doi.org/10.3390/ma13010065
105. Tkaczyk A et al (2021) Daphnia magna model in the toxicity assessment of pharmaceuticals: a review. Sci Total Environ 763:143038. https://doi.org/10.1016/j.scitotenv.2020.143038
106. Seydi E, Tabbati Y, Pourahmad J (2020) Toxicity of atenolol and propranolol on rat heart mitochondria. Drug Res 70:151–157. https://doi.org/10.1055/a-1112-7032
107. Diniz MS et al (2015) Ecotoxicity of ketoprofen, diclofenac, atenolol and their photolysis byproducts in zebrafish (Danio rerio). Sci Tot Environ 505:282–289. https://doi.org/10.1016/j.scitotenv.2014.09.103
108. Foran CM et al (2004) Reproductive assessment of Japanese Medaka (*Oryzias latipes*) following a four-week fluoxetine (SSRI) exposure. Arch Environ Contam Toxicol 46:511–517. https://doi.org/10.1007/s00244-003-3042-5
109. Henry TB, Black MC (2008) Acute and chronic toxicity of fluoxetine (selective serotonin reuptake inhibitor) in western Mosquitofish. Arch Environ Contam Toxicol 54:325–330. https://doi.org/10.1007/s00244-007-9018-0

110. Wan J et al (2015) Effect of erythromycin exposure on the growth, antioxidant system and photosynthesis of Microcystis flos-aquae. J Hazar Mater 283:778–786. https://doi.org/10.1016/j.jhazmat.2014.10.026
111. Nie XP et al (2013) Toxic effects of erythromycin, ciprofloxacin and sulfamethoxazole exposure to the antioxidant system in *Pseudokirchneriella subcapitata*. Environ Poll 172:23–32. https://doi.org/10.1016/j.envpol.2012.08.013
112. Liu B et al (2011) Toxic effects of erythromycin, ciprofloxacin and sulfamethoxazole on photosynthetic apparatus in *Selenastrum capricornutum*. Ecotoxicol Environ Safety 74:1027–1035. https://doi.org/10.1016/j.ecoenv.2011.01.022
113. Rangasamy B et al (2018) Developmental toxicity and biological responses of zebrafish (Danio rerio) exposed to anti-inflammatory drug ketoprofen. Chemosphere 213:423–433. https://doi.org/10.1016/j.chemosphere.2018.09.013
114. Chen X et al (2021) Occurrence and risk assessment of pharmaceuticals and personal care products (PPCPs) against COVID-19 in lakes and WWTP-river-estuary system in Wuhan, China. Sci Tot Environ 792:148352. https://doi.org/10.1016/j.scitotenv.2021.148352
115. Shiraki K, Daikoku T (2020) Since January 2020 Elsevier has created a COVID-19 resource centre with free information in English and Mandarin on the novel coronavirus COVID-19. The COVID-19 resource centre is hosted on Elsevier Connect, the company's public news and information. Pharmacol Therapeutics 2(9):107512
116. Azuma T et al (2017) Fate of new three anti-influenza drugs and one prodrug in the water environment. Chemosphere 169:550–557. https://doi.org/10.1016/j.chemosphere.2016.11.102
117. Hassanipour S et al (2021) The efficacy and safety of Favipiravir in treatment of COVID-19: a systematic review and meta-analysis of clinical trials. Scien Rep 11:1–11. https://doi.org/10.1038/s41598-021-90551-6
118. Kuroda K et al (2021) Predicted occurrence, ecotoxicological risk and environmentally acquired resistance of antiviral drugs associated with COVID-19 in environmental waters. Sci Tot Environ 776:145740. https://doi.org/10.1016/j.scitotenv.2021.145740
119. Tong S et al (2020) Ribavirin therapy for severe COVID-19: a retrospective cohort study. Int J Antimicrobial Agents 56:1–5. https://doi.org/10.1016/j.ijantimicag.2020.106114
120. Kasonga TK et al (2021) Endocrine-disruptive chemicals as contaminants of emerging concern in wastewater and surface water: a review. J Environ Manage 277:111485. https://doi.org/10.1016/j.jenvman.2020.111485
121. Joseph L et al (2019) Removal of contaminants of emerging concern by metal-organic framework nanoadsorbents: a review. Chem Eng J 369:928–946. https://doi.org/10.1016/j.cej.2019.03.173
122. Kaur L (2020) Role of phytoremediation strategies in removal of heavy metals. https://doi.org/10.1007/978-981-32-9771-5_13

Chapter 2
Techniques for the Detection, Quantifications, and Identification of Pharmaceutically Active Compounds and Their Removal Mechanisms

2.1 Introduction

The presence of pharmaceutically active compounds (PhACs) in water bodies has created global concerns considering their possible environmental and health impacts. For instance, the appearance of bacterial antibiotic resistance (BARs) is an example of such consequences that has caused major concerns in healthcare organizations [1, 2]. With the increasing trends in PhAC consumption, especially under pandemic conditions (e.g., COVID-19), the development of more complicated resistance mechanisms is expected by microorganisms. PhACs can also have toxic effects on terrestrial or aquatic microorganisms and hence can threaten the ecological balance [3, 4].

As discussed in Chap. 1, PhACs can be detected at different concentrations in the receiving environments according to their patterns of use, their molecular structure and resistance to the environment, and the efficiency of the current wastewater treatment facilities. Various technologies have been developed to address these pollutants, which can be divided into biological (e.g., anaerobic digestion (AD), microbial fuel cells (MFC), and constructed wetlands (CW)) and physico-chemical treatments (e.g., adsorption, membrane technologies, and advanced oxidation processes (AOPs)). However, the potential effects of PhACs in the environment, as well as the efficiency of the treatment methods applied, are highly dependent on the environmental concentrations of various types of PhACs. Hence, the first step to controlling the trace of the PhACs in the environment is to determine the concentration of the pollutants using precise and reliable analytical methods. Such techniques are also required to evaluate the efficiency of the developed treatment methods.

It is worth mentioning that other useful methods, such as UV–VIS spectroscopy,[1] which have been considered rapid, simple, and precise methods for the detection

[1] UV–VIS spectrophotometry measures the absorbance of light by an analyte at a certain wavelength to determine the concentration of the analyte. Conventional UV–VIS spectrophotometers normally contain two different lamps for the emitting of light. The first lamp is made of deuterium, which provides wavelengths from 190 to 400 nm. In addition to the deuterium lamp, the UV–VIS

M. Kamali et al., *Advanced Wastewater Treatment Technologies for the Removal of Pharmaceutically Active Compounds*, Green Energy and Technology,
https://doi.org/10.1007/978-3-031-20806-5_2

of pollutants and the determination of their concentrations, especially for colored compounds such as dyes [6, 7], normally fail to separate different pollutants. Hence, such methods can only be used where previous knowledge is available regarding the composition of the effluents (e.g., synthetic wastewaters). However, in most environmental conditions, there are possibilities of having pollutants with similar optical behavior, which can negatively affect the accuracy of the conventional analytical methods.

Furthermore, there is a need to detect and quantify the reaction intermediates and the by-products of the decomposition of PhACs when (waste)water treatment methods are applied for the removal of such compounds. Regardless of the concentration of the parent pollutants, the type and the concentration of the decomposition product have determinant effects on the toxicity of the effluents and the probable environmental and health issues [8]. In this regard, there is a need to couple various analytical techniques to determine the degree of mineralization as well as the content of the decomposition products. It is evident that a sustainable method for the removal of PhACs should also result in effluents that are safe to be discharged into the environment.

The present chapter aims to introduce the methods that can be used for the detection and quantification of PhACs and their decomposition products and the existing gaps to be addressed by future studies. Furthermore, methods that can be used for the investigation of the mechanisms involved in the removal of PhACs using various physico-chemical and biological treatment methods are introduced and discussed in this chapter.

2.2 Detection and Quantification Techniques

Sample preparation is the initial step for the detection and analysis of PhACs, especially while working with real samples where relatively low concentrations of such compounds are expected. Proper extraction and cleaning of the samples can ensure the accuracy of the results. Procedures are also needed to remove the impurities in the sample with the potential of interfering with the analysis. A schematic of the water sample preparation for the measurements of the PhAC concentrations is illustrated in Fig. 2.1. High-performance liquid chromatography (HPLC) is a popular technique that has been widely used for the identification of PhAC concentrations.[2]

---

spectrophotometers normally contain a tungsten lamp, which provides wavelengths from 300 to 900 nm. By combining these two lamps both UV and visible light can be provided [5].

[2] In HPLC, the pressurized solvents containing the water sample is passed through a column packed with an adsorbent (i.e., granuls of silica or polymers). The components in the solvent react differently with the adsorbent, allowing separation of the components due to their difference in flow rates. To quantify the desired component a detector is used which generates a signal proportional to the quantity of component exiting the column [9].

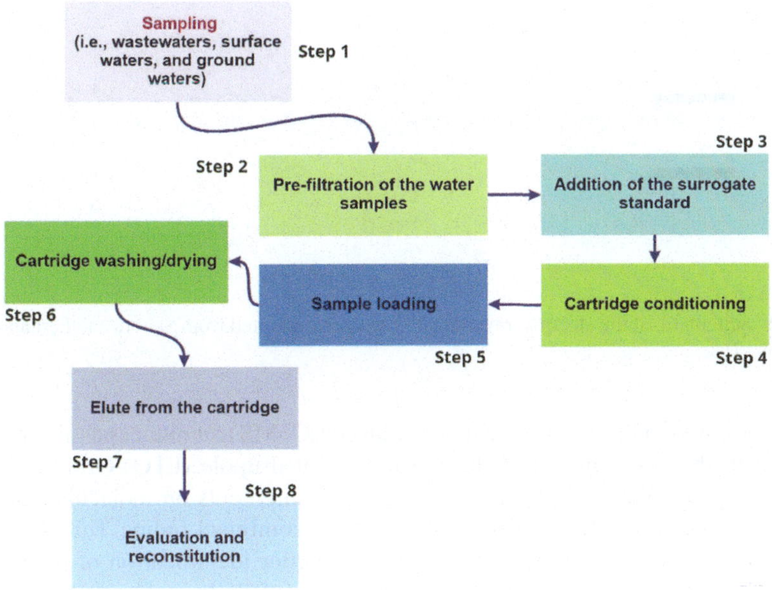

Fig. 2.1 Schematic of the steps required for sample preparation for HPLC analysis. Solid-phase extraction (SPE, Step 4) is an important task that requires a precise selection of the adsorbent in the cartridges

Other methods, such as gas chromatography with mass spectrometry (GC–MS[3]) or GC with tandem MS (GC–MS/MS) and liquid chromatography with mass spectrometry (LC–MS) or LC with tandem MS (LC–MS/MS), have also been well developed and used in recent years for the detection and quantification of PhACs. LC–MS/MS, however, is the most preferred technique when compared to other analytical techniques for PhACs since this technique allows the precise separation of coeluted compounds with different products but the same molecular weights. MS/MS is also a highly useful technique to increase the analytical selectivity of the applied method, which is an essential need when analyzing samples from complex matrices.[4] There have also been efforts to enhance the sensitivity of the analytical techniques, especially to detect unknown organic compounds. For instance electrospray ionization (ESI) is a technique that is used in mass spectrometry and provides the possibility to identify structurally important fragment patterns [11]. This technique generates ions by using a high-voltage electrospray to produce aerosols (Fig. 2.2).

---

[3] Mass spectroscopy (MS) is based on ionization of the molecules in the gas phase, their separation according to their mass to charge (m/z) ratios, and detecting the separated ionizied molecules (see [10].

[4] Lichrospher® 100 RP-18 column (250 × 4 mm, 5 μm particle size), Phenomenex® Synergy™ HYDRO-RP column (C18, polar endcapped; 50 × 2.00 mm, 4 μm particle size), and Titan C18 (3.0 × 100 mm, 1.9 μm particle size) are the widely used columns.

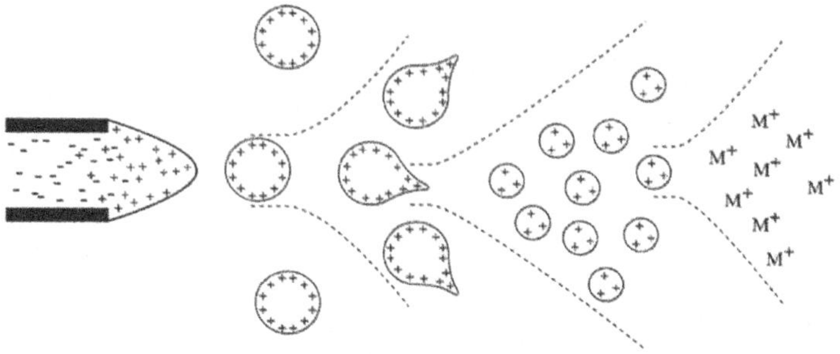

Fig. 2.2 Schematic of the electrospray ionization process, adopted from Sahora and Fernández-del Castillo [12]

The progress in the development of advanced LC–MS techniques has also recently resulted in the development of the linear trap quadrupole (LTQ)-Orbitrap, which enables high-resolution/high mass accuracy measurements on molecular ions [13, 14]. There are also advanced methods, such as combined online Turbulent Flow Chromatography (TFC)-LTQ-Orbitrap, for the better identification of pharmaceutical degradation products [14]. Quadrupole time-of-flight mass spectrometry Q-TOF–MS is also a hybrid technique that combines quadrupole technologies with a time-of-flight mass analyzer.

2.3 Techniques for Identification of the Removal Mechanisms

PhACs can be removed under various physical, chemical, or biological pathways. This section explores the most important techniques that have been developed to identify the mechanism involved in this process for the removal of PhACs.

2.3.1 Adsorptive Removal

Adsorption is the most widely known process in which adsorbates (i.e., organic compounds or inorganic compounds) are attached to specific adsorbents. There are various parameters that determine the efficiency of the adsorption process, such as the type and properties of the adsorbent and adsorbate, the pKa value of the PhACs, and the operating pH. Various techniques have been developed for the characterization of adsorbents, including X-ray diffraction (to identify the composition and crystalline phases of the materials), gas adsorption for the determination of the specific surface area and porosity of the materials (using Brunauer–Emmett–Teller (BET) theory), and scanning electron microscopy (SEM, to explore the surface methodology of the porous structure of the adsorbents). As discussed in Chap. 8, the presence of surface functional groups is also an essential asset for adsorbents that can potentially promote the adsorption of pollutants. Techniques arc also available for the identification of such functional groups, such as Fourier transform infrared spectroscopy (FTIR).[5] Raman spectroscopy is also used to identify the surface bonds in adsorbents to characterize materials regarding the presence of functional groups as essential elements in the adsorption of pollutants [16, 17].

The efficiency of the adsorption process is generally expressed by the adsorption kinetics, which evaluates the rate of the adhesion of the adsorbates onto the adsorbents. For PhACs, the adsorption process normally obeys the pseudofirst-order model (Eqs. 2.1 and 2.2).

$$q(t) = q_e\left(1 - e^{-kt}\right) \tag{2.1}$$

$$q(t) = \frac{K_2 q_e 2t}{1 + K_2 q_e t}. \tag{2.2}$$

These equations, $q(t)$ represents the adsorbed compounds onto the adsorbent at the given time (t), expressed by mg/g. Additionally, q_x is the mg/g of the solute adsorbed by the adsorbent at equilibrium. Furthermore, K_1 and K_2 describe the rate constants of the pseudofirst-order and pseudosecond-order models, respectively, (g/mg/min) [18].

Various adsorption and desorption isotherms, such as Freundlich, linear, and Langmuir isotherms, can also be used to describe the adsorption process equilibrium [19]. The relationship between the sorption of the PhACs and the equilibrium concentration of these compounds in the liquid phase can be calculated empirically using Eq. 2.3.

$$q = K_{\mathrm{f}} C_{\mathrm{eq}}^{1/n} \tag{2.3}$$

In this equation, q_{eq} is the amount of PhACs adsorbed at equilibrium, K_{f} is the Freundlich adsorption coefficient, and C_{eq} is the equilibrium concentration in the liquid phase. This isotherm can also be linearized in the logarithmic form, as mentioned in Eq. 2.4.

$$\log q_{\mathrm{ep}} = \log K_{\mathrm{f}} + \frac{1}{n} \log C_{\mathrm{eq}} \tag{2.4}$$

In the linear isotherm (Eq. 2.5), the constant $1/n$ of the Freundlich model (Eq. 2.3) approximates unity. This model is generally used when no specific bonding between the adsorbate and the adsorbent occurs.

[5] A technique which is generally employed to obtain the infrared (IF) spectrum of adsorption or emission of materials in various states (i.e., solid, liquid, or gas) [15].

$$q = K_f C_{\mathrm{eq}} \tag{2.5}$$

Finally, Langmuir describes the monolayer sorption of PhACs onto the surface of the adsorbent with a finite number of identical sites without surface discussion (Eq. 2.6).

$$q_{\mathrm{ep}} = \frac{QbC_{\mathrm{eq}}}{1 + bC_{\mathrm{eq}}} \tag{2.6}$$

In addition to the adsorption kinetics and equilibrium, the thermodynamic study can also aid in obtaining a better understanding of the nature of the adsorption process. In this regard, thermodynamic parameters, including Gibb's energy (ΔG, kJ/mol), enthalpy (ΔH, kJ/mol), and entropy (ΔS, J/(mol·K)), are generally calculated according to Eqs. 2.7 and 2.8 [19].

$$\ln K = \frac{-\Delta G}{RT} \tag{2.7}$$

$$\Delta G = \Delta H - T\Delta S \tag{2.8}$$

When $\Delta G < 0$, the adsorption process is favorable. While chemosorption is the dominant mechanism under $\Delta H > 0$, physical adsorption, chemical adsorption, or a combination of both can occur under $\Delta H < 0$ (which represents an exothermic condition). Further randomness of the solid/solution interface can also be concluded by $\Delta H > 0$ values. This is because the translational energy lost by the adsorbate is less than that gained by the displaced solvent molecules, such as water.

Various mechanisms, such as ion exchange, hydrogen bonds, π–π interactions, hydrophobic interactions, and electrostatic interactions (see Chap. 8), can be involved in the adsorption process. However, it is normally difficult to identify the contribution of each adsorption mechanism to the overall removal of PhACs. Hence, the overall free energy of adsorption, ΔG_{ads}, is generally considered representative of various adsorption mechanisms (Eq. 2.9). It is worth mentioning that adsorption occurs when ΔG_{ads} is negative.

$$\Delta G_{\mathrm{ads}} = \Delta G_{\mathrm{elec}} + \Delta G_{\mathrm{hydro}} + G_{\mathrm{H\text{-}bond}} + \Delta G_{\pi-\pi\ \mathrm{EDA}} + \ldots \tag{2.9}$$

It has also been well documented that adsorption is the first step of the chemical transformation of PhACs when catalytic treatment techniques such as photocatalysis are applied for the removal of such compounds. Furthermore, adsorption of the PhACs may occur by the microbial communities and remain in the waste sludge from the biological treatment methods. Hence, it would be of high importance to detect the mechanisms involved in the adsorption of PhACs.

FTIR is also a useful technique to detect the chemical bonds present on the surface of materials, including adsorbents, after the adhesion of adsorbates. There are reports of the successful detection of PhAC adsorption mechanisms, as indicated in Fig. 2.3.

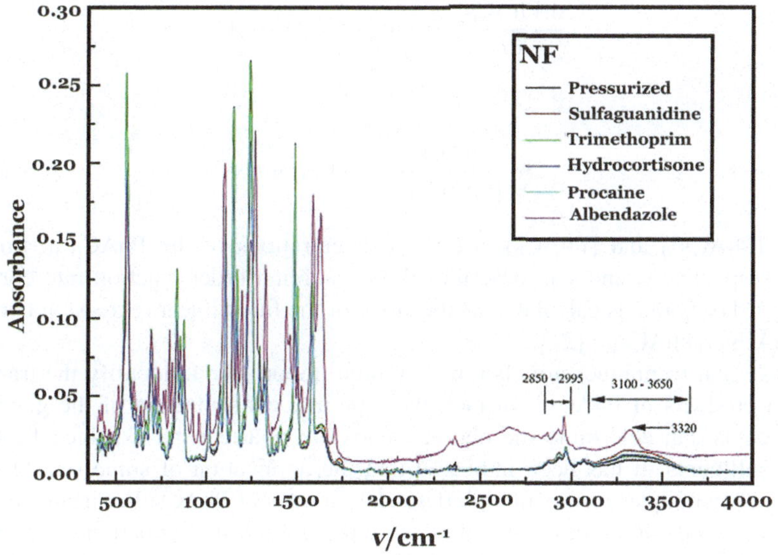

Fig. 2.3 FTIR spectra for the adsorption of albendazole using reverse osmosis (RO)/nanofiltration (NF) membranes. The rise in the baseline in the range of 3100–3650 cm^{-1} is an indication of the H-bonding process. Additionally, a direct H-bond with nitrogen can be observed at 3320 cm^{-1}. Finally, the carbonyl group and the bending of the methyl group of albendazole can be seen at 1620 cm^{-1} and between 800 and 1000 1632 cm^{-1}, respectively, adopted from Dolar et al. [20]

This figure can provide clues to the adsorption of albendazole under mechanisms including hydrogen bonds and $\pi-\pi$ interactions.

Other advanced techniques, such as X-ray photoelectron spectroscopy (XPS), can also be used to study the adsorption process of PhACs [21]. As a quantitative spectroscopic technique, XPS can provide the advantage of not only the determination of the elements covering the surface of the materials but also the bonds between these elements. For instance, significant adsorption of tetracyclines and sulfamethazine on reduced graphene oxides, identified by the presence of N 1s and S 2p peaks in the XPS spectra, is an example of the application of XPS for the identification of the adsorbed compounds discussed by Song et al. [22].

2.3.2 Advanced Oxidation Processes

Advanced oxidation processes are defined as techniques that result in the generation of active species (ASs), including radical or nonradical agents that can attack and decompose complex organic compounds in the content of polluted (waste)waters [23]. The efficiency of AOPs is normally described by the pseudofirst-order model fitted to the reaction kinetics of PhAC removal (Eqs. 2.10 and 2.11).

$$\frac{d[\mathrm{PhACs}]}{dt} = -k_{\mathrm{obs}} \times [\mathrm{PhACs}]_t \tag{2.10}$$

Or:

$$\ln \frac{[\mathrm{PhACs}]_t}{[\mathrm{PhACs}]_0} = -k_{\mathrm{obs}} \times t \tag{2.11}$$

where $[\mathrm{PhACs}]_0$ and $[\mathrm{PhACs}]_t$ are the concentrations of the PhACs at times 0 and t, respectively, and k_{obs} describes the pseudofirst-order reaction rate constant (min^{-1}). This value is calculated as the slope of the fitted linear regression between $\ln([\mathrm{PhACs}]_t/[\mathrm{PhACs}]_0)$ [24].

Analytical techniques are also of very high importance to identify the transformation products of PhACs.[6] In fact, the type and concentration of the generated by-products can determine the degree of toxicity that can be expected from the treated effluents. It has been indicated that the application of some (waste)water treatment techniques results in high degradation rates of PhACs. For instance, most advanced oxidation processes (AOPs) can represent a fast elimination of PhACs in relatively short reaction times [25, 26]. However, the degradation products of some treatment techniques can be of high recalcitrance and are difficult to further degrade using such processes. In this regard, there is a need to assess the additive, synergetic, and antagonistic effects of such degradation products when present in the treated effluents.

It has also been demonstrated that the type and effects of the decomposition products of PhACs can be influenced by the initial concentrations of the parent compounds and the efficiency of the applied (waste)water treatment methods. For instance, the degradation pathways and the generation of by-products when AOPs are applied can also be directly related to the type and amount of oxidation agents produced in the medium. As an example, for fluoroquinolone antibiotics (such as ciprofloxacin, norfloxacin, and lomefloxacin), the degradation products differ in the presence or absence of hydroxyl radicals. It has been indicated that in such compounds, hydroxyl radicals attach to the quinolone ring. Additionally, the piperazine ring of such compounds can be readily oxidized in the presence of molecular ozone. The bioaccumulation and toxicity of the degradation products are also highly dependent on their solubility in the water medium. Those compounds with higher solubility in (i.e., polar or nonpolar) medium can be readily absorbed by living organisms due to their higher bioavailability, causing toxic effects to the membrane, cell organs, or nuclei of the cells [27, 28].

The advanced analytical techniques discussed above can also be efficiently used to identify the transformation products of the PhACs. For instance, sulfamethoxazole (SML) is one of the most widely used pharmaceuticals and can be frequently found in water bodies (for instance, up to 50 μg/L in countries such as Mozambique and Kenya [29]). Hence, it can be considered a model for investigating the degradation of

[6] Especially when different wastewater treatment methods are applied for the removal of PhACs from the polluted (waste)waters.

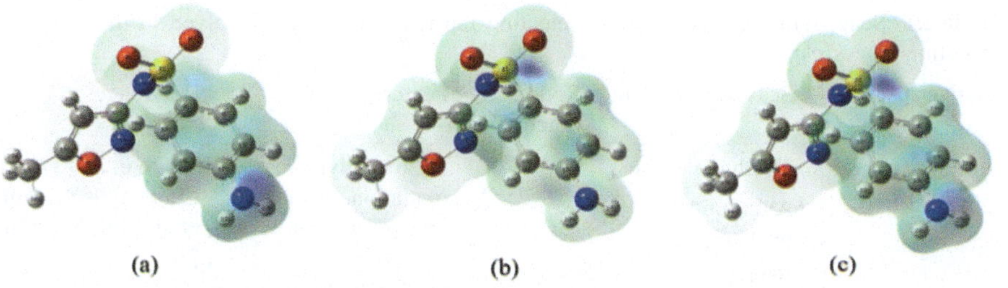

Fig. 2.4 Mapped electron density isosurface of sulfamethoxazole ($\rho = 0.01$ a.u. **a** f − (r); **b** f + (r); **c** f0(r), mapped using the Fukui function, adapted from Luo et al. [30]

products using various AOPs. There are some predictions regarding the reaction sites and the degradation products of these pharmaceuticals. In these molecules, N-7 has been identified with the highest likelihood for electrophilic reactions, indicated as the dark blue region in Fig. 2.4 [30]. Identification of the degradation using advanced techniques such as UHPLC-HRMS/MS allows for confirming such discussions.

Various techniques have been developed to identify the mechanisms involved in various types of AOPs for the removal of various organic pollutants. Detection of the type and share of the active agents that are generated and contribute to the degradation of organic pollutants can be performed using the scavenging agents that are introduced to the reaction medium to consume certain types of ASs, resulting in a drop in the efficiency of the organic compounds. Several scavenging agents have been introduced and employed thus far for the determination of the share of various ASs, as listed in Table 2.1. Notably, other ASs, such as peroxymonosulfate radicals $\left(SO_5^{\cdot -}\right)$, chlorine radicals ($Cl^{\cdot}$), chlorine-free radicals $\left(Cl_2^{\cdot -}\right)$, carbonate radicals $\left(CO_3^{\cdot -}\right)$, bicarbonate radicals $\left(HCO_3^{\cdot -}\right)$, nitrate radicals $\left(NO_3^{\cdot -}\right)$, and phosphate radicals $\left(PO_4^{\cdot 2-}\right)$, , can be involved in the decomposition of PhACs with different redox potentials (i.e., 1.1, 2.03, 2.0, 1.59, 1.7, 2.3–2.5, and > 2, respectively) and can be involved in the degradation of recalcitrant organic compounds, including PhACs [31–36].

Electron spin resonance (ESR) is another technique that has been used for the identification of the mechanisms involved in advanced oxidation processes for the removal of organic pollutants. ESR was first developed to investigate magnetic substances that contain one or more unpaired electrons, including transition metal ions and radicals [54]. The progress in this field later resulted in the development of electron paramagnetic resonance (EPR) [55]. Currently, these techniques are being widely used to detect various types of radicals (i.e., hydroxyl, sulfate, and superoxide radicals) generated in the reaction medium, which have a very short lifetime [56] (Fig. 2.5).

EPR is principally based on the high-rate addition of free radicals to spin-trapping agents (STAs) (i.e., nitrones or nitroso). As a result, paramagnetic nitroxides with a higher degree of stability[7] are formed, which can be detected and recorded. The

[7] ($t_{1/2}$(spin-adducts) = seconds – hours).

Table 2.1 Scavenging agents reported detecting the reactive species involved in the advanced oxidation processes

Active species	Scavenging agent	Redox potential	Featured AOPs	References
Protons (h^+)	Triethanolamine	–	Photocatalytic oxidation, wet air oxidation, Fenton oxidation, electrochemical oxidation, ozonation	Chadi et al. [37], Liu et al. [38], Yan et al. [39], Ng et al. [40]
	Ammonium oxalate			
	EDTA			
	Potassium iodine			
Superoxide radicals[8] ($O_2^{\cdot-}$)	Benzoquinone	2.4	Fenton oxidation, ultrasonic oxidation, wet air oxidation, ozonation, gamma-ray/electron beam radiation	Teel and Watts [42]
	Chloroform			
Hydroxyl radicals ($HO^{\cdot}$)	Tertiary butanol	2.7	Fenton oxidation, photocatalytic oxidation, electrochemical oxidation, wet air oxidation, gamma-ray/electron beam radiation	Bilanda et al. [43], Rastogi et al. [44], Raja et al. [45]; Yu et al. [46], Zhang et al. [36]
	Methanol			
	Isopropanol			
	Ethanol			
	N-butanol			
Electron (e^-)	Ethanol	−2.9	Photocatalytic oxidation, electrooxidation, photolysis, electrochemical oxidation, gamma-ray/electron beam radiation, etc.	Raja et al. [45], Zhang et al. [24], Yan et al. [39], Ramacharyulu et al. [47], Schwarz [48], Hasty et al. [49], Yi et al. [50]
	Potassium dichromate			
	Silver nitrate ($AgNO_3$)			
Singlet oxygen ($^1O_2$)	Sodium azide	2.2	Photocatalytic processes, electrooxidations, PS activation, etc.	Shang et al. [51], Zhou et al. [52]
	Furfuryl alcohol			
Iodate radicals ($IO_3^{\cdot}$)	Phenol	1.6	Iodate-assisted AOPs such as iodate activation with catalytic materials	Zhang et al. [24]
Sulfate radicals ($SO_4^{\cdot-}$)	Ethanol	2.5–3.1	Activation of persulfate using photocatalytic oxydation	Zhang et al. [24], Clifton and Huie [53]
	Methanol			
	1-Octanol			

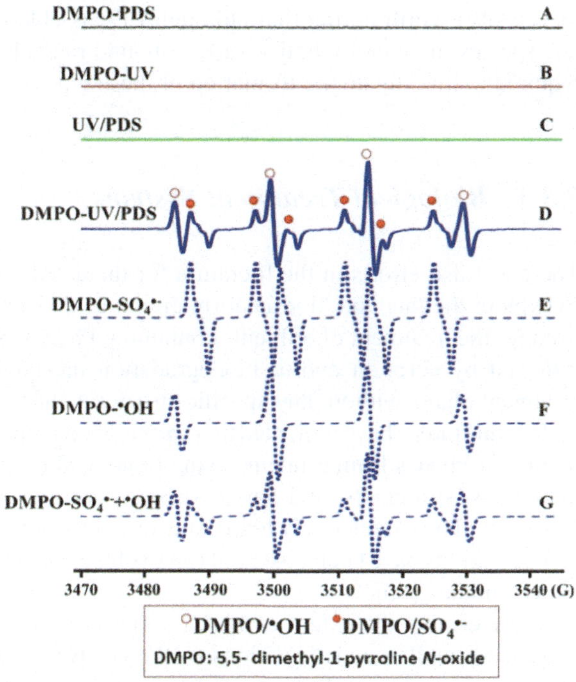

Fig. 2.5 ESR spectra obtained from UV photolysis of peroxydisulfate (PDS) (**a**), without UV (**b**), without spin-trapping agents (**c**), and with UV, PDS, and spin-trapping agents (**d**), indicating the formation of hydroxyl and sulfate radicals (E–G), adopted from Gao et al. [56]

respective radicals are identified based on the characteristics of the peaks in these spectra.

To date, over 100 STAs (nitrones) have been assessed.[9] In addition, the entries from approximately 10,000 experiments involving the application of 20 STAs have been provided in the Spin Trap Database.[10] In this regard, classic STAs such as 5,5-dimethyl-1-pyrroline N-oxide (DMPO) and *N-tert*-butylmethanimine *N*-oxide have received popularity for the detection of hydroxyl radicals and sulfate radicals [57, 58]. The 2,2,6,6-Tetramethylpiperidine (TEMP) is also used for trapping singlet oxygen [59].

Although this technique can help to identify the types of radicals involved in the AOPs for the removal of organic compounds, there is a risk of relying on the curve integral area to determine the concentration of the radicals. This is because various parameters affect the intensity of EPR spectra, such as the sensitivity of the instrument, as well as the sample preparation and the time interval between the sampling and the analysis.

---

[8] Artepillin C with a rate constant of 7.44 × 107 can be used for the protonated form of the superoxie (HO_2•) [41].

[9] Analysis of spectra achieved from various STAs indicate that the cyclic compounds with an α−H and a tertiary α′−C atom can form more stable nitroxides resulting in a better-quality spectrum [57].

[10] The entries as well as the updates can be found at: Spin Trap Database (nih.gov).

It is also worth noting that although EPR is able to efficiently detect free reactive species, it cannot identify surface-bound radicals. Hence, other techniques are required to investigate the formation of these types of active radicals.

2.3.3 *Biological Treatment Systems*

There are also efforts in the literature for the development of analytical techniques to explore the fate and biotransformation mechanism when biological methods are used for the treatment of effluents containing PhACs. Such compounds can be either removed by sorption and/or biodegradation mechanisms or discharged from the treatment plants without any specific structural changes.

Recent studies have indicated that sorption is an important mechanism for the elimination of PhACs from effluents using biological treatment methods. In this regard, antibiotics, stimulants, and antidepressants are present in sludge from wastewater treatment plants at relatively high concentrations (up to q[11] = 232 mg/kg), while low concentrations have been observed (up to 686 μg/kg) for diuretics, antianxieties, and anticoagulants [60].

To determine the extent of PhAC sorption by microorganisms, a preseparation stage is normally needed. Various methods have been introduced to separate these compounds, such as ultrasonic solvent extraction (USE) [61]. According to this method, strong solvents such as methanol are added to the samples after filtration or centrifugation, and extraction is performed by introducing ultrasonic irradiation. Analytical techniques such as HPLC can then be applied for the analysis of the concentration of the extracted compounds. Various sorption and desorption isotherms, such as Freundlich, linear, and Langmuir isotherms, which have already been described in 2.3.1, can be used to describe the equilibrium of PhAC sorption on biomass [62]. There is also a need to distinguish and discriminate between the sorption and biodegradation of PhACs when biological treatment techniques are used for the removal of these compounds. Inactivation of the biomass can lead to identifying the share of sorption in the overall removal efficiency of biological treatment methods. This can be performed by the addition of some chemicals, such as azide. Blocking electron transfer is the main mechanism involved in this process, which limits energy production in cells by inhibiting adenosine triphosphate (ATP) synthesis. This stops the energy-intensive activities inside the cells of the microorganisms [63–65].

Specific microorganisms have been reported for the biodegradation of PhACs. Table 2.2 provides some detailed information about such microorganisms and the respective PhACs they can eliminate.

It is also well known that efficient electron transfer between microorganisms is a determinant parameter for the efficiency of biological processes (such as anaerobic digestion) for the elimination of pollutants, including PhACs [93]. In addition to

---

[11] $q = (C_0 - C_e)V/XV$, where C_0 and C_e are the initial and the residual concentrations of the PhACs at a specific moment, V is the solution volume (L), and X is the concentration (mg/L) of the sludge.

Table 2.2 Specific microorganisms reported for the biodegradation of PhACs

PhACs	Molecular structure	Microorganism	References
Carbamazepine	$C_{15}H_{12}N_2O$	*Trametes versicolor*	Rodríguez-Rodríguez et al. [66]
Ibuprofen	$C_{13}H_{18}O_2$	*Phanerochaete chrysosporium*	Li et al. [67]
		Trametes versicolor	Marco-Urrea et al. [68]
Ciprofloxacin	$C_{17}H_{18}FN_3O_3$	*Pleurotus ostreatus*	Singh et al. [69]
		Thermus sp.	Pan et al. [70]
		Labrys portucalensis F11	Amorim et al. [71]
Naproxen	$C_{14}H_{14}O_3$	*Trametes versicolor*	Rodríguez-Rodríguez et al. [66]
		Phanerochaete chrysosporium	Li et al. [67]
Clofibric acid	$C_{10}H_{11}ClO_3$	*Trametes versicolor*	Marco-Urrea et al. [68]
Cefalexin	$C_{16}H_{17}N_3O_4S$	*Pseudomonas* sp. *Strain CE22*	Lin et al. [72]
Diclofenac	$C_{14}H_{11}Cl_2NO_2$	*Trametes versicolor*	Rodríguez-Rodríguez et al. [66]
		Actinoplanes sp.	Osorio-Lozada et al. [73]
		A consortium of Alcaligenes faecalis, Staphylococcus aureus, Staphylococcus haemolyticus, and Proteus mirabilis	Murshid and Dhakshinamoorthy [74]
Sulfamethazine	$C_{12}H_{14}N_4O_2S$	*Microbacterium* sp.	Topp et al. [75], Billet et al. [76]
Mefenamic acid	$C_{15}H_{15}NO_2$	*A consortium of Alcaligenes faecalis, Staphylococcus aureus, Staphylococcus haemolyticus, and Proteus mirabilis*	Murshid and Dhakshinamoorthy [74]
		Phanerochaete sordida	Hata et al. [77]
Sulfamethoxazole	$C_{10}H_{11}N_3O_3S$	*Pseudomonas psychrophila HA-4*	Jiang et al. [78, 79]
Etonogestrel	$C_{22}H_{28}O_2$	*Cunninghamella blakesleeana*	Baydoun et al. [80]

(continued)

Table 2.2 (continued)

PhACs	Molecular structure	Microorganism	References
Indomethacin	$C_{19}H_{16}ClNO_4$	*unninghamella blakesleeana*	Zhang et al. [81]
Clofibric acid	$C_{10}H_{11}ClO_3$	*Streptomyces MIUG 4.89*	Popa Ungureanu et al. [82]
Norfloxacin	$C_{16}H_{18}FN_3O_3$	*Microbacterium* sp.	Kim et al. [83, 84]
		Labrys portucalensis F11	Amorim et al. [71]
Ketoprofen	$C_{16}H_{14}O_3$	*Raoultella ornithinolytica B6*	Ismail et al. [85]
		Pseudomonas aeruginosa	Tajani et al. [86]
		Pseudomonas sp. *S34*	Kim et al. [87]
Flurbiprofen	$C_{15}H_{13}FO_2$	*Streptomyces* sp.	Bright et al. [88]
Trimethoprim	$C_{14}H_{18}N_4O_3$	*Trametes versicolor*	Alharbi et al. [89]
6-Dehydroprogesterone	$C_{21}H_{28}O_2$	*Aspergillus niger*	Ahmad et al. [90]
		Gibberella fujikuroi	Ahmad et al. [90]
Atenolol	$C_{14}H_{22}N_2O_3$	*Trametes versicolor*	Marco-Urrea et al. [91]
Propanolol	$C_{16}H_{21}NO_2$	*Trametes versicolor*	Marco-Urrea et al. [91]
17 alpha-Ethynylestradiol (EE2)	$C_{20}H_{24}O_2$	*Rhodococcus zopfii*	Menashe et al. [92]
		Pseudomonas putida	Menashe et al. [92]
Ofloxacin	$C_{18}H_{20}FN_3O_4$	*Labrys portucalensis F11*	Amorim et al. [71]

the direct electron transfer between the microorganisms and electrically conductive pili, electron transfer by conductive materials present in the medium has been sown as the main mechanism for electron transfer in biological treatment systems [94]. Hence, the properties of the materials, such as electrical conductivity, are of essential importance. The specific surface area of the conductive materials added to the media is also of high significance due to its role in providing active sites for the adhesion and colonization of the microbial communities [95].

There are also analytical techniques, such as quantitative PCR[12] (qPCR) and DNA/RNA sequencing, which can be used to study the microbial communities (i.e., bacteria, archaea, and fungi) in the medium. Such studies can also aid in investigating the diversity of microbial communities and their population dynamics [96]. The qPCR (which utilizes fluorescent labeling) detects and quantifies nucleic

[12] Polymerase chain reaction (PCR) is a qualitative technique which allows acquiring the "presence or absence" results. while qPCR is used for the quantitative analysis. In fact, qPCR, quantifies the amount of DNA amplified in each cycle through repeating steps including: denaturation, annealing and elongation.

acids and can be used to quantify RNA transcripts (RT–qPCR). This is normally performed through the reverse transcription of RNA transcripts into complementary DNA (cDNA) followed by qPCR analysis.

However, qRT–PCR can only be used to detect known sequences. In contrast, RNA-Seq assists in detecting both known and novel transcripts. In this technique, next-generation sequencing (NGS) is used to identify the presence and quantify the RNA in cells. RNA-Seq provides possibilities to study the alternative gene spliced transcripts, the post-transcriptional modifications, gene fusion, mutations/SNPs, and changes in gene expression over time. It can also be used to identify the differences in gene expression between different groups [97]. Notably, in biological treatment processes such as CWs, the removal of PhACs is governed by a combination of various processes, including adsorption (either to the plants or to the base materials), biodegradation, photodegradation, and phytoremediation [98]. Hence, different analytical techniques may be required to identify the removal and transformation mechanisms of such emerging pollutants.

In addition to the detection and quantification of the mother pollutants, as well as degradation products, there are techniques that can help to better understand the efficiency of the (waste)water treatment methods applied for the removal of PhACs. For instance, techniques are available to identify the extent of mineralization of pharmaceutical compounds and degradation products formed using various biological and physico-chemical treatment methods. In this regard, the chemical oxygen demand (COD) and total organic carbon (TOC) content of the effluents after treatment can lead to conclusions about the efficiency of the applied AOPs for the removal of organic compounds. COD and TOC are currently the basis of the standards for the discharge of effluents that originate from industrial and nonindustrial activities in many countries all over the world [99]. However, such variables are not able to demonstrate the toxic nature of the treated effluents. To overcome these issues, advanced analytical methods discussed in Sect. 2.1 (such as LC–MS) have been well developed in recent years to assist in identifying the by-products of the treatments employed for the removal of pharmaceutical compounds.

It is also evident that effluents containing recalcitrant and nonbiodegradable compounds represent a high COD and a low biological oxygen demand (BOD_5) [100]. Hence, the biodegradability index (defined as BOD_5/COD) can be considered an important indirect measure indicating the degree of complexity or toxicity of the degradation products formed in the medium. In this regard, few studies have aimed to measure the biodegradability of pharmaceutical-containing effluents before and after treatment with AOPs. For instance, sonolysis at 520 kHz has been indicated as an efficient method for increasing the biodegradability of fluoroquinolone antibiotic (such as ciprofloxacin containing effluents from 0.06 to 0.60, 0.17, and 0.18 depending on the pH (3, 7, and 10, respectively) after 120 min of reaction time [101].

2.3.4 Toxicity Studies

Toxicity tests are also essential to validate the effectiveness of the (waste)water treatment methods applied for the removal of PhACs. In this regard, various assays have been developed that can be implemented for the assessment of the toxic nature of effluents containing these compounds before and after the treatment process.

Respirometry is among the most widely used methods to this end and involves measuring the rate at which the available substances in the medium are consumed by the biomass. There are various types of respirometric equipment ranging from simple bottles that are controlled manually to sophisticated instruments equipped with automated sensors and recording accessories [102]. The presence of toxic substances can impair the activity of microorganisms, which leads to a drop in the oxygen uptake rate (OUR) [24] (Fig. 2.6).

In addition to indirect toxicity studies such as respirometry, various direct toxicity assays have been developed to explore the toxic effects of effluents on indicator organisms (such as *Daphina magna*). Table 2.3 presents various toxicity assays used for testing the toxicity of the PhACs containing effluents before and after treatment with various biological and physico-chemical (waste)water treatment methods [104].

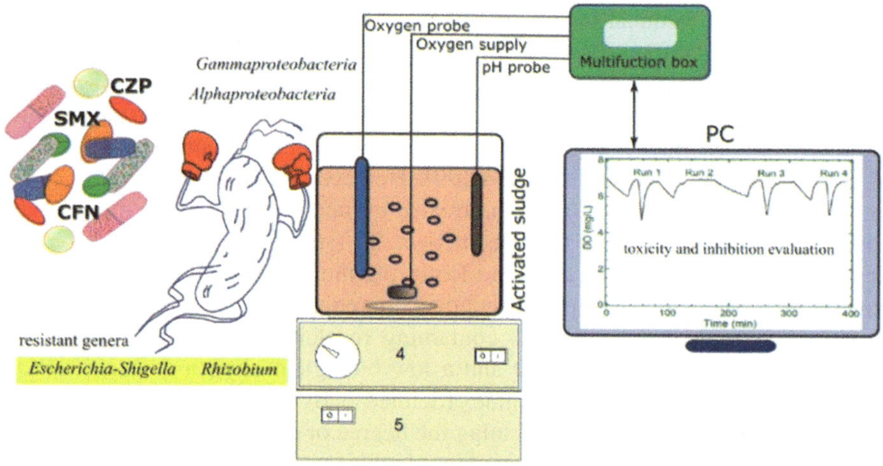

Fig. 2.6 Schematic illustration of an automated respirometric system used by Vasiliadou et al. [103] for the study of the toxicity of the PhACs

2.4 Further Reading

Table 2.4 contains items from the recent literature that the reader can consult for more detailed information regarding analytical methods for the detection and quantification of the PhACs and their decomposition products when (waste)water treatment technologies are applied for the removal of these compounds.

2.5 Summary

Various methods have been developed thus far to remove pharmaceutically active compounds from the content of polluted (waste)waters and to prevent the possible

Table 2.3 Summary of direct toxicity studies on pharmaceutical-containing effluent protocols and the observed results and remarks

Wastewater/applied method	Toxicity assay	Remarks	Referencs
Secondary sedimentation effluent/ferrate(VI)	Embryo toxicity assay (zebrafish embryo model)	Complete removal of the toxicity	Jiang et al. [105]
Hospital wastewater/solar photo-Fenton with Fe^{3+}-EDDS[a]	Silico (Q)SAR	Reduction in toxicity after treatment with the oxidation method	Cuervo Lumbaque et al. [106]
Pharmaceutical industrial/NA[b]	Dehydrogenase activity (DHA) and bioluminescent bacteria (*Vibrio qinghaiensis*) tests	Confirming a direct relationship between the toxicity and COD, TSS, TS, and TN	Ma et al. [107]
Municipal effluents/NA	Lethality and sublethal effects on *Hydra attenuata*, hydranth number, attachment, and ability to ingest prey	The five most common pharmaceuticals (ibuprofen, naproxen, bezafibrate, gemfibrozil, and carbamazepine) were found to be the most toxic to *H. attenuata* with and toxicity threshold as low as 320 μg/L. based	Quinn et al. [108]

(continued)

Table 2.3 (continued)

Wastewater/applied method	Toxicity assay	Remarks	Referencs
Equalization tank wastewater/anaerobic–aerobic treatment	Acute toxicity tests to *Vibrio fischeri* (*V. fischeri*) and *Daphnia magna* (D. magna)	The applied treatment resulted in a substantial reduction in the toxicity of the effluents Antibiotics, amoxicillin, cephalexin, ammonia nitrogen, and total phosphorus were considered as the main causes of the toxicity in the untreated effluents	Hu et al. [109]
Hospital wastewater/solar photocatalysis using TiO_2 nanomaterials	Bioluminescence inhibition in the *Vibrio fischeri* bacteria using DeltaTox® II toxicity analyzer	Reduction in the toxicity of the effluents after using the AOP	Pino-Sandoval et al. [110]
Hospital wastewater/biological treatment using the white-rot fungi, specifically *Trametes Versicolor*	*Daphnia magna* immobilization test and seed germination test on *Lactuca sativa*	Reduction in the toxicity of the effluents after treatment with the biological agents	Tormo-Budowski et al. [111]

[a]Ethylenediamine-N,N′-disuccinic acid
[b]No treatment was applied, and the study focused on the analysis of the toxicity of the effluents

Table 2.4 Further reading suggestions for more detailed coverage of the literature on various analytical techniques for the detection and quantification of the PhACs and their decomposition products

References	Subject
Williams [10]	Fundamentals of the mass spectrometry
Zhou [112]	HPLC for analysis of the PhACs
Blitz et al. [97]	Applications of deep sequencing
Mainardis et al. [104]	Microalgae strains used for the respirometric studies

environmental and health effects of their release, bioaccumulation, and possible biomagnification in the environment. The detection and quantification of PhACs are an essential step to determine the efficiency of the developed methods. Additionally, it would be of very high importance to elucidate the products that result from the decomposition of pharmaceuticals, their persistency, and possible effects when released into aquatic or terrestrial environments compared to those of the parent compounds. This chapter discussed the progress and the current status of analytical techniques developed to address the mentioned needs and the future perspectives in this regard.

References

1. Abushaheen MA et al (2020) Antimicrobial resistance, mechanisms and its clinical significance. Dis Mon 66:100971. https://doi.org/10.1016/j.disamonth.2020.100971
2. Zhou G et al (2015) The three bacterial lines of defense against antimicrobial agents. Int J Mol Sci 16:21711–21733. https://doi.org/10.3390/ijms160921711
3. Baillie TA (2008) Metabolism and toxicity of drugs. Two decades of progress in industrial drug metabolism. Chem Res Toxicol 21:129–137. https://doi.org/10.1021/tx7002273
4. Khan HK, Rehman MYA, Malik RN (2020) Fate and toxicity of pharmaceuticals in water environment: an insight on their occurrence in South Asia. J Environ Manage 271:111030. https://doi.org/10.1016/j.jenvman.2020.111030
5. Skoog DA, Holler FJ, Crouch SR (2007) Principles of instrumental analysis, 6th ed. Thomson Brooks/Cole
6. Ma Y et al (2018) Magnetic lignin-based carbon nanoparticles and the adsorption for removal of methyl orange. Coll Surf A: Physicochem Eng Aspects 559:226–234. https://doi.org/10.1016/j.colsurfa.2018.09.054
7. Ying Chin L et al (2015) Immobilization of nano-sized TiO_2 on glass plate for the removal of methyl orange and methylene blue. In: ICGSCE, pp 105–113. https://doi.org/10.1007/978-981-287-505-1_13
8. Trovó AG et al (2012) Paracetamol degradation intermediates and toxicity during photo-Fenton treatment using different iron species. Water Res 46:5374–5380. https://doi.org/10.1016/j.watres.2012.07.015
9. Gerber F et al (2004) Practical aspects of fast reversed-phase high-performance liquid chromatography using 3 μm particle packed columns and monolithic columns in pharmaceutical development and production working under current good manufacturing practice. J Chromatogr A 1036(2):127–133. https://doi.org/10.1016/j.chroma.2004.02.056
10. Williams DA (1997) Chapter 3: Fundamentals of mass spectrometry. Pharmacochemistry Library, 26, pp 19–45. https://doi.org/10.1016/S0165-7208(97)80150-X
11. da Silva FMA et al (2017) Positive electrospray ionization ion trap mass spectrometry and ab initio computational studies of the multi-pathway fragmentation of oxoaporphine alkaloids. Int J Mass Spectrom 418:30–36. https://doi.org/10.1016/j.ijms.2016.12.004
12. Sahora K, Fernández-del Castillo C (2015) Surgical management of mucinous cystic neoplastic lesions of the pancreas. In: Pancreatic cancer, cystic neoplasms and endocrine tumors: diagnosis and management, vol 24, pp 243–248. https://doi.org/10.1002/9781118307816.ch34
13. Abramović BF et al (2021) Experimental and computational study of hydrolysis and photolysis of antibiotic ceftriaxone: degradation kinetics, pathways, and toxicity. Sci Total Environ 768:144991. https://doi.org/10.1016/j.scitotenv.2021.144991
14. Llorca M et al (2015) Identification of new transformation products during enzymatic treatment of tetracycline and erythromycin antibiotics at laboratory scale by an on-line turbulent flow liquid-chromatography coupled to a high resolution mass spectrometer LTQ-Orbitrap. Chemosphere 119:90–98. https://doi.org/10.1016/j.chemosphere.2014.05.072
15. Somsen GW, Gooijer C, Brinkman UAT (1999) Liquid chromatography–Fourier-transform infrared spectrometry. J Chromatogr 856:213–242. https://doi.org/10.1016/B978-0-12-409547-2.14514-7

16. Balasubramani K, Sivarajasekar N, Naushad M (2020) Effective adsorption of antidiabetic pharmaceutical (metformin) from aqueous medium using graphene oxide nanoparticles: equilibrium and statistical modelling. J Mol Liq 301:112426. https://doi.org/10.1016/j.molliq.2019.112426
17. Maged A et al (2021) New mechanistic insight into rapid adsorption of pharmaceuticals from water utilizing activated biochar. Environ Res 202(July):111693. https://doi.org/10.1016/j.envres.2021.111693
18. Ho YS, McKay G (1998) A Comparison of chemisorption kinetic models applied to pollutant removal on various sorbents. Process Saf Environ Prot 76:332–340. https://doi.org/10.1205/095758298529696
19. De Andrade JR et al (2018) Adsorption of pharmaceuticals from water and wastewater using nonconventional low-cost materials: a review. Ind Eng Chem Res 57:3103–3127. https://doi.org/10.1021/acs.iecr.7b05137
20. Dolar D et al (2017) Adsorption of hydrophilic and hydrophobic pharmaceuticals on RO/NF membranes: identification of interactions using FTIR. J Appl Polym Sci 134:17–21. https://doi.org/10.1002/app.44426
21. Veclani D, Tolazzi M, Melchior A (2020) Molecular interpretation of pharmaceuticals' adsorption on carbon nanomaterials: theory meets experiments. Processes 8:1–39. https://doi.org/10.3390/PR8060642
22. Song W et al (2016) Experimental and theoretical evidence for competitive interactions of tetracycline and sulfamethazine with reduced graphene oxides. Environ Sci Nano 3:1318–1326. https://doi.org/10.1039/c6en00306k
23. Deniere E et al (2018) Advanced oxidation of pharmaceuticals by the ozone-activated peroxymonosulfate process: the role of different oxidative species. J Hazard Mater 360:204–213. https://doi.org/10.1016/j.jhazmat.2018.07.071
24. Zhang X et al (2022) Kinetics and mechanisms of the carbamazepine degradation in aqueous media using novel iodate-assisted photochemical and photocatalytic systems. Sci Total Environ 825:153871. https://doi.org/10.1016/j.scitotenv.2022.153871
25. Oller I, Malato S (2021) Photo-Fenton applied to the removal of pharmaceutical and other pollutants of emerging concern. Curr Opin Green Sustain Chem 29:100458. https://doi.org/10.1016/j.cogsc.2021.100458
26. Talwar S, Verma AK, Sangal VK (2021) Synergistic degradation employing photocatalysis and photo-Fenton process of real industrial pharmaceutical effluent utilizing the Iron-Titanium dioxide composite. Process Saf Environ Prot 146:564–576. https://doi.org/10.1016/j.psep.2020.11.029
27. MacKay D, Fraser A (2000) Bioaccumulation of persistent organic chemicals: mechanisms and models. Environ Pollut 110:375–391. https://doi.org/10.1016/S0269-7491(00)00162-7
28. Turner A, Williamson I (2005) Octanol-water partitioning of chemical constituents in river water and treated sewage effluent. Water Res 39:4325–4334. https://doi.org/10.1016/j.watres.2005.08.009
29. Avramiotis E et al (2021) Oxidation of sulfamethoxazole by rice husk biochar-activated persulfate. Catalysts 11:850. https://doi.org/10.3390/catal11070850
30. Luo T et al (2019) Sulfamethoxazole degradation by an Fe(ii)-activated persulfate process: insight into the reactive sites, product identification and degradation pathways. Environ Sci Process Impacts 21:1560–1569. https://doi.org/10.1039/c9em00254e
31. Beitz T, Bechmann W, Mitzner R (1998) Investigations of reactions of selected azaarenes with radicals in water. 1. Hydroxyl and sulfate radicals. J Phys Chem A 102:6760–6765. https://doi.org/10.1021/jp980654i
32. Czapski G (1999) Acidity of the carbonate radical. J Phys Chem A 103:3447–3450. https://doi.org/10.1021/jp984769y
33. Huie RE, Clifton CL, Neta P (1991) 'Electron transfer reaction rates and equilibria of the carbonate and sulfate radical anions. Int J Rad Appl Instrum 38:477–481. https://doi.org/10.1016/1359-0197(91)90065-A

34. Maruthamuthu P, Neta P (1977) Reactions of phosphate radicals with organic compounds. J Phys Chem:1622–1625
35. Shields BJ, Doyle AG (2016) Direct C(sp^3)-H cross coupling enabled by catalytic generation of chlorine radicals. J Am Chem Soc 138:12719–12722. https://doi.org/10.1021/jacs.6b08397
36. Zhang T, Zhu H, Croué JP (2013) Production of sulfate radical from peroxymonosulfate induced by a magnetically separable $CuFe_2O_4$ spinel in water: efficiency, stability, and mechanism. Environ Sci Technol 47:2784–2791. https://doi.org/10.1021/es304721g
37. Chadi NE, Merouani S, Hamdaoui O, Bouhelassa M, Ashokkumar M (2019) H_2O_2/periodate (IO_4-): a novel advanced oxidation technology for the degradation of refractory organic pollutants. Environ Sci Water Res Technol 5:1113–1123.https://doi.org/10.1039/c9ew00147f
38. Liu W et al (2019) Visible-light-driven photocatalytic degradation of diclofenac by carbon quantum dots modified porous g-C3N4: mechanisms, degradation pathway and DFT calculation. Water Res 151:8–19. https://doi.org/10.1016/j.watres.2018.11.084
39. Yan J et al (2018) Activation $CuFe_2O_4$ by hydroxylamine for oxidation of antibiotic sulfamethoxazole. Environ Sci Technol 52:14302–14310. https://doi.org/10.1021/acs.est.8b03340
40. Ng J, Wang X, Sun DD (2011) One-pot hydrothermal synthesis of a hierarchical nanofungus-like anatase TiO_2 thin film for photocatalytic oxidation of bisphenol A. Appl Catal B 110:260–272. https://doi.org/10.1016/j.apcatb.2011.09.011
41. Boulebd H et al (2021) Insights into the mechanisms and kinetics of the hydroperoxyl radical scavenging activity of Artepillin C. New J Chem 45:7774–7780. https://doi.org/10.1039/d1nj00666e
42. Teel AL, Watts RJ (2002) Degradation of carbon tetrachloride by modified Fenton's reagent. J Hazard Mater 94(2):179–189. https://doi.org/10.1016/S0304-3894(02)00068-7
43. Bilanda DC et al (2018) *Allablanckia floribunda* hypotensive activity on ethanol induced hypertension in rats. J Phytopharmacol 7:146–151. https://doi.org/10.31254/phyto.2018.7208
44. Rastogi A, Al-Abed SR, Dionysiou DD (2009) Sulfate radical-based ferrous-peroxymonosulfate oxidative system for PCBs degradation in aqueous and sediment systems. Appl Catal B 85:171–179. https://doi.org/10.1016/j.apcatb.2008.07.010
45. Raja A et al (2020) Visible active reduced graphene oxide-$BiVO_4$-ZnO ternary photocatalyst for efficient removal of ciprofloxacin. Sep Purif Technol 233(June 2019):115996. https://doi.org/10.1016/j.seppur.2019.115996
46. Yu X et al (2020) Advanced oxidation of benzalkonium chloride in aqueous media under ozone and ozone/UV systems—degradation kinetics and toxicity evaluation. Chem Eng J (Nov):127431. https://doi.org/10.1016/j.cej.2020.127431
47. Ramacharyulu PVRK et al (2018) Mechanistic insights into 4-nitrophenol degradation and benzyl alcohol oxidation pathways over MgO/g-C_3N_4 model catalyst systems. Catal Sci Technol 8:2825–2834. https://doi.org/10.1039/c8cy00431e
48. Schwarz HA (1981) Free radicals generated by radiolysis of aqueous solutions. J Chem Educ 58:101–105. https://doi.org/10.1021/ed058p101
49. Hasty N et al (1972) Role of azide in singlet oxygen reactions: reaction of azide with singlet oxygen. Tetrahedron Lett 13:49–52. https://doi.org/10.1016/S0040-4039(01)84236-2
50. Yi Q et al (2019) Singlet oxygen triggered by superoxide radicals in a molybdenum Cocatalytic fenton reaction with enhanced REDOX activity in the environment. Environ Sci Technol 53:9725–9733. https://doi.org/10.1021/acs.est.9b01676
51. Shang K et al (2022) Degradation of sulfamethoxazole (SMX) by water falling film DBD plasma/persulfate: reactive species identification and their role in SMX degradation. Chem Eng J 431:133916. https://doi.org/10.1016/j.cej.2021.133916
52. Zhou X et al (2021) Regulating activation pathway of Cu/persulfate through the incorporation of unreducible metal oxides: pivotal role of surface oxygen vacancies. Appl Catal B 286:119914. https://doi.org/10.1016/j.apcatb.2021.119914
53. Clifton CL, Huie RE (1993) Rate constants for some hydrogen abstraction reactions of the carbonate radical. Int J Chem Kinet 25:199–203. https://doi.org/10.1002/kin.550250308

54. Ôsawa Y, Saitô N (1968) Electron spin resonance studies on anthocyanins. Phytochemistry 7:1189–1195. https://doi.org/10.1016/S0031-9422(00)88269-2
55. Kempe S, Metz H, Mäder K (2010) Application of Electron Paramagnetic Resonance (EPR) spectroscopy and imaging in drug delivery research—chances and challenges. Eur J Pharm Biopharm 74:55–66. https://doi.org/10.1016/j.ejpb.2009.08.007
56. Gao HY et al (2020) First direct and unequivocal electron spin resonance spin-trapping evidence for pH-dependent production of hydroxyl radicals from sulfate radicals. Environ Sci Technol 54:14046–14056. https://doi.org/10.1021/acs.est.0c04410
57. Scott MJ, Billiar TR, Stoyanovsky DA (2016) N-tert-butylmethanimine N-oxide is an efficient spin-trapping probe for EPR analysis of glutathione thiyl radical. Sci Rep 6:1–11. https://doi.org/10.1038/srep38773
58. Guan YH et al (2011) Influence of pH on the formation of sulfate and hydroxyl radicals in the UV/Peroxymonosulfate system. Environ Sci Technol 45:9308–9314. https://doi.org/10.1021/es2017363
59. Hideg É, Spetea C, Vass I (1994) Singlet oxygen and free radical production during acceptor- and donor-side-induced photoinhibition. Studies with spin trapping EPR spectroscopy. BBA—Bioenergetics 1186:143–152. https://doi.org/10.1016/0005-2728(94)90173-2
60. Sellier A, Khaska S, Le Gal La Salle C (2022) Assessment of the occurrence of 455 pharmaceutical compounds in sludge according to their physical and chemical properties: a review. J Hazar Mater 426:128104. https://doi.org/10.1016/j.jhazmat.2021.128104
61. Salgado R et al (2012) Biodegradation of clofibric acid and identification of its metabolites. J Hazard Mater 241–242:182–189. https://doi.org/10.1016/j.jhazmat.2012.09.029
62. Yang SF et al (2011) Sorption and biodegradation of sulfonamide antibiotics by activated sludge: experimental assessment using batch data obtained under aerobic conditions. Water Res 45:3389–3397. https://doi.org/10.1016/j.watres.2011.03.052
63. Lin AYC et al (2010) Potential for biodegradation and sorption of acetaminophen, caffeine, propranolol and acebutolol in lab-scale aqueous environments. J Hazard Mater 183(1–3):242–250. https://doi.org/10.1016/j.jhazmat.2010.07.017
64. Rattier M et al (2012) Investigating the role of adsorption and biodegradation in the removal of organic micropollutants during biological activated carbon filtration of treated wastewater. J Water Reuse Desalin 2:127–139. https://doi.org/10.2166/wrd.2012.012
65. Yamamoto H et al (2009) Persistence and partitioning of eight selected pharmaceuticals in the aquatic environment: laboratory photolysis, biodegradation, and sorption experiments. Water Res 43(2):351–362. https://doi.org/10.1016/j.watres.2008.10.039
66. Rodríguez-Rodríguez CE, Marco-Urrea E, Caminal G (2010) Degradation of naproxen and carbamazepine in spiked sludge by slurry and solid-phase Trametes versicolor systems. Biores Technol 101:2259–2266. https://doi.org/10.1016/j.biortech.2009.11.089
67. Li X et al (2015) Removal of carbamazepine and naproxen by immobilized Phanerochaete chrysosporium under non-sterile condition. New Biotechnol 32:282–289. https://doi.org/10.1016/j.nbt.2015.01.003
68. Marco-Urrea E et al (2009) Ability of white-rot fungi to remove selected pharmaceuticals and identification of degradation products of ibuprofen by Trametes versicolor. Chemosphere 74:765–772. https://doi.org/10.1016/j.chemosphere.2008.10.040
69. Singh SK, Khajuria R, Kaur L (2017) Biodegradation of ciprofloxacin by white rot fungus *Pleurotus ostreatus*. 3 Biotech 7:1–8. https://doi.org/10.1007/s13205-017-0684-y
70. Pan L et al (2018) Study of ciprofloxacin biodegradation by a *Thermus* sp. isolated from pharmaceutical sludge. J Hazar Mater 343:59–67. https://doi.org/10.1016/j.jhazmat.2017.09.009
71. Amorim CL et al (2014) Biodegradation of ofloxacin, norfloxacin, and ciprofloxacin as single and mixed substrates by *Labrys portucalensis* F11. Appl Microbiol Biotechnol 98:3181–3190. https://doi.org/10.1007/s00253-013-5333-8
72. Lin B et al (2015) Characterization of cefalexin degradation capabilities of two Pseudomonas strains isolated from activated sludge. J Hazard Mater 282:158–164. https://doi.org/10.1016/j.jhazmat.2014.06.080

73. Osorio-Lozada A et al (2008) Biosynthesis of drug metabolites using microbes in hollow fiber cartridge reactors: case study of diclofenac metabolism by actinoplanes species. Drug Metab Dispos 36:234–240. https://doi.org/10.1124/dmd.107.019323
74. Murshid S, Dhakshinamoorthy GP (2019) Biodegradation of Sodium Diclofenac and Mefenamic acid: kinetic studies, identification of metabolites and analysis of enzyme activity. Int Biodeterior Biodegradation 144:104756. https://doi.org/10.1016/j.ibiod.2019.104756
75. Topp E et al (2013) Accelerated biodegradation of veterinary antibiotics in agricultural soil following long-term exposure, and isolation of a sulfamethazine-degrading *Microbacterium* sp. J Environ Qual 42:173–178. https://doi.org/10.2134/jeq2012.0162
76. Billet L et al (2021) Antibiotrophy: key function for antibiotic-resistant bacteria to colonize soils—case of sulfamethazine-degrading *Microbacterium* sp. C448. Front Microbiol 12:1–13. https://doi.org/10.3389/fmicb.2021.643087
77. Hata T et al (2010) Removal of diclofenac and mefenamic acid by the white rot fungus *Phanerochaete sordida* YK-624 and identification of their metabolites after fungal transformation. Biodegradation 21:681–689. https://doi.org/10.1007/s10532-010-9334-3
78. Jiang B et al (2012) Genome sequence of a cold-adaptable sulfamethoxazole-degrading bacterium, *Pseudomonas psychrophila* HA-4. J Bacteriol 194:5721–5721. https://doi.org/10.1128/JB.01377-12
79. Jiang B et al (2014) Biodegradation and metabolic pathway of sulfamethoxazole by *Pseudomonas psychrophila* HA-4, a newly isolated cold-adapted sulfamethoxazole-degrading bacterium. Appl Microbiol Biotechnol 98:4671–4681. https://doi.org/10.1007/s00253-013-5488-3
80. Baydoun E et al (2016) Microbial transformation of contraceptive drug etonogestrel into new metabolites with *Cunninghamella blakesleeana* and *Cunninghamella echinulata*. Steroids 115(016):56–61. https://doi.org/10.1016/j.steroids.2016.08.003
81. Zhang P et al (2006) Biotransformation of indomethacin by the fungus *Cunninghamella blakesleeana*. Acta Pharmacol Sin 27:1097–1102. https://doi.org/10.1111/j.1745-7254.2006.00350.x
82. Popa Ungureanu C, Favier L, Bahrim G (2016) Screening of soil bacteria as potential agents for drugs biodegradation: a case study with clofibric acid. J Chem Technol Biotechnol 91:1646–1653. https://doi.org/10.1002/jctb.4935
83. Kim DW et al (2011) Modification of norfloxacin by a *Microbacterium* sp. strain isolated from a wastewater treatment plant. Appl Environ Microbiol 77(17):6100–6108. https://doi.org/10.1128/AEM.00545-11
84. Kim DW et al (2013) Identification of the enzyme responsible for N-Acetylation of norfloxacin by *Microbacterium* sp. strain 4N2-2. Appl Environ Microbiol 79:314–321. https://doi.org/10.1128/AEM.02347-12
85. Ismail MM et al (2016) Biodegradation of ketoprofen using a microalgal–bacterial consortium. Biotech Lett 38:1493–1502. https://doi.org/10.1007/s10529-016-2145-9
86. Tajani AS et al (2021) Anti-quorum sensing potential of ketoprofen and its derivatives against *Pseudomonas aeruginosa*: insights to in silico and in vitro studies. Arch Microbiol 203(8):5123–5132. https://doi.org/10.1007/s00203-021-02499-w
87. Kim GJ et al (2002) Screening, production and properties of a stereospecific esterase from *Pseudomonas* sp. S34 with high selectivity to (S)-ketoprofen ethyl ester. J Mole Catalysis—B Enzymatic 17:29–38. https://doi.org/10.1016/S1381-1177(01)00077-7
88. Bright TV et al (2011) Bacterial production of hydroxylated and amidated metabolites of flurbiprofen. J Mol Catal B Enzym 72:116–121. https://doi.org/10.1016/j.molcatb.2011.05.008
89. Alharbi SK et al (2019) Degradation of diclofenac, trimethoprim, carbamazepine, and sulfamethoxazole by laccase from Trametes versicolor: transformation products and toxicity of treated effluent. Biocatal Biotransform 37:399–408. https://doi.org/10.1080/10242422.2019.1580268
90. Ahmad MS et al (2016) Biotransformation of 6-dehydroprogesterone with *Aspergillus niger* and *Gibberella fujikuroi*. Steroids 112:62–67. https://doi.org/10.1016/j.steroids.2016.04.008

91. Marco-Urrea E et al (2010) Oxidation of atenolol, propranolol, carbamazepine and clofibric acid by a biological Fenton-like system mediated by the white-rot fungus Trametes versicolor. Water Res 44:521–532. https://doi.org/10.1016/j.watres.2009.09.049
92. Menashe O et al (2020) Biodegradation of the endocrine-disrupting chemical 17α-ethynylestradiol (EE2) by *Rhodococcus zopfii* and *Pseudomonas putida* encapsulated in small bioreactor platform (SBP) capsules. Appl Sci (Switzerland) 10:336. https://doi.org/10.3390/app10010336
93. Zhu X et al (2017) Effects and mechanisms of biochar-microbe interactions in soil improvement and pollution remediation: a review. Environ Pollut 227:98–115. https://doi.org/10.1016/j.envpol.2017.04.032
94. Pan J et al (2019) Enhanced methane production and syntrophic connection between microorganisms during semi-continuous anaerobic digestion of chicken manure by adding biochar. J Clean Prod 240:118178. https://doi.org/10.1016/j.jclepro.2019.118178
95. Baniamerian H et al (2019) Application of nano-structured materials in anaerobic digestion: current status and perspectives. Chemosphere 229:188–199. https://doi.org/10.1016/j.chemosphere.2019.04.193
96. Gallardo-Altamirano MJ et al (2021) Insights into the removal of pharmaceutically active compounds from sewage sludge by two-stage mesophilic anaerobic digestion. Sci Total Environ 789:147869. https://doi.org/10.1016/j.scitotenv.2021.147869
97. Blitz IL et al (2014) Applications of deep sequencing to developmental systems. In: Principles of developmental genetics, 2nd ed. Elsevier Inc. https://doi.org/10.1016/B978-0-12-405945-0.00002-8
98. Mannina G (2017) Erratum: energy recovery from immobilised cells of *Scenedesmus obliquus* after wastewater treatment (Lecture notes in civil engineering, 4, 2017, pp 266–271, https://doi.org/10.1007/978-3-319-58421-8_42), (Lecture notes in civil engineering, https://doi.org/10.1007/978-3-319-58421-8_116)
99. Ren G et al (2020) A novel stacked flow-through electro-Fenton reactor as decentralized system for the simultaneous removal of pollutants (COD, NH3-N and TP) and disinfection from domestic sewage containing chloride ions. Chem Eng J 387:124037. https://doi.org/10.1016/j.cej.2020.124037
100. Kamali M, Khodaparast Z (2015) Review on recent developments on pulp and paper mill wastewater treatment. Ecotoxicol Environ Saf 114:326–342. https://doi.org/10.1016/j.ecoenv.2014.05.005
101. De Bel E et al (2009) Influence of pH on the sonolysis of ciprofloxacin: biodegradability, ecotoxicity and antibiotic activity of its degradation products. Chemosphere 77:291–295. https://doi.org/10.1016/j.chemosphere.2009.07.033
102. Khashij M, Mehralian M, Goodarzvand Chegini Z (2020) Degradation of acetaminophen (ACT) by ozone/persulfate oxidation process: experimental and degradation pathways. Pigm Resin Technol 49:363–368. https://doi.org/10.1108/PRT-11-2019-0107
103. Vasiliadou IA et al (2018) Toxicity assessment of pharmaceutical compounds on mixed culture from activated sludge using respirometric technique: the role of microbial community structure. Sci Total Environ 630:809–819. https://doi.org/10.1016/j.scitotenv.2018.02.095
104. Mainardis M et al (2021) Respirometry tests in wastewater treatment: why and how? a critical review. Sci Tot Environ 793:148607. https://doi.org/10.1016/j.scitotenv.2021.148607
105. Jiang JQ et al (2013) Pharmaceutical removal from wastewater by ferrate(VI) and preliminary effluent toxicity assessments by the zebrafish embryo model. Microchem J 110:239–245. https://doi.org/10.1016/j.microc.2013.04.002
106. Cuervo Lumbaque E et al (2021) Removal of pharmaceuticals in hospital wastewater by solar photo-Fenton with Fe^{3+}-EDDS using a pilot raceway pond reactor: transformation products and in silico toxicity assessment. Microchem J 164:106014. https://doi.org/10.1016/j.microc.2021.106014
107. Ma K et al (2016) Toxicity evaluation of wastewater collected at different treatment stages from a pharmaceutical industrial park wastewater treatment plant. Chemosphere 158:163–170. https://doi.org/10.1016/j.chemosphere.2016.05.052

108. Quinn B, Gagné F, Blaise C (2008) An investigation into the acute and chronic toxicity of eleven pharmaceuticals (and their solvents) found in wastewater effluent on the cnidarian, Hydra attenuata. Sci Tot Environ 389:306–314. https://doi.org/10.1016/j.scitotenv.2007.08.038
109. Hu Y et al (2022) Residual β-lactam antibiotics and ecotoxicity to *Vibrio fischeri*, *Daphnia magna* of pharmaceutical wastewater in the treatment process. J Hazard Mater 425:127840
110. Pino-Sandoval DA et al (2022) Solar photocatalysis for degradation of pharmaceuticals in hospital wastewater: influence of the type of catalyst aqueous matrix, and toxicity evaluation. Water, Air Soil Poll 233:14. https://doi.org/10.1007/s11270-021-05484-7
111. Tormo-Budowski R et al (2021) Removal of pharmaceuticals and ecotoxicological changes in wastewater using Trametes versicolor: a comparison of fungal stirred tank and trickle-bed bioreactors. Chem Eng J 410:128210. https://doi.org/10.1016/j.cej.2020.128210
112. Zhou LZ (2005) Handbook of pharmaceutical analysis by HPLC. Separation science and technology. http://www.sciencedirect.com/science/article/pii/S0149639505800637

Chapter 3
Removal of Pharmaceutically Active Compounds in Water Bodies—Science History and Research Hotspots

3.1 Introduction

The presence of pharmaceutically active compounds (PhACs) in water bodies has become a concern for the scientific community, considering their potential impacts on the ecosystem balance, living organisms, and human health [1, 2]. This has triggered huge efforts all over the world to develop efficient technologies to deal with these compounds and to prevent increasing their environmental concentrations. Such technologies can be divided into biological and physico-chemical treatments, which have demonstrated different efficiencies for the removal of PhACs. In this stage, it would be of high importance to explore the research statistics in this field, the scientific progress, the current trends, and the existing gaps for future studies [3].

Scientometric analysis has been frequently employed to map scientific progress in numerous fields [4–6]. Such studies provide an overall view of science philosophy by exploring science history, the contributions of various parties (i.e., countries, organizations, and authors), and past and existing trends in a specific scientific area [4–8]. The state-of-the-art information provided by scientometric studies can aid in identifying the existing gaps and recommendations for future studies to overcome the barriers to the implementation of science and technology [9]. Various tools and algorithms have been used for scientometric analysis, such as CiteSpace [10], VOS viewer [11], Histcite [12], Pajek [13], and ScientoPy [6], to visualize bibliometric networks. A combination of these tools can also be used to present a more detailed analysis of the literature.

This chapter aims to provide an overview of the literature using scientometric indicators. A combination of CiteSpace and ScientoPy toll has been employed to reach this goal. The results of this chapter have been used as a basis for further discussions in the following chapters. The outcomes can also provide robust insights for decision-makers to approach more sustainable technologies for the elimination of PhACs and for research bodies to orient their efforts to address the existing issues in the development and implementation of viable (waste)water treatment technologies.

M. Kamali et al., *Advanced Wastewater Treatment Technologies for the Removal of Pharmaceutically Active Compounds*, Green Energy and Technology,
https://doi.org/10.1007/978-3-031-20806-5_3

Table 3.1 The keywords used for the advanced search in WoS for the technologies developed thus far for the treatment of pharmaceutically active compounds[1]

Set	Search field	Keywords	Results[2]
#1	Title	treat*[3] or removal or degrad* or adsor* or absor* or elimin* or puri* or desalin* or *decontamin*	2,921,208
#2	Topic[4]	pharmaceutical*	233,526
#3	Title	*water* or Aqu* or synthetic or effluent* or solution	2,153,884
#4	Summary	Operator: AND (#1 AND #3 AND #4)	6196

3.2 Methodology

To achieve the goals of this chapter, an advanced search was conducted in the Web of Science (WOS) core collection database using the set of keywords mentioned in Table 3.1. The keywords were selected based on a primary literature review, and a set of keywords was designed to cover all the possibilities in terms of the documents published on the development and application of various methods for the elimination of PhACs.

Various databases, including Google Scholar, Scopus, and WoS, can be used for retrieving scientific documents in a specific scientific area [14]. The selection of the WoS for further analysis was based on the fact that this database covers all the valid documents published with a high degree of reliability [14].

Analysis of the retrieved documents included the following steps:

(a) Screening of the published documents to polish the databank and to ensure the accuracy and relevancy of the retrieved documents.
(b) Creation of the marked list in WoS to export the retrieved documents with the required format for analysis using ScientoPy (Tab delimited file) and CiteSpace (6.0. R1) (Plain text).
(c) Analysis of the exported database regarding the scientometric criteria, including the "document status", "keyword analysis", "countries/organizations", "authors", and "publication sources". Some analysis was also performed directly using WoS, including the "publication year", "authors", and "subject categories".

Discussions have also been provided in this chapter to identify the current trends and future perspectives of wastewater treatment methods for the elimination of such compounds.

[1] Analysis was performed on March 1, 2022.

[2] The number of the published documents in WoS with this set of keywords.

[3] The sign "*" is to include all the possible continuations.

[4] Covering the keywords which appear in the title, abstract, and keyword of the published documents.

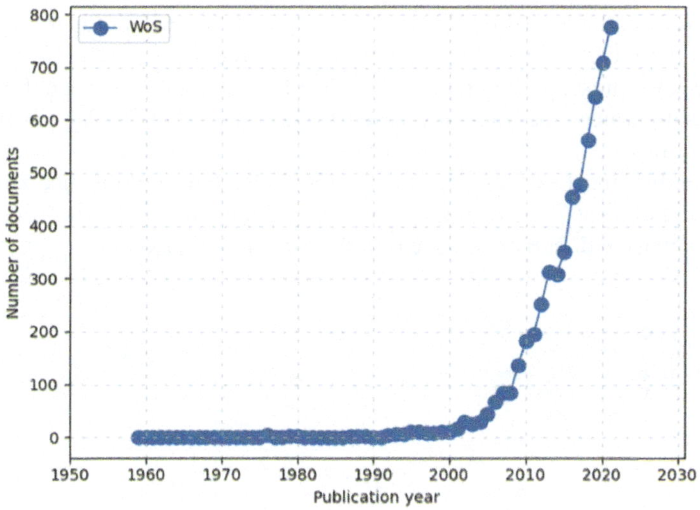

Fig. 3.1 The number of published documents per year on the wastewater treatment method for the elimination of PhACs. As seen in this figure, publications in this field have been initiated since the 1950s and have accelerated since 2000. There has also been a sharp increase in the number of publications in this field in recent years

3.3 Results and Discussion

3.3.1 Research Statistics

The number of published documents in this field is indicated in Fig. 3.1. As seen in this figure, publications in WoS on the application of various methods for the removal of PhACs have been initiated since the 1950s. The initial publications in this field mainly focus on the treatment of wastewater from the pharmaceutical industry [15]. As can be observed in this figure, the main milestone in the number of publications occurred around 2005, and a sharp increase in this parameter can be observed, especially after 2010, when it reached approximately 800 documents in 2021. The figure also represents an exponential growth trend, indicating that the research in this field has not yet been saturated [16], and there is still an increasing need for research in this area.

Figure 3.2 indicates that articles share 84.9% of all the published documents in this scientific field. Other publication types, including reviews, proceeding papers, book chapters, and data papers, also shared 7.4%, 6.4%, 1%, and > 1% of all the documents, respectively. It is also worth mentioning that the review papers share a relatively high percentage of all the documents published in this literature.[5] This is probably due to the need for critical conclusions regarding the applicability of various

[5] In most cases, this value has been observed less than 2% (see, e.g., [3, 17]).

(waste)water treatment technologies to deal with PhACs to minimize the possible impacts of these compounds on the environment and human health [18, 19].

Figure 3.3 represents the contribution of different countries in publications on wastewater treatment methods developed to deal with PhACs. As can be clearly observed in this figure, China, with 1143 documents, is the leading country in this regard. Notably, most of these documents (56%) have been published since 2018. The United States (667 documents), Spain (628 documents), India (461 documents), and Germany (346 documents) have occupied the next ranks.

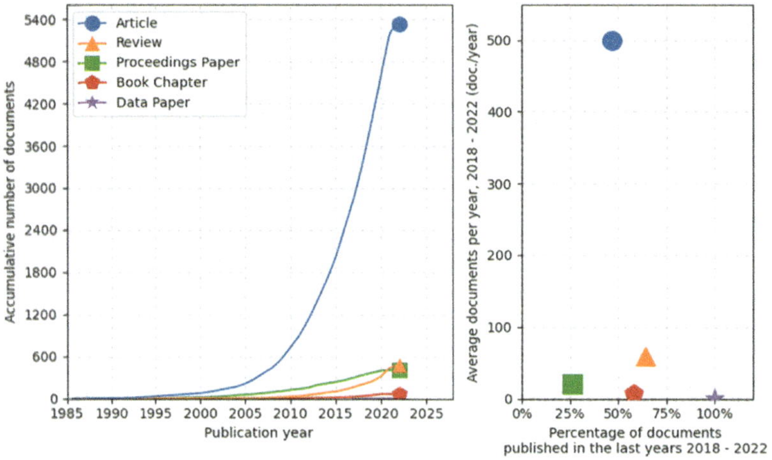

Fig. 3.2 Various types of documents (and their relative shares) published on the removal of PhACs and the respective evaluation trends. The analysis was performed using the ScientoPy tool

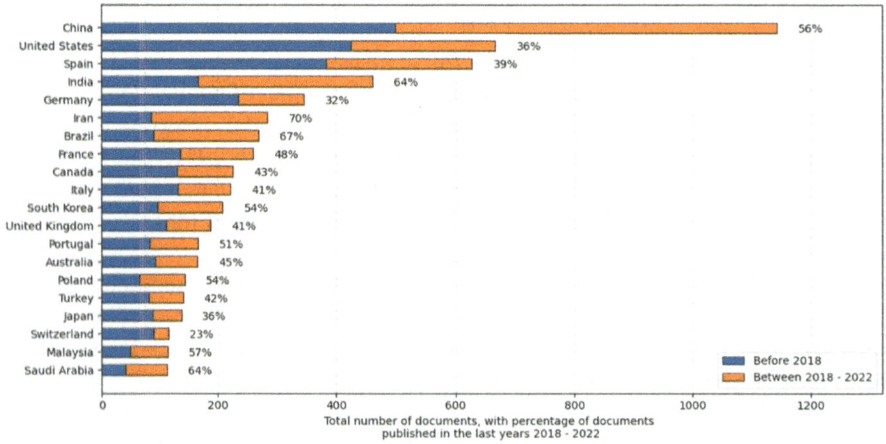

Fig. 3.3 Contribution of various countries to publications on wastewater treatment methods for PhACs. The analysis was performed using the ScientoPy tool

Notably, the number of published documents in China accelerated after 2018, as shown in Fig. 3.3, with 56% of all the documents published by this country on this subject. Very fast progress in the number of documents can also be observed in countries such as Iran, Brazil, and India by 70%, 67%, and 64% of all the documents published in these countries, respectively. This is probably because such countries have recently faced severe environmental issues related to the generation and discharge of large amounts of polluted (waste)waters and the need for the development of efficient and cost-effective treatment technologies in these areas of the world [20]. It has also been previously discussed in the literature that the rapid progress of China in recent years is due to the successful development plans that have been implemented in this country, such as "special economic zones of the People's Republic of China" [21] and "economic and technological development zones" [22]. Attracting foreign investments has also been among the goals established by China to accelerate the development of science and technology in this country.

Figure 3.4 demonstrates the contribution of various countries by focusing on the existing collaborations among various countries that are active in the publications on the development of (waste)water treatment methods for the removal of PhACs. The network contains both nodes and links. Each node refers to a country, and the links visually connect the countries to illustrate the pattern of cooperation among various countries [7]. As seen in this figure, countries such as China, the US, and Spain have already established a very strong collaboration with other countries to develop science in this field, but countries such as Japan, Sweden, and Greece are still not in such strong connections with other countries all over the world. Hence, it can be assumed that establishing strong collaborations with pioneer countries can considerably help them implement sustainable technologies to prevent water pollution by PhACs.

There are also institutions all over the world that have actively participated in the production of scientific documents on (waste)water treatment technologies for the removal of PhACs, as listed in Fig. 3.5. According to this figure, 5 Chinese institutions are among the most active bodies in this area, which demonstrates the importance of generating science and knowledge in this area in rapidly developing countries such as China. Four Spanish institutions have also been placed among the top 15 in this field, followed by 2 institutions in Portugal and Australia.

Regarding the publishing sources, the analysis was performed on the retrieved documents using the ScientoPy tool. The results are presented in Fig. 3.6. According to the results achieved, "Science of the Total Environment", "Water Research", and "Chemosphere", with 373, 364, and 300 documents, respectively, are the most active journals for the application of various (waste)water treatments for the elimination of PhACs.

According to the data presented in Table 3.2, "Water research" has received the highest number of citations for the published documents, with an average citation per item (ACI) of approximately 118. It is also worth noting that "Environmental Science and Technology" has received the highest ACI value of approximately 129, which can be considered an indicator for the quality of the published documents based on the attention of the scientific community they have attracted. Hence, ACI

Fig. 3.4 The contributions of various countries all over the world and their cooperation in the production of scientific documents on the application of (waste)water treatment technologies for the removal of PhACs were analyzed using the CiteSpace tool

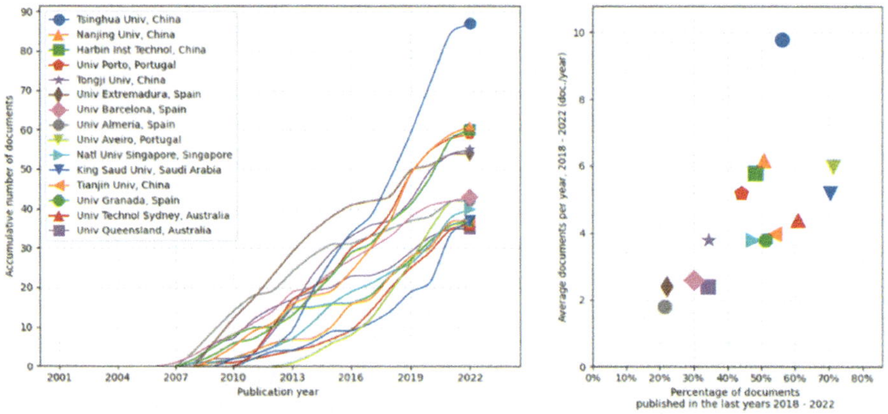

Fig. 3.5 Contribution of various institutions throughout the world to the production of scientific documents on wastewater treatment technologies for the elimination of PhACs. The analysis was performed using the ScientoPy tool

should also be considered an important parameter when analyzing the contribution of a journal for the publication of scientific documents.

The contribution of the authors in the relevant publications was also analyzed using ScientoPy and CiteSpace tools. Figure 3.7 represents the contribution of the authors in terms of the frequency of the documents they have coauthored. As seen

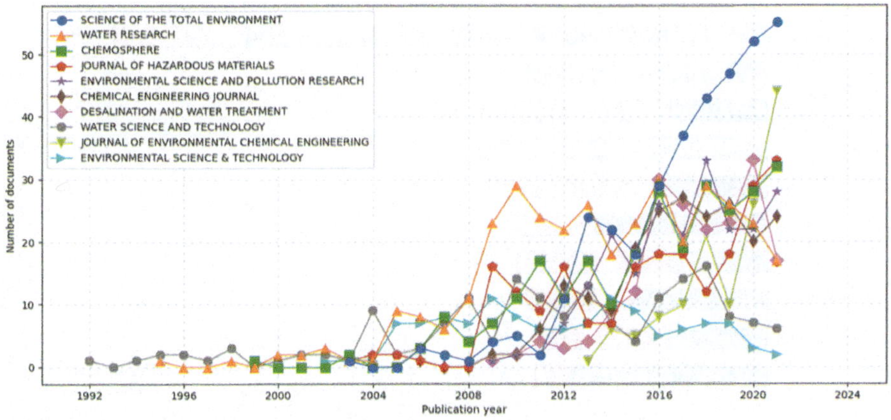

Fig. 3.6 Analysis of the sources active in publishing the scientific documents on the development of (waste)water treatment methods for the removal of PhACs. The analysis was performed using the ScientoPy tool on the data retrieved from WoS

Table 3.2 Contribution of the scientific journals to the publication of scientific documents on the application of various (waste)water treatment methods for the removal of PhACs

Rank	Source		Frequency	Time cited	ACI[a]
	Full title	ISSN			
1	Science of the Total Environment	0048-9697	373	23,368	65.65
2	Water Research	0043-1354	364	42,870	117.77
3	Chemosphere	0045-6535	300	15,184	50.61
4	Environmental Science and Pollution Research	09441344	244	4778	19.58
5	Journal of Hazardous Materials	03043894	228	15,052	66.02
6	Chemical Engineering Journal	13858947	215	13,425	62.44
7	Desalination and Water Treatment	19443986	187	1692	9.05
8	Water Science and Technology	02731223	172	3837	22.31
9	Environmental Science and Technology	10643389	121	15,586	128.81
10	Journal of Environmental Chemical Engineering	22133437	114	1383	12.13

[a]Average citations per item

in this figure, Barcelo, D.[6] with 78 documents (see, e.g., Mir-Tutusaus et al. [23],

[6] CSIC—Instituto de Diagnostico Ambiental y Estudios del Agua (IDAEA), Barcelona, Spain.

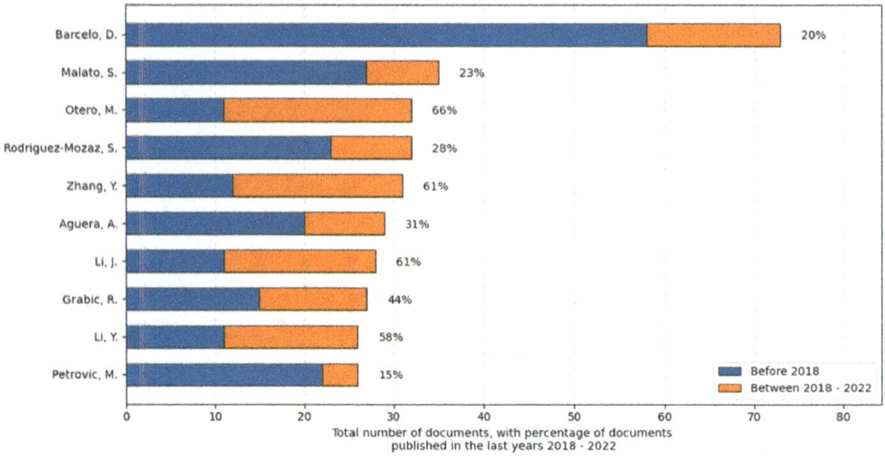

Fig. 3.7 Contribution of authors in publications on wastewater treatment methods for PhACs. The figure also includes the number of published documents since 2018. The analysis was performed using the ScientoPy tool

Malato, S.[7] with 35 documents (see, e.g., Mir-Tutusaus et al. [23], and Rodriguez-Mozaz, S.[8] (see, e.g., Cibati et al. [24] have been recognized as the most active authors for publications on (waste)water treatment methods for PhACs. Barcelo, D., has received 8502 citations for the published documents, with a relatively high average citation per item of 109 ($h = 46$). Furthermore, Otero, M. [25] has been the most active author since 2018, with 66% of all the published documents by this author (Fig. 3.7).

3.3.2 *Research Trends and Hotspots*

The research trends and hotspots in this field were identified using variables such as category (subject) analysis, keyword analysis, and exploring the "Hot Papers" and "Highly Cited Documents" in WoS. The results of the subject analysis are shown in Fig. 3.8.

As shown in Fig. 3.8, most of the studies in this field can be classified under the categories of "Environmental Science", "Environmental Engineering", "water Resources", and "Chemical Engineering" with 3153, 1878, 1204, and 1184 documents, respectively. Other categories, such as "Materials Science" and "Toxicology", can also be seen in this figure, which indicates the multidisciplinary nature of the research related to the removal of the PhACs from the (waste)waters.

[7] Centro de Investigaciones Energeticas, Medioambientales Tecnologicas Ctra Senes km 4, Almeira, Spain.

[8] Universidade de Aveiro, Aveiro, Portugal.

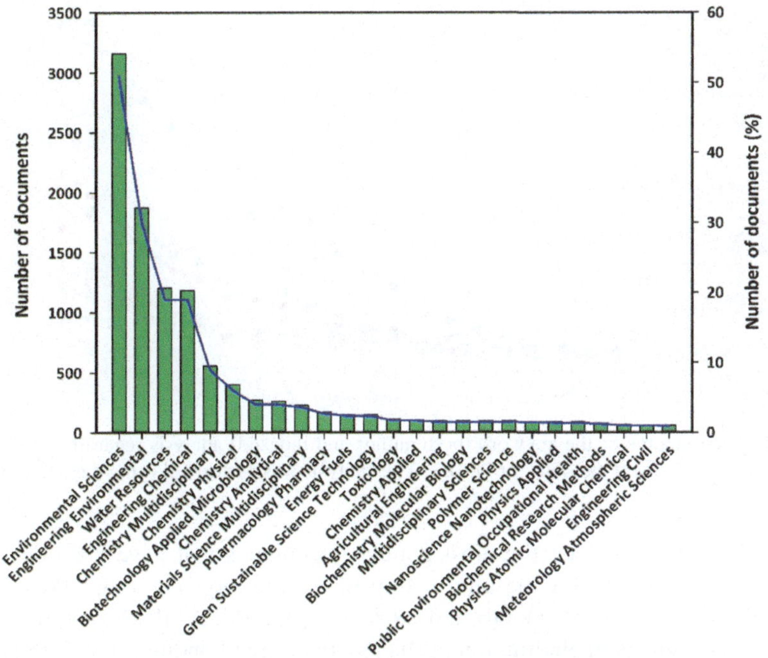

Fig. 3.8 The outcome of the category analysis regarding publications on (waste)water treatment methods for PhACs. The analysis was performed using WoS (retrieved 22/03/2022)

Figure 3.9 represents the keywords that have been used by the relevant publications on the application of various (waste)water treatment technologies for the removal of PhACs. In addition to the general keywords related to the PhACs such as "Pharmaceutical", "Personal Care Products", and "Wastewater", there are keywords that have been frequently used in the relevant publications regarding the methods developed for the removal of PhACs from the (waste)waters. "Fate" is among the most widely used keywords reflecting the concerns that exist among the scientific communities to explore the final destination of PhACs and the mechanisms involved in their behavior and possible transformation in the environment [26]. Existing studies have discussed the possibility of direct toxic effects of PhACs on living organisms, as well as the generation of antibiotic-resistant bacteria (ARBs) or antibiotic resistance genes (ARGs) [27–30]. This is mainly because most conventional (waste)water treatment technologies have not been designed for the elimination of PhACs, ABRs, or ARGs. To minimize such risks, there have been efforts to optimize various biological methods, such as activated sludge (AS) [31] and anaerobic digestion (AD) [32], and their state-of-the-art generations, such as constructed wetlands (CWs) [33] and microbial fuel cells (MFCs) [34]. The applicability of such methods and the recent progress toward their optimization toward the efficient removal of PhACs have been discussed in various chapters of the present book.

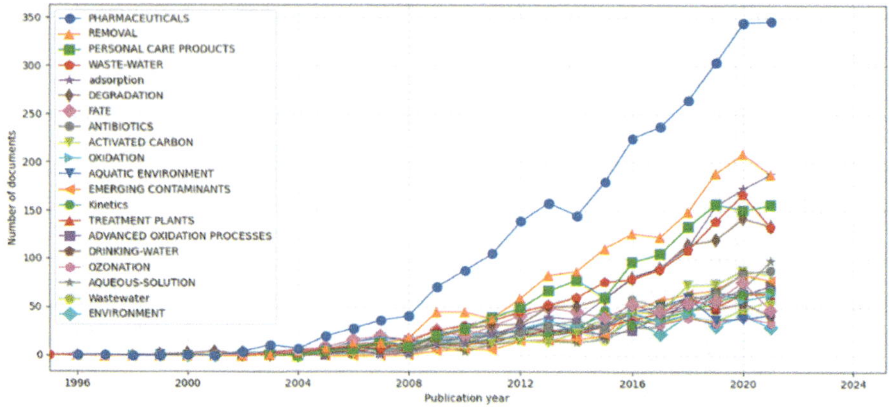

Fig. 3.9 The outcome of the keyword (both author and indexed) analysis regarding publications on wastewater treatment methods for PhACs. The analysis was performed using the ScientoPy tool

"Adsorption" is among the most popular techniques that have been developed and employed rapidly for the elimination of various organics (e.g., PhACs) and inorganics (such as heavy metals and nutrients) [35, 36]. As discussed in Chap. 9, various mechanisms, including ion exchange, hydrogen bonding, $\pi-\pi$ interactions, hydrophobic interactions, and electrostatic interactions, have been detected for the application of various adsorbents for the elimination of PhACs [37, 38]. The latest trend in this field is to develop and apply low-cost and efficient adsorbents such as clay-based materials (e.g., bentonite [39]), carbonaceous materials such as activated carbon (which appears among the most frequently used keywords, Fig. 3.9), and, more recently, biochar, which can provide large specific surface areas for the attraction of adsorbates [40]. Optimization of the properties of such novel adsorbents by, for instance, introducing surface functional groups, is a current trend in the literature for promoting adsorption processes for the removal of PhACs [41–43]. There have also been trends in the literature for the development of chemical-based treatment methods such as advanced oxidation processes (AOPs) for the efficient removal of PhACs from the containing (waste)waters. AOPs are based on the generation of active species (including radicals, superoxides, singlet oxygen, etc.) which can attack and decompose complex organic compounds. Ozonation [44], activation of oxidation agents such as persulfate, iodine, and chlorine [45, 46], light-assisted H-AOPs[9] [47], and those based on the application of electricity [48], heat [49], ultrasound [50], and microwaves [51] are examples of AOPs that have been used thus far for the removal of PhACs. Analysis of the keywords was also performed using the CiteSpace tool to identify the trends and hotspots in this field. The results are presented in Fig. 3.10. According to this figure, research in this field was initiated with the

[9] Including photolysis, and the photocatalytic processes using either ultraviolet or visible-light irradiation.

identification of the "mechanisms" involved in the removal of PhACs at the beginning of the 1990s. "Biosorption", as the main mechanism involved in the removal of PhACs using biological treatment methods [52, 53], appeared in 1998 to describe the behavior of some PhACs when introduced into biological wastewater treatment systems. Attention has also been paid after the 2000s to physico-chemical methods for the removal of such compounds, which resulted in keywords such as "adsorption" in 2001, "membrane" separation in 2003, "photolysis", and "ozonation" after 2004, as well as novel AOPs such as "Fenton oxidation" in 2015, "catalysis" since 2018, and more recently advanced materials for the adsorption of PhACs such as "graphene-based materials" highlighted in the literature after 2019. Figure 3.10 also demonstrates the current trends in this field, including in-depth studies for the identification of the "mechanisms" involved in the removal of PhACs, especially using various types of AOPs. Sustainable and low-cost approaches, such as the application of "visible light" for the activation of novel catalysts (such as graphitic carbon nitride and g-C_3N_4) [54, 55], are also among the current trends in this field. There have also been attempts to develop "facile" and cost-effective methods for the fabrication of efficient photocatalysts for the degradation of such compounds.

Table 3.3 also represents the topics of the "Hot Papers" published in WoS on the development of (waste)water treatments to eliminate PhACs. This table also confirms that the development of efficient adsorbents and catalysts is at the top of the attention of the scientific community.

There has also been a trend in the literature for the development of "sustainable" wastewater treatment technologies for the removal of PhACs to fulfill the United Nations Sustainable Development Goals (SDGs) [62].[10] To identify the respective documents, an advanced search in WoS was also performed by the inclusion of the keyword related to sustainability (i.e., "sustainab*" in Topics) with those mentioned in Table 3.4. According to the results achieved, 231 documents concerned the sustainability aspects of wastewater treatment methods for the elimination of PhACs. The results of the precise analysis of these documents are presented in Table 2.3.

3.4 Further Reading

Table 3.5 contains items from the recent literature that the reader can consult for more detailed information regarding the history and science background of various technologies for the removal of PhACs.

[10] Especially SDG 6: clean water and sanitation.

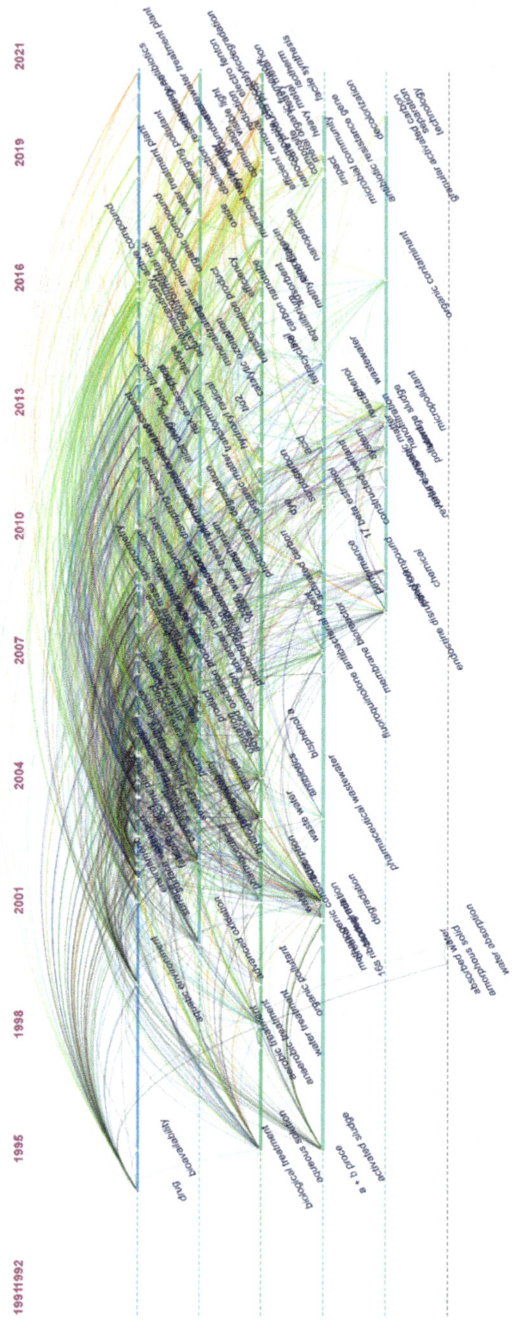

Fig. 3.10 The timeline of the evolution of the keywords in the scientific documents published on the application of various (waste)water treatment methods for the removal of PhACs. The analysis was performed using CiteSpace on the data retrieved from WoS (22/3/2022)

Table 3.3 The topics of the "Hot Papers" in WoS, published on the removal of PhACs from the polluted (waste)waters

Authors	Topic	Citations[11]
Oba et al. [56]	Adsorption of PhACs (ibuprofen) using efficient and low-cost adsorbents	31
Tran et al. [57]	Adsorption of PhACs with carbonaceous materials (spherical biochar)	81
Ahmed et al. [58]	Physico-chemical, biological, and combined treatments for the removal of PhACs	29
Giannakis et al. [59]	Sulfate radical-based Advanced Oxidation Processes for the degradation of PhACs	147
Wang et al. [60]	Presence and fate of PhACs and antibiotic-resistant genes (ARGs) and antibiotic-resistant bacteria (ARB) in municipal wastewater treatment plants	134
Isari et al. [61]	Visible-light photocatalysis and sono-photocatalysis for the degradation of PhACs	67

3.5 Summary

The present chapter has aimed to explore the scientific progress in the development of (waste)water treatment technologies for the elimination of PhACs. Various scientometric criteria, such as publication history, publication types, contributions (i.e., countries, organizations, journals, and authors), and trend analysis (i.e., using the evolution of the keywords as well as exploring hot- and highly cited papers), were employed to map the progress of science and technology in this field. The results indicated that a total of 6197 documents have been published on this topic since the 1950s, with an increasing trend after the 2000s. The study also indicated that the development of low-cost and sustainable materials to be used as adsorbents for the removal of PhACs or as catalysts (e.g., visible-light active photocatalysts) is among the current trends in this field. There are also efforts in progress to identify the mechanisms involved in the degradation of PhACs using novel AOPs and the fate of the PhACs and their degradation products to the environment.

[11] Based on the citations received in WoS.

Table 3.4 The summary of the documents concerned the sustainability aspects in wastewater treatment methods for the elimination of PhACs

Technology	Remarks/References
Adsorption	– Waste-derived materials such as biochar [63], hydrochar [64], and activated carbon [65] are highly recommended for the efficient and cost-effective removal of PhACs – Clay-based materials are low-cost and efficient adsorbents to satisfy the sustainability considerations [66–68]
Advanced oxidation processes (gamma irradiation, GA)	– Gamma irradiation is a sustainable technology for the elimination of PhACs, especially when combined with nanofiltration [69] – Strategies such as immobilization of the biological strong oxidizing agents (i.e., enzymes) on nanostructured supports [70] can offer highly efficient and low-cost techniques for the degradation of PhACs, especially using novel techniques such as 3D printing [71] – Biosynthesis approaches can considerably lead to the production of sustainable nanocatalysts for the removal of PhACs [72] – Sulfate-based radicals (such as UV-PDS[12]) are low-cost and efficient for the decomposition of PhACs [73] – Magnetization creates the possibility of recovery and reuse of photocatalytic materials [74] – Visible-light active materials are currently under the spotlight for the degradation of a wide range of organic pollutants, especially PhACs [75] – Combination of AOPs with solar irradiation (such as solar-Fenton systems [76]) can enhance the efficiency and reduce the costs of these methods toward the sustainable removal of PhACs
Biological treatments	– Bioaugmentation can be considered as a sustainable biological treatments satisfying the economic and environmental considerations [77] – Copelletization, as a sustainable method, allows complete harvesting of single algae cells for the efficient removal of PhACs [78]
	– Membrane bioreactors appear among the most sustainable treatment processes for the elimination of PhACs [79–81] – The current review highlights microalgae as a promising and sustainable approach to efficiently biotransform or bioadsorb PPCPs – Application of microalgae is considered a promising and sustainable technique for the biodegradation of PhACs [62]

Table 3.5 Further reading suggestions for more detailed coverage of the literature on various technologies for the removal of pharmaceuticals in (waste)waters

References	Item	Subject
Zima-Kulisiewicz and Delgado [82]	Section 5	Evolution and the impact of the publications on the effects of the PhACs in the environment
Barcellos et al. [20]	Table 3	Opportunities for the developing countries for the management of PhACs
Guo et al. [83]	Full text	Current trends in the application of biological wastewater treatment technologies for the removal of antibiotics

References

1. Chang T et al (2018) Toxic/hazardous substances and environmental engineering effect of hydraulic retention time on electricity generation using a solid plain-graphite plate microbial fuel cell anoxic/oxic process for treating pharmaceutical sewage. J Environ Sci Health, Part A 53(13):1185–1197. https://doi.org/10.1080/10934529.2018.1530338
2. Mramba AS et al (2020) A review on electrochemical degradation and biopolymer adsorption treatments for toxic compounds in pharmaceutical effluents. Electroanalysis 32(12):2615–2634. https://doi.org/10.1002/elan.202060454
3. Zandi S et al (2019) Industrial biowastes treatment using membrane bioreactors (MBRs)—a scientometric study. J Environ Manage 247:462–473. https://doi.org/10.1016/j.jenvman.2019.06.066
4. Darko A et al (2019) A scientometric analysis and visualization of global green building research. Build Environ 149(Nov 2018):501–511. https://doi.org/10.1016/J.BUILDENV.2018.12.059
5. Mostafaie A et al (2021) A scientometric study on industrial effluent and sludge toxicity. Toxics 9:176
6. Ruiz-Rosero J, Ramirez-Gonzalez G, Viveros-Delgado J (2019) Software survey: ScientoPy, a scientometric tool for topics trend analysis in scientific publications. Scientometrics 121:1165–1188. https://doi.org/10.1007/s11192-019-03213-w
7. Olawumi TO, Chan DWM (2018) A scientometric review of global research on sustainability and sustainable development. J Clean Prod 183:231–250. https://doi.org/10.1016/j.jclepro.2018.02.162
8. Tsay M, Lai C (2018) A scientometric study of heat transfer journal literature from 1900 to 2017. Int Commun Heat Mass Transfer 98:258–264. https://doi.org/10.1016/j.icheatmasstransfer.2018.09.006
9. Bernabò N et al (2017) Scientometric study of the effects of exposure to non-ionizing electromagnetic fields on fertility: a contribution to understanding the reasons of partial failure. PLoS ONE 12:1–17. https://doi.org/10.1371/journal.pone.0187890
10. Chen C (2017) CiteSpace: a practical guide for mapping scientific literature
11. Shomoossi N et al (2019) Understanding the research process and historical trends in English for medical purposes using scientometrics and co-occurrence analysis. Acta Facultatis Medicae Naissensis 36:235–247. https://doi.org/10.5937/afmnai1903236S
12. Ren M et al (2021) Visualizing the knowledge domain of pulsed light technology in the food field: a scientometrics review. Innov Food Sci Emerg Technol 74:102823. https://doi.org/10.1016/j.ifset.2021.102823

[12] A combination of ultraviolet irradiation and proxydisulfate for the generation of active species in the medium.

13. Cobo MJ et al (2011) Science mapping software tools: review, analysis, and cooperative study among tools. J Am Soc Inform Sci Technol 62:1382–1402. https://doi.org/10.1002/asi.21525
14. Falagas ME et al (2008) Comparison of PubMed, Scopus, Web of Science, and Google Scholar: strengths and weaknesses. FASEB J 22:338–342. https://doi.org/10.1096/fj.07-9492lsf
15. Koppernock F (1968) Reinigung stark wechselnder Abwässer der pharmazeutischen Chemie bei der Merck AG. Chem Ing Tec 40:263–268. https://doi.org/10.1002/cite.330400603
16. Gandia RM et al (2019) Scientometric and bibliometric review. Autonom Veh 39:9–28. https://doi.org/10.1080/01441647.2018.1518937
17. Khalaj M et al (2020) Green synthesis of nanomaterials—a scientometric assessment. J Clean Prod 267:122036. https://doi.org/10.1016/j.jclepro.2020.122036
18. Abushaheen MA et al (2020) Antimicrobial resistance, mechanisms and its clinical significance. Dis Mon 66:100971. https://doi.org/10.1016/j.disamonth.2020.100971
19. Zhou G et al (2015) The three bacterial lines of defense against antimicrobial agents. Int J Mol Sci 16:21711–21733. https://doi.org/10.3390/ijms160921711
20. da Barcellos DS, Procopiuck M, Bollmann HA (2022) Management of pharmaceutical micropollutants discharged in urban waters: 30 years of systematic review looking at opportunities for developing countries. Sci Tot Environ 809:151128. https://doi.org/10.1016/j.scitotenv.2021.151128
21. Crane B et al (2018) China's special economic zones: an analysis of policy to reduce regional disparities. Region Stud Region Sci 5:98–107. https://doi.org/10.1080/21681376.2018.1430612
22. Zhao Y et al (2008) Simulation and evaluation on the eco-industrial system of Changchun economic and technological development zone, China. Environ Monit Assess 139:339–349. https://doi.org/10.1007/s10661-007-9840-x
23. Mir-Tutusaus JA et al (2021) Prospects on coupling UV/H_2O_2 with activated sludge or a fungal treatment for the removal of pharmaceutically active compounds in real hospital wastewater. Sci Tot Environ 773:145374. https://doi.org/10.1016/j.scitotenv.2021.145374
24. Cibati A et al (2022) Unravelling the performance of UV/H_2O_2 on the removal of pharmaceuticals in real industrial, hospital, grey and urban wastewaters. Chemosphere 290:133315. https://doi.org/10.1016/j.chemosphere.2021.133315
25. Rocha LS et al (2021) Sustainable and recoverable waste-based magnetic nanocomposites used for the removal of pharmaceuticals from wastewater. Chem Eng J 426:129974. https://doi.org/10.1016/j.cej.2021.129974
26. Krzeminski P et al (2019) Performance of secondary wastewater treatment methods for the removal of contaminants of emerging concern implicated in crop uptake and antibiotic resistance spread: a review. Sci Total Environ 648:1052–1081. https://doi.org/10.1016/j.scitotenv.2018.08.130
27. Ben W et al (2017) Distribution of antibiotic resistance in the effluents of ten municipal wastewater treatment plants in China and the effect of treatment processes. Chemosphere 172:392–398. https://doi.org/10.1016/j.chemosphere.2017.01.041
28. Li J et al (2016) Occurrence and removal of antibiotics and the corresponding resistance genes in wastewater treatment plants: effluents' influence to downstream water environment. Environ Sci Pollut Res 23:6826–6835. https://doi.org/10.1007/s11356-015-5916-2
29. Prata JC et al (2018) Influence of microplastics on the toxicity of the pharmaceuticals procainamide and doxycycline on the marine microalgae *Tetraselmis chuii*. Aquat Toxicol 197:143–152. https://doi.org/10.1016/j.aquatox.2018.02.015
30. Tkaczyk A et al (2021) *Daphnia magna* model in the toxicity assessment of pharmaceuticals: a review. Sci Total Environ 763:143038. https://doi.org/10.1016/j.scitotenv.2020.143038
31. Kim D, Nguyen LN, Oh S (2020) 'Ecological impact of the antibiotic ciprofloxacin on microbial community of aerobic activated sludge. Environ Geochem Health 42:1531–1541. https://doi.org/10.1007/s10653-019-00392-6
32. Aziz A et al (2022) Anaerobic digestion in the elimination of antibiotics and antibiotic-resistant genes from the environment—a comprehensive review. J Environ Chem Eng 10:106423. https://doi.org/10.1016/j.jece.2021.106423

33. Vo HNP et al (2019) Removal and monitoring acetaminophen-contaminated hospital wastewater by vertical flow constructed wetland and peroxidase enzymes. J Environ Manage 250:109526. https://doi.org/10.1016/j.jenvman.2019.109526
34. Uddin MJ, Jeong YK, Lee W (2021) Microbial fuel cells for bioelectricity generation through reduction of hexavalent chromium in wastewater: a review. Int J Hydrogen Energy 46:11458–11481. https://doi.org/10.1016/j.ijhydene.2020.06.134
35. Jain SN et al (2020) Batch and continuous studies for adsorption of anionic dye onto waste tea residue: kinetic, equilibrium, breakthrough and reusability studies. J Clean Prod 252:119778. https://doi.org/10.1016/j.jclepro.2019.119778
36. Naushad M et al (2019) Adsorption kinetics, isotherm and reusability studies for the removal of cationic dye from aqueous medium using arginine modified activated carbon. J Mol Liq 293:111442. https://doi.org/10.1016/j.molliq.2019.111442
37. Bernal V, Giraldo L, Moreno-Piraján JC (2020) Adsorption of pharmaceutical aromatic pollutants on heat-treated activated carbons: effect of carbonaceous structure and the adsorbent-adsorbate interactions. ACS Omega 5:15247–15256. https://doi.org/10.1021/acsomega.0c01288
38. Feng J et al (2015) Effect of hydroxyl group of carboxylic acids on the adsorption of Acid Red G and Methylene Blue on TiO_2. Chem Eng J 269:316–322. https://doi.org/10.1016/j.cej.2015.01.109
39. Ain QU et al (2020) Superior dye degradation and adsorption capability of polydopamine modified Fe_3O_4-pillared bentonite composite. J Hazard Mater 397:122758. https://doi.org/10.1016/j.jhazmat.2020.122758
40. Pauletto PS, Bandosz TJ (2022) Activated carbon versus metal-organic frameworks: a review of their PFAS adsorption performance. J Hazard Mater 425:127810. https://doi.org/10.1016/j.jhazmat.2021.127810
41. Afzal MZ et al (2018) Enhancement of ciprofloxacin sorption on chitosan/biochar hydrogel beads. Sci Total Environ 639:560–569. https://doi.org/10.1016/j.scitotenv.2018.05.129
42. Liu H, Xu G, Li G (2021) Preparation of porous biochar based on pharmaceutical sludge activated by NaOH and its application in the adsorption of tetracycline. J Colloid Interface Sci 587:271–278. https://doi.org/10.1016/j.jcis.2020.12.014
43. Lonappan L et al (2018) Adsorption of diclofenac onto different biochar microparticles: dataset—characterization and dosage of biochar. Data Brief 16:460–465. https://doi.org/10.1016/j.dib.2017.10.041
44. Umar M et al (2013) Application of ozone for the removal of bisphenol A from water and wastewater—a review. Chemosphere 90:2197–2207. https://doi.org/10.1016/j.chemosphere.2012.09.090
45. Clarizia L et al (2017) Homogeneous photo-Fenton processes at near neutral pH: a review. Appl Catal B 209:358–371. https://doi.org/10.1016/j.apcatb.2017.03.011
46. Wei Y, Li G, Wang B (2011) Research on harbor oily wastewater treatment by Fenton oxidation. Adv Mater Res 322:164–168. https://doi.org/10.4028/www.scientific.net/AMR.322.164
47. Abrile MG et al (2021) 'Degradation and mineralization of the emerging pharmaceutical pollutant sildenafil by ozone and UV radiation using response surface methodology. Environ Sci Pollut Res 28:23868–23886. https://doi.org/10.1007/s11356-020-11717-9
48. Tian L et al (2022) Mineralization of cyanides via a novel Electro-Fenton system generating $\cdot OH$ and $\cdot O^{2-}$. Water Res 209:117890
49. Deng J et al (2013) Thermally activated persulfate (TAP) oxidation of antiepileptic drug carbamazepine in water. Chem Eng J 228:765–771. https://doi.org/10.1016/j.cej.2013.05.044
50. De Bel E et al (2009) Influence of pH on the sonolysis of ciprofloxacin: biodegradability, ecotoxicity and antibiotic activity of its degradation products. Chemosphere 77:291–295. https://doi.org/10.1016/j.chemosphere.2009.07.033
51. Miljevic B et al (2014) To sonicate or not to sonicate PM filters: reactive oxygen species generation upon ultrasonic irradiation. Aerosol Sci Technol 48:1276–1284. https://doi.org/10.1080/02786826.2014.981330

52. Azizan NAZ, Yuzir A, Abdullah N (2021) Pharmaceutical compounds in anaerobic digestion: a review on the removals and effect to the process performance. J Environ Chem Eng 9:105926. https://doi.org/10.1016/j.jece.2021.105926
53. Stasinakis AS (2012) Review on the fate of emerging contaminants during sludge anaerobic digestion. Biores Technol 121:432–440. https://doi.org/10.1016/j.biortech.2012.06.074
54. Garg R, Gupta R, Bansal A (2021) Synthesis of g-C_3N_4/ZnO nanocomposite for photocatalytic degradation of a refractory organic endocrine disrupter. Mater Today: Proc 44:855–859. https://doi.org/10.1016/j.matpr.2020.10.787
55. Shen M et al (2021) gCN-P: a coupled g-C_3N_4/persulfate system for photocatalytic degradation of organic pollutants under simulated sunlight. Environ Sci Poll Res:0123456789. https://doi.org/10.1007/s11356-021-17540-0
56. Oba SN et al (2021) Removal of ibuprofen from aqueous media by adsorption: a comprehensive review. Sci Total Environ 780:146608. https://doi.org/10.1016/j.scitotenv.2021.146608
57. Tran HN et al (2020) Innovative spherical biochar for pharmaceutical removal from water: insight into adsorption mechanism. J Hazard Mater 394:122255. https://doi.org/10.1016/j.jhazmat.2020.122255
58. Ahmed SF et al (2021) Recent developments in physical, biological, chemical, and hybrid treatment techniques for removing emerging contaminants from wastewater. J Hazard Mater 416:125912. https://doi.org/10.1016/j.jhazmat.2021.125912
59. Giannakis S, Lin KYA, Ghanbari F (2021) A review of the recent advances on the treatment of industrial wastewaters by sulfate radical-based advanced oxidation processes (SR-AOPs). Chem Eng J 406:127083. https://doi.org/10.1016/j.cej.2020.127083
60. Wang J et al (2020) Occurrence and fate of antibiotics, antibiotic resistant genes (ARGs) and antibiotic resistant bacteria (ARB) in municipal wastewater treatment plant: an overview. Sci Total Environ 744:140997. https://doi.org/10.1016/j.scitotenv.2020.140997
61. Isari AA et al (2020) N, Cu co-doped TiO_2@functionalized SWCNT photocatalyst coupled with ultrasound and visible-light: an effective sono-photocatalysis process for pharmaceutical wastewaters treatment. Chem Eng J 392:123685. https://doi.org/10.1016/j.cej.2019.123685
62. Hena S, Gutierrez L, Croué JP (2021) Removal of pharmaceutical and personal care products (PPCPs) from wastewater using microalgae: a review. J Hazard Mater 403:124041. https://doi.org/10.1016/j.jhazmat.2020.124041
63. Krasucka P et al (2021) Engineered biochar—a sustainable solution for the removal of antibiotics from water. Chem Eng J 405:126926. https://doi.org/10.1016/j.cej.2020.126926
64. Román S et al (2018) Towards sustainable micro-pollutants' removal from wastewaters: caffeine solubility, self-diffusion and adsorption studies from aqueous solutions into hydrochars. Mol Phys 116:2129–2141. https://doi.org/10.1080/00268976.2018.1487597
65. Joly M et al (2013) Ice nucleation activity of bacteria isolated from cloud water. Atmos Environ 70:392–400. https://doi.org/10.1016/j.atmosenv.2013.01.027
66. Chauhan M, Saini VK, Suthar S (2020) Removal of pharmaceuticals and personal care products (PPCPs) from water by adsorption on aluminum pillared clay. J Porous Mater 27(2):383–393. https://doi.org/10.1007/s10934-019-00817-8
67. Levard C et al (2021) Silica-clay nanocomposites for the removal of antibiotics in the water usage cycle. Environ Sci Pollut Res 28:7564–7573. https://doi.org/10.1007/s11356-020-11076-5
68. Yadav P, Yadav A, Labhasetwar PK (2022) Sustainable adsorptive removal of antibiotics from aqueous streams using Fe_3O_4-functionalized MIL101(Fe) chitosan composite beads. Environ Sci Poll Res 101.https://doi.org/10.1007/s11356-021-18385-3
69. Ghazouani S et al (2022) Removal of tramadol hydrochloride, an emerging pollutant, from aqueous solution using gamma irradiation combined by nanofiltration. Process Saf Environ Prot 159:442–451. https://doi.org/10.1016/j.psep.2022.01.005
70. Pylypchuk IV et al (2020) Removal of diclofenac, paracetamol, and carbamazepine from model aqueous solutions by magnetic sol–gel encapsulated horseradish peroxidase and lignin peroxidase composites. Nanomaterials 10:282. https://doi.org/10.3390/nano10020282

71. Xu X et al (2022) A customizable 3D printed device for enzymatic removal of drugs in water. Water Res 208:117861. https://doi.org/10.1016/j.watres.2021.117861
72. He F et al (2021) Reducing the impact of antibiotics in wastewaters: Increased removal of mitoxantrone from wastewater by biosynthesized manganese nanoparticles. J Clean Prod 293:126207. https://doi.org/10.1016/j.jclepro.2021.126207
73. Sbardella L et al (2020) Integrated assessment of sulfate-based AOPs for pharmaceutical active compound removal from wastewater. J Clean Prod 260:121014. https://doi.org/10.1016/j.jclepro.2020.121014
74. Fawzi Suleiman Khasawneh O, Palaniandy P (2021) Removal of organic pollutants from water by Fe_2O_3/TiO_2 based photocatalytic degradation: a review. Environ Technol Inno 21:101230. https://doi.org/10.1016/j.eti.2020.101230
75. Majumdar A, Pal A (2020) Recent advancements in visible-light-assisted photocatalytic removal of aqueous pharmaceutical pollutants. Clean Technol Environ Policy. https://doi.org/10.1007/s10098-019-01766-1
76. Gallego-Schmid A et al (2019) Environmental assessment of solar photo-Fenton processes in combination with nanofiltration for the removal of micro-contaminants from real wastewaters. Sci Tot Environ 650:2210–2220. https://doi.org/10.1016/j.scitotenv.2018.09.361
77. Benner J et al (2013) Is biological treatment a viable alternative for micropollutant removal in drinking water treatment processes? Water Res 47:5955–5976. https://doi.org/10.1016/j.watres.2013.07.015
78. Hom-Diaz A et al (2017) Performance of a microalgal photobioreactor treating toilet wastewater: Pharmaceutically active compound removal and biomass harvesting. Sci Total Environ 592:1–11. https://doi.org/10.1016/j.scitotenv.2017.02.224
79. Goswami L et al (2018) Membrane bioreactor and integrated membrane bioreactor systems for micropollutant removal from wastewater: a review. J Water Process Eng 26:314–328. https://doi.org/10.1016/j.jwpe.2018.10.024
80. Melvin SD, Leusch FDL (2016) Removal of trace organic contaminants from domestic wastewater: a meta-analysis comparison of sewage treatment technologies. Environ Int 92–93:183–188. https://doi.org/10.1016/j.envint.2016.03.031
81. Wang Y (2021) Anaerobic membrane bioreactors for trace organic contaminants with wastewater treatment: a review. IOP Conf Ser: Earth Environ Sci 691:012002. https://doi.org/10.1088/1755-1315/691/1/012002
82. Zima-Kulisiewicz BE, Delgado A (2009) Synergetic microorganismic convection generated by Opercularia asymmetrica ciliates living in a colony as effective fluid transport on the micro-scale. J Biomech 42:2255–2262. https://doi.org/10.1016/j.jbiomech.2009.06.018
83. Guo WQ, Yin RL, Zhou XJ (2013) Current trends for biological antibiotic pharmaceutical wastewater treatment. Adv Mater Res 726–731:2140–2145. https://doi.org/10.4028/www.scientific.net/AMR.726-731.2140

Chapter 4
Pharmaceutically Active Compounds in Activated Sludge Systems—Presence, Fate, and Removal Efficiency

4.1 Introduction

Activated sludge (AS) is the most widely used biological wastewater treatment process in large-scale wastewater treatment plants (WWTPs). Such systems normally perform under hydraulic retention times ranging from 4 to 14 h. In this process, aerobic microorganisms are employed to mineralize the organic compounds present in the composition of wastewater into the final degradation products (i.e., water and carbon dioxide) or less complex intermediates.

AS systems are normally considered very efficient for the removal of a wide range of biodegradable organic compounds. However, the presence of contaminants of emerging concern (CECs), especially pharmaceutically active compounds (PhACs), has raised concerns regarding the quality of the final discharged effluents. This is mainly due to the inefficiency of AS processes to deal with some CECs and the increasing pattern of the consumption and release of such compounds, especially pharmaceutical compounds. The yearly consumption of pharmaceuticals is estimated at 15 g per capita on average. This value is much higher in industrialized countries (50–150 g per capita). For instance, the global consumption of carbamazepine is estimated to be 1014 tons per year [1].

Various routes have been identified for the release of PhACs into water bodies, such as effluents from the pharmaceutical industries, unaltered excretion in feces and urine, household disposal of these compounds, and their release from the veterinary and the medicines used in aquaculture. The fate of these compounds in WWTPs is highly dependent on parameters such as type, concentration, probable toxic effects on AS bacteria, biological degradation rate constant (K_{biol}), and mechanisms involved in their interaction with AS microorganisms. There are estimations regarding the concentrations of PhACs in the influents and effluents of WWTPs. As an example, Nakada et al. [2] indicated that aspirin (7300 ng/L) is the most abundant pharmaceutical in 5 municipal WWTPs in Tokyo, followed by crotamiton (921 ng/L), ibuprofen (669 ng/L), triclosan (511 ng/L), and diethyltoluamide (503 ng/L). Variables such

M. Kamali et al., *Advanced Wastewater Treatment Technologies for the Removal of Pharmaceutically Active Compounds*, Green Energy and Technology,
https://doi.org/10.1007/978-3-031-20806-5_4

as the temperature and the solid retention time (SRT) can determine the microbial composition in such systems. For instance, studies have revealed that the microbial communities present in AS systems are highly sensitive to water temperature, and higher diversities can be normally expected in activated sludge during summer [3, 4]. Mechanisms including abiotic transformations, sorption to biomass (or suspended solids), and biodegradation can play roles in the removal of PhACs in AS treatment plants, depending on the type of PhACs and the composition of the sludge in these systems. This can potentially determine the need for any subsequent treatment to remove the residual PhACs from the effluents treated by activated sludge systems.

This chapter has aimed to explore the presence, fate, and removal of different types of PhACs in activated sludge systems and the mechanisms involved in their elimination or transformation. Additional literature, as well as further research opportunities, has also been provided in the present chapter to direct future studies aimed at minimizing the potential risks from the release of PhACs into the environment.

4.1.1 Effects of Pharmaceutically Active Compounds on Aerobic Microorganisms

According to the literature, exposure to environmentally relevant concentrations of some PhACs can influence the composition of AS microbial communities. For instance, ciprofloxacin can be found in sewage effluents at concentrations ranging from 50 to 500 μg/L. Kim et al. [5, 6] demonstrated that the diversity of microorganisms can be reduced (by approximately 20%) in the presence of ciprofloxacin (500 μg/L) in effluents. They also discussed that *Rhodobacteraceae* and *Nakamurellaceae,* which are widely found in large-scale WWTPs, are the most influenced microorganisms (> tenfold) in the presence of CIP. PhACs such as ibuprofen, naproxen, ketoprofen, diclofenac, and clofibric acid can also reduce the microbial diversity only under higher concentrations (i.e., > 50 mg/L) [7]. It has also been indicated that PhACs such as caffeine, sulfamethoxazole, and carbamazepine can inhibit the growth of microorganisms and reduce microbial diversity (especially for sulfamethoxazole and carbamazepine). It has also been indicated that acclimatization can result in microbial communities being resistant to the toxic effects induced by PhACs. A recent study demonstrated that after acclimatization with pharmaceuticals, multiresistant genera such as *Escherichia–Shigella* are abundant in the composition of the activated sludge [8]. In contrast, quantitative image analysis (QIA) has indicated that ibuprofen favors the growth of aggregated microbial communities rather than filamentous bacteria [9]. Hence, AS is considered a promising technology to deal with such a biodegradable compound.

4.2 Biodegradation of PhACs With AS

The biotransformation of PhACs is normally expressed by K_{biol}. However, this value has been calculated only for a limited number of pharmaceuticals, mainly using pseudo first-order kinetics. According to relevant studies, pharmaceuticals such as carbamazepine and naproxen have low K_{biol} values, meaning that they cannot be efficiently biodegraded. A recent study revealed that ciprofloxacin at concentrations ranging from 50 to 500 μg/L is poorly removed in AS (> 20%) [5, 6]. In conclusion, this compound is released from AS processes without any efficient treatment/transformation. However, for compounds such as ibuprofen, diclofenac, and ketoprofen, moderate to high biodegradability (20–90%) has been observed according to the calculated K_{biol} values [1]. A study performed in Southern England also concluded that PhACs, including ibuprofen, paracetamol, salbutamol, and mefenamic acid, can be effectively eliminated (up to 90%) using AS processes [10]. Ibuprofen and paracetamol in the range 0.4–1 mg/L were also efficiently removed (> 90%) according to the results achieved by [9]. There are also observations that sulfamethoxazole can be biodegraded using AS systems after the acclimatization of the microbial communities [11]. The latter study revealed that this pharmaceutical can be consumed by microorganisms as a source of both carbon and nitrogen. In another recent study aimed at the identification of the concentration of some PhACs in the influents and effluents of WWTPs in Tokyo, it was reported that compounds including aspirin, ibuprofen, and thymol can be efficiently degraded during the primary and secondary treatment processes. However, poor removals were observed for amide-type pharmaceuticals, ketoprofen, and naproxen due to their low degree of hydrophobicity [2]. The authors also stated that crotamiton was the most abundant pharmaceutical in the effluents of the WWTPs due to its recalcitrant structure. It has also been concluded that the age of sludge is a determinant parameter determining the biodegradation of PhACs. In this regard, aerobic sludge with an age of over 7 days can be used more efficiently for the biodegradation of PhACs, especially for compounds such as naproxen and ketoprofen. This is due to the enrichment of the sludge with the heterotrophic bacterial community [12].

According to the recent literature, PhACs can be degraded via two main routes: (a) the nitrification process (autotrophic biodegradation) and (b) the COD removal process (heterotrophic biodegradation). A combination of these two processes can also contribute to the biodegradation of PhACs. Figure 4.1 illustrates the main removal routes of some widely used PhACs. According to this figure, cefalexin is mainly degraded via adsorption, hydrolysis, nitrification, and COD degradation routes. Additionally, norfloxacin and ibuprofen are mainly biodegraded through the COD degradation process in AS systems. As is evident in Fig. 4.1, COD biodegradation (heterotrophic biodegradation) is the main mechanism for the removal of biodegradable PhACs such as norfloxacin, sulfamethoxazole, sulfamethazine, ibuprofen, and cephalexin.

It is also observed that the degradation of PhACs can be influenced by the AS operating parameters, such as HRT and SRT. The increase in SRT can result in an

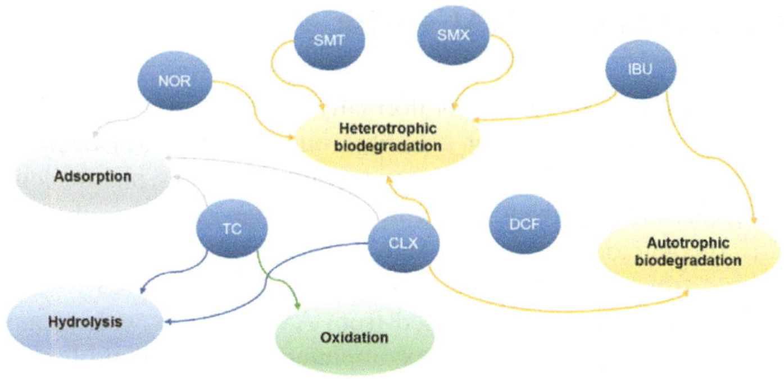

Fig. 4.1 Main removal routes of some widely used PHACs in the AS treatment process, reprinted with permission from Peng et al. [13]. According to this figure, norfloxacin, sulfamethazine, sulfamethoxazole, ibuprofen, and cephalexin are biodegraded mainly under the COD biodegradation process. Nitrification can also contribute to the degradation of ibuprofen and cephalexin. Low degradation efficiencies (approximately 10%) can also be expected for some PhACs, such as cephalexin and tetracycline, under the hydrolysis route

improvement in the degradation of moderately biodegradable pharmaceuticals. For instance, in a recent study, the increase in the SRT (to 20 days) increased diclofenac biodegradation (up to 25%) [14]. However, nonbiodegradable compounds such as roxithromycin and erythromycin were not removed even under very long SRTs. The increase in HRT can also increase the contact time between the microorganisms and the pharmaceutical compounds, leading to an increase in their removal efficiency. However, this strategy can increase the operating costs due to the higher energy demand per unit of treated water.

It can also be expected that by increasing the biomass dosage (expressed as g volatile suspended solids per liter of effluents (gVSS/L)), more pharmaceutical degrading microorganisms will be available in the medium. However, there is a need for more studies to demonstrate the role of biomass concentration and its optimum level on the biodegradation of PhACs. It is also of high importance to have a conclusion about the effects of the season on the biodegradation of the PhACs using AS systems. In this regard, a recent study to study the biotransformation rate of some types of pharmaceuticals under summer and winter conditions concluded that acetaminophen, metformin, metoprolol, terbutaline, and phenazone (ranked based on their biotransformation rates) can be biodegraded using the activated sludge process and that their degradation kinetics are influenced by the season. In fact, no significant changes in the microbial community structure were identified as a function of the season (Fig. 4.2) [3].

There is also a need for further studies concerning the effects of other influencing parameters, such as the presence of nutrients/salts, pH, and different concentrations of PhACs, on their biodegradability using activated sludge systems.

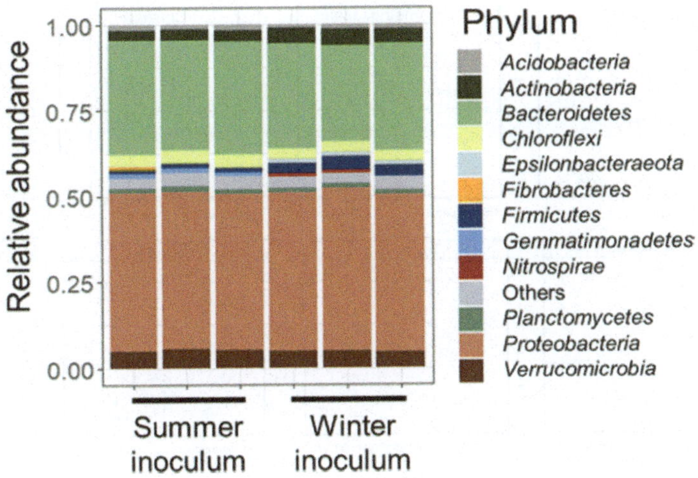

Fig. 4.2 Abundance of the most important microbial phylum as a function of the season (summer and winter), reprinted with permission from van Bergen et al. [3]

4.3 Adsorption of PhACs With AS

Adsorption is also a mechanism for the removal of some PhACs from wastewater streams using an AS system. There are a limited number of studies in the literature evaluating the adsorption potential (expressed as the adsorption coefficient (K_d)) of some specific types of PhACs. For instance, Martínez-Alcalá et al. [1] observed higher K_d values for pharmaceuticals such as diclofenac and naproxen. According to Fig. 4.3 [13], tetracycline can also be efficiently removed by the adsorption route, while a relatively low degree of adsorption can be expected for compounds such as sulfamethazine, sulfamethoxazole, and ibuprofen.

The interaction of other pharmaceuticals with activated sludge (especially in sewage treatment plants) has also been studied and discussed recently. For ciprofloxacin, it has been indicated that adsorption is the dominant mechanism for its removal from effluents, contributing to more than 90% of the removed pharmaceutical. This can be due to the difference between the sludge surface charge and ciprofloxacin charge, which can result in electrostatic interactions between ciprofloxacin and AS. In a recent study, [15] demonstrated that the presence of ciprofloxacin in biological reactors containing AS results in (a) a reduction in the concentration of soluble protein in the medium as well as a remarkable increase in extracellular proteins. The adsorption of ciprofloxacin by extracellular proteins plays an important role in the removal of this pharmaceutical from the medium. Remarkable adsorption of PhACs has also been observed by Suarez et al. [16]. They studied the main removal mechanisms of 16 pharmaceutical and personal care products (PPCPs), including medicines, fragrances, and hormones, using an activated sludge wastewater treatment process with various concentrations of PPCPS ranging

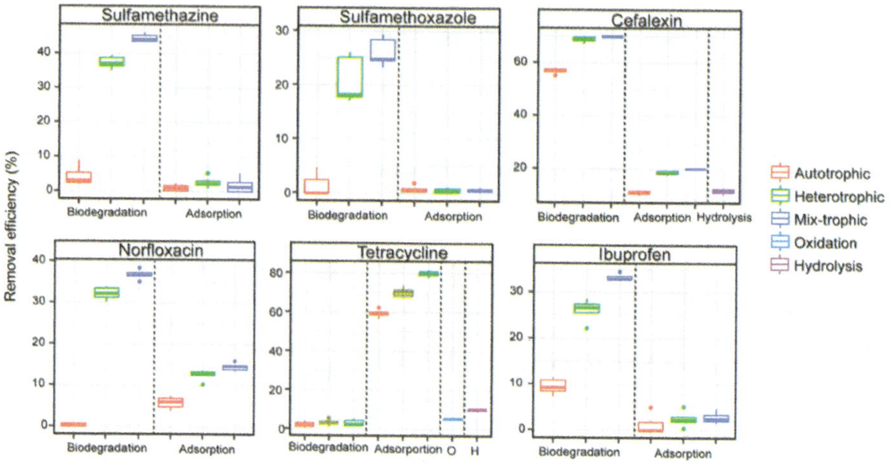

Fig. 4.3 Various mechanisms involved in the removal of some PhACs. Tetracycline is efficiently removed by adsorption, while a relatively low degree of adsorption has been observed for compounds such as sulfamethazine, sulfamethoxazole, and ibuprofen, reprinted with permission from Peng et al. [13]

from 10 to 40 μg/L. They observed lower adsorption efficiencies for carbamazepine, diazepam, and diclofenac. For compounds such as fragrances, fluoxetine, ibuprofen, naproxen, and natural estrogens, adsorption efficiencies above 80% were achieved. It was also concluded in their research that volatilization can play a role in the removal of the fragrance celestolide (up to 45%).

The study of the concentration of a specific compound pharmaceutical compound in water and sludge phases can also be used to gain a better understanding of the mechanisms involved in the adsorption of PhACs into activated sludge. This can be performed by measuring the water-sludge partition coefficients (k_p). At neutral pH values, the k_p values of most of the pharmaceuticals and estrogens (e.g., benzophenone, estrone, 17β-estradio, and 17α-ethynilestradiol) are normally very low, indicating that most of these compounds remain soluble in the water phase. Under acidic pH values, the k_p values of the acidic pharmaceutical compounds increase, indicating that lower pH values can favor the adsorption of such PhACs [17].

There are also clues indicating that the adsorption of the pharmaceuticals can be promoted (but to a lesser extent) by the adsorption of the organic matter present in the effluents, rather than the activated sludge. For instance, a study indicated that enrofloxacin and tetracycline, as common veterinary drugs, at initial concentrations of 100 μg/L can be removed (i.e., 68% and 77% for enrofloxacin and tetracycline, respectively, after 10 days) following adsorption to both sludge particles and organic particulate matter [18]. However, it should be stated that most wastewater treatment plants based on the activated sludge process work under much shorter hydraulic and solid retention times. Hence, less removal of such compounds under real treatment conditions can be expected, which may be the cause of their release into the environment and create probable subsequent environmental issues. The elimination of other

PhACs, such as ibuprofen and ketoprofen, can also be highly affected by the HRT and SRT of the activated sludge process [16].

In addition to the existing organic matter, the presence of elements such as iron can also influence the adsorption of PhACs. Enhanced adsorptive removal of ciprofloxacin in the presence of iron salts is an example of this phenomenon [19]. In fact, the iron salt dosage and pH can determine the extent of ciprofloxacin adsorption and removal from the effluents, in addition to the electrostatic interactions between ciprofloxacin and the sludge particles. From a mechanistic point of view, iron can establish complexes with pharmaceuticals, resulting in their adsorptive removal from the medium.

Although the adsorption process can partially aid in removing some types of PhACs, the management and disposal of the generated sludge can bring environmental concern because pharmaceuticals can accumulate and biomagnify in the food chain and cause ecological and health problems (Fig. 4.4) [20]. In this regard, there is a need for more investigations on the fate of sludge from wastewater treatment plants treating effluents containing PhACs and possible related environmental and health issues.

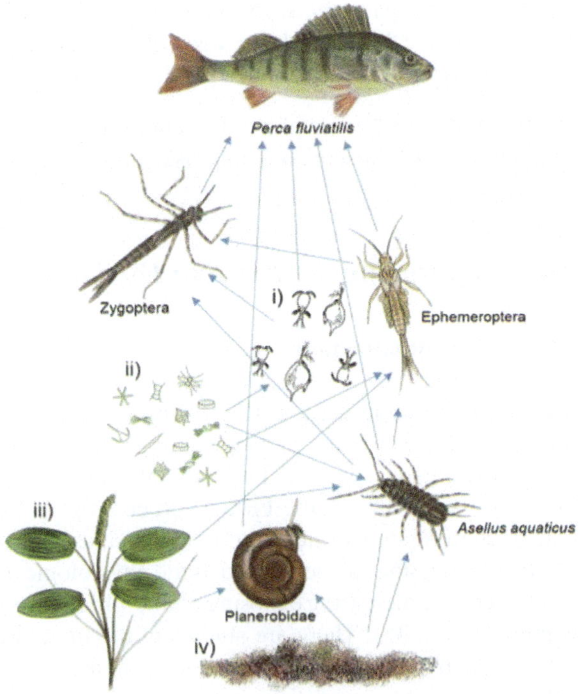

Fig. 4.4 Schematic of a food web showing the possible movement, bioaccumulation, and biomagnification of PhACs, reprinted with permission from [21]

4.4 Modifications in AS Processes for the Efficient Removal of PhACs

As discussed before in this chapter, activated sludge processes are effective methods for the treatment of a wide range of effluents [14, 22]. According to the relevant sustainability criteria [23, 24], activated sludge technologies are cost-effective candidates for the treatment of low- and medium-strength effluents from various origins. They are also relatively easy to implement with a high degree of stability and social acceptance.

However, they normally fail to efficiently remove complex organic compounds such as most PhACs. Table 4.1 evaluates the performance of the AS processes for the treatment of pharmaceuticals. As shown in this table, AS systems are currently struggling with sustainability issues such as the following:

- Difficulties in the efficient treatment of PhACs.
- Process stability due to the possible toxic effects of some types of PhACs on the microbial communities in the content of activated sludge.
- Generation of solid wastes contaminated by the PhACs adsorbed by either the microorganisms or the suspended solids in the effluents.
- Retaining the PhACs in the treated effluents and the possible effects of their release into the environment.

Hence, there is a crucial need to apply modifications in conventional AS-based technologies to reduce the ongoing risks of the release of PhACs in the environment.

4.4.1 Upgrading the Existing Facilities

4.4.1.1 Acclimatization

Most of the existing wastewater treatment plants are based on AS systems. Therefore, it would be an economic idea to upgrade the existing facilities to enhance their performance to deal with the PhACs. In this regard, strategies including acclimatization processes with specific PhACs, process optimization, and modifications in the reactor configurations can be recommended, as discussed in this section.

Acclimatization can be defined as the physiological adaptation process that occurs by the stepwise increasing exposure of microbial communities to harsh environmental conditions [44, 45]. There are studies in the literature on the successful acclimatization of activated sludge to phenolic effluents. An example of these studies is the enhanced phenol degradation at a maximum rate of 0.12 g phenol/g VSS/h under room temperature and near-neutral pH conditions [25]. There are also recent reports on the effects of the acclimatization process on the promotion of the performance of conventional activated sludge processes for the biodegradation of some PhACs. For instance, efficient degradation of triclosan (0.5, 1, and 2 mg/L) in an activated

Table 4.1 Evaluation of the performance of conventional AS systems to deal with pharmaceutical compounds

Criteria	Description	Evaluation of AS performance	Relevant literature
Treatment efficiency	The performance of the treatment system to treat the effluents considering the treatment process variables	Moderate to low and selective	[19]
Ease of implementation	The complexity level of the treatment system in terms of the equipment and expertise needed for the treatment process	Relatively easy to operate	[26]
Combination possibility	The potential of the method to be combined with other treatment methods	Can be combined with various biological and physico-chemical methods	[27, 28]
Process stability	Resistance of the treatment system against the failures and the possibility of recovery after failures	Stable to treat the effluents containing biodegradable organic compounds	[29]
Health and safety risks	The associated occupational and health risks with the implementation of the treatment method	Relatively low; the possibility of exposure to the biological agents	[30, 31]
Generation of solid wastes	Solid wastes (e.g., sludge generation) generated by the treatment method	High; solid waste management plans are needed	[32–34]
Release of chemical substances	The possibility of the release of additives (mainly chemicals) used in the treatment process into the environment	Low (no chemicals are normally used in the conventional AS processes)	[26]
CO_2 emission	The amount of carbon dioxide emitted from the treatment process or from the treatment facilities	Low to moderate (bacterial activities can result in the release of some extent of carbon dioxide to the environment)	[35]
Water reuse potential	Quality of the treated water to be reused in various applications	Moderate to low; due to the presence of nonremovable PhACs in the treated water	[18, 36, 37]

(continued)

Table 4.1 (continued)

Criteria	Description	Evaluation of AS performance	Relevant literature
Potential to recover by-products	Possibility of recovery of the by-products including materials and energy	Moderate; the sludge can be contaminated by the adsorbed PhACs which make it difficult to be recovered and reused in land applications. However, there is a possibility for the production of carbonaceous materials using methods such as the pyrolysis process	[38, 39]
Initial investments	The initial investments required for the treatment process including infrastructures, land area, equipment, certificates, etc.	High for the effluents containing PhACs due to the need for high HRTs which necessitate larger reactors and infrastructures	[14, 40, 41]
Maintenance costs	Overall expenses needed for the maintenance of the facilities required for the treatment process	High; the need for higher HRT can also increase the maintenance costs of the treatment of each volume unit of the effluents	[40, 41]
Odor impact	Odor which is normally generated by the performance of the employed treatment technology	High. Additionally, efficient treatment of some pharmaceuticals necessitates higher SRTs which may affect the odor from the activated sludge process (more studies are required)	[42]
Noise impact	Noise produced by the performance of the treatment process	Low to moderate	[26]
Visual impact	The influence of the treatment infrastructures on the local visual properties	High; this is especially because of the need for larger treatment infrastructures due to the need for higher HRTs	[26]

(continued)

Table 4.1 (continued)

Criteria	Description	Evaluation of AS performance	Relevant literature
Public acceptance	The overall perception of the public community's about the impact of the applied treatment method on their routine life	Low; this can be because the low efficiency of the conventional activated sludge processes can result in the presence of these compounds in the treated water which can deteriorate human health and environmental safety	[43]

The criteria and their descriptions are based on Kamali et al. [25]

sludge system (up to 97%) has been reported after the acclimatization process, as reported by Stasinakis et al. [46]. They concluded that triclosan is rapidly adsorbed by suspended solids and undergoes subsequent biodegradation. It is worth mentioning that the presence of triclosan in nonacclimatized AS systems can deteriorate the removal of ammonia and nitrification capacity of the system. However, the acclimatization of biomass resulted in the recovery of the nitrification capacity, and the system was able to biodegrade TCS even at an initial concentration of 2 mg/L. There are also few reports on the effectiveness of such strategies, especially under pandemic conditions, in which the consumption of some specific drugs increases considerably. For instance, a report on the effect of adaptation[1] of the AS microorganisms during the 2009–2010 influenza pandemic indicated that no adaptation happens to the antivirals and antibiotics (e.g., tamiflus, oseltamivir carboxylate (OC)) that are consumed more under these conditions [47]. This is an alert for the uncontrolled release of drugs that have been increasingly used during COVID-19 pandemic conditions, especially azithromycin, chloroquine, hydroxychloroquine, ivermectin, and dexamethasone, as well as antivirals such as remdesivir and favipiravir [48].

The application of additives can also be considered a possible way to enhance the performance of the existing activated sludge systems. There are some reports on the application of inexpensive materials such as biochar for the enhancement of microbial communities during the acclimatization process using model pollutants. This can be achieved mainly by the promoted colonization of the microorganisms that use biochar as a support and shelter [49]. Such a strategy can be used for the promotion of the existing AS systems for the efficient degradation of pharmaceutically active compounds. Additives such as biochar can also be efficiently used for promoting the adsorption of nonbiodegradable compounds from effluents during the activated sludge process. For instance, in a recent study, various types of biochars were used for the efficient adsorption of ciprofloxacin (up to 94%). Intraparticle diffusion, π–π electron–donor–acceptor interactions, and hydrophobic and electrostatic

[1] Heritable modification in the function or structure of the microorganisms which can enhance their performance under a stressful environment.

interactions have also been identified as the main mechanisms for the adsorption of this pharmaceutical [5, 6].

There are also other additive materials, such as carbon nanotubes, clay-based materials (such as montmorillonite), graphene, and activated carbon, which have demonstrated high potential for the removal of pharmaceuticals [50–53]. However, their applicability to enhance the performance of AS processes needs to be assessed because some materials, such as graphene and graphene oxide, can have antibacterial and toxic effects [54–56].

4.4.1.2 Configuration Modifications

Applying modifications in the configuration of the existing AS systems can also be considered a way to enhance the potential of these technologies to deal with PhACs. There are various possibilities for such upgrading purposes. Converting conventional activated sludge systems to moving-bed biofilm reactors (MBBRs) is an alternative that can potentially enhance the potential of AS processes to deal with highly polluted effluents and reduce the HRT of the system, which can help reduce the overall treatment costs (see Table 4.1). Figure 4.5 illustrates a schematic of the conversion of a conventional activated sludge process to an MBBR system, as reported by Falletti and Conte [57].

As a mature technology, MBBR allows growing the attached microorganisms to the plastic carriers, which can enhance the resistance of the system against harsh environmental conditions. There are studies on the enhanced performance of MBBRs for the removal of PhACs compared to conventional activated sludge processes [58]. This is due to the effectiveness of the compact biofilm for the degradation of compounds with relatively low biodegradability. In this regard, more efficient technologies, such

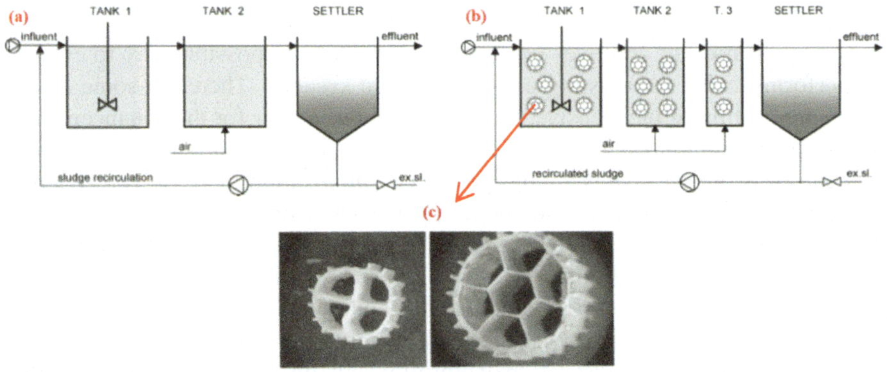

Fig. 4.5 Upgrading of a conventional activated sludge process (**a**) to an MBBR system (**b**) using microbial carriers (**c**), adopted from Falletti and Conte [57]

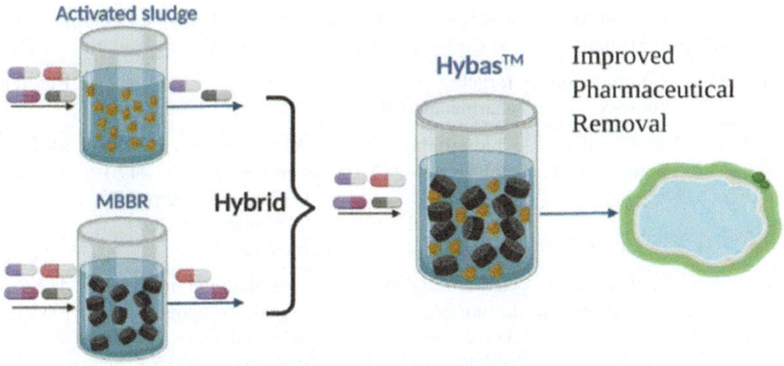

Fig. 4.6 Integration of conventional activated sludge systems with MBBRs (innovative Hybas™ pilot-scale system) for the efficient degradation of pharmaceuticals, reprinted with permission from Tang et al. [59]

as integrated fixed-film activated sludge (IFAS),[2] can be proposed for upgrading conventional AS systems. Such systems are especially beneficial for the degradation of pharmaceuticals because they can provide sludge of different ages: low-age sludge (suspended flocs) and aged sludge (sludge carriers). This can potentially facilitate nitrification at lower SRTs[3] (Fig. 4.6).

4.4.1.3 Combination with Other Technologies

The lack of efficiency of conventional activated sludge processes for the removal of pharmaceuticals (and other emerging pollutants) can also be addressed by combining these technologies with efficient (mainly physico-chemical) treatments.

The application of tertiary treatments can be considered in this regard as an option for polishing treated effluents using activated sludge systems. Advanced oxidation processes (AOPs) are considered promising candidates for coupling with conventional AS systems. AOPs are based on the generation of powerful radicals that can attack and decompose organic pollutants, including recalcitrant and nonbiodegradable organic compounds. There are various types of AOPs, such as ozonation, electrooxidation, photocatalysis, Fenton oxidation, and catalysis, as well as their combinations [60, 61]. Based on the type and operating conditions, various degrees of degradation have been observed for recalcitrant organic compounds. There are reports of the successful combination of some AOPs with activated sludge processes for the efficient degradation of pharmaceuticals. Table 4.2 summarizes some of the studied combinations and their overall effects on the removal of pharmaceuticals.

[2] In such systems, the flocs of the microorganisms and the attached biofilms of MBBR carriers are integrated.

[3] COD and nitrogen removal are both involved in the degradation of pharmaceuticals.

Table 4.2 Summary of some recent studies on the combination of physico-chemical treatment techniques with conventional activated sludge processes for the efficient degradation of PhACs

Physico-chemical treatment	Combination strategy	Target pharmaceuticals	Remarks	References
Sonophotolysis	AOP as the pretreatment	Chloramphenicol, diclofenac, salicylic acid, and paracetamol	AS alone: 65–73% TOC removal (HRT = 24 h) Combined system: 98% TOC and 99% COD removals	Mowla et al. [62]
UV/H_2O_2	AOP as the post-treatment	Carbamazepine, acetaminophen, diclofenac, sulfamethoxazole, and 17-α-ethinylestradiol	Complete removal of organic compounds was achieved under a UV intensity of 130.5 kJ/m^2 and an H_2O_2 dosage of 200 mg/L	da Silva et al. [63]
UV/H_2O_2	AOP as the post-treatment	22 different types of pharmaceuticals	All the pharmaceuticals were efficiently removed (93–95%) using the combined method	Mir-Tutusaus et al. [64]
UV/H_2O_2	AOP as the pretreatment	22 different types of pharmaceuticals	The combined method represented a moderate performance (83%) for the overall removal of pharmaceuticals	Mir-Tutusaus et al. [64]
Electro-Fenton (EF)	EF as the pretreatment	Trimethoprim	Complete removal of the pharmaceutical using the EF technology (0.69 mM Fe^{2+}, 466 mA, and 30 min) Improvement in biodegradability from 0.11 to 0.52 by the pretreatment Up to 90% mineralization using the subsequent AC process	Mansour et al. [65]
Adsorption	Cotreatment	Ciprofloxacin	Mechanisms including diffusion in macropores, π–π electron-donor–acceptor interactions, and electrostatic attraction are involved in the removal of PhACs in such combined systems	[5, 6]

Table 4.3 Further reading suggestions for more detailed coverage of the literature on the fate and removal of pharmaceutically active compounds using activated sludge processes

Reference	Item	Subject
Alfonso-Muniozguren et al. [66]	Section 2.1	Activated sludge processes for the elimination of PhACs from the containing effluents

4.5 Further Reading

Table 4.3 contains items from the recent literature that the reader can consult for more detailed information regarding the fate and removal of pharmaceutically active compounds using activated sludge processes.

4.6 Summary

The activated sludge (AS) process has been widely used for the treatment of effluents from various industrial and nonindustrial (e.g., municipal) origins. Hence, their performance for the removal of PhACs is very important to estimate the amount of PhACs that can be released into the environment. Biodegradation and adsorption are the main mechanisms identified for the removal of PhACs using activated sludge processes. However, relatively low removals have been observed for most of the PhACs using conventional AS systems. Hence, there is a need to apply modifications in such facilities to enable them to deal with PhACs. Strategies such as acclimatization (especially in the presence of cheap and efficient materials such as biochar), upgrading the reactor configuration, and combination with other physicochemical treatment technologies (such as post-treatment with advanced oxidation processes) can be adopted to efficiently remove PhACs from the effluents.

References

1. Martínez-Alcalá I, Guillén-Navarro JM, Fernández-López C (2017) Pharmaceutical biological degradation, sorption and mass balance determination in a conventional activated-sludge wastewater treatment plant from Murcia, Spain. Chem Eng J 316:332–340. https://doi.org/10.1016/j.cej.2017.01.048
2. Nakada N et al (2006) Pharmaceutical chemicals and endocrine disrupters in municipal wastewater in Tokyo and their removal during activated sludge treatment. Water Res 40:3297–3303. https://doi.org/10.1016/j.watres.2006.06.039
3. van Bergen TJHM et al (2021) Do initial concentration and activated sludge seasonality affect pharmaceutical biotransformation rate constants? Appl Microbiol Biotechnol 105:6515–6527. https://doi.org/10.1007/s00253-021-11475-9
4. Saleem M, Bukhari AA, Al-Malack MH (2003) Seasonal variations in the bacterial population in an activated sludge system. J Environ Eng Sci 2:155–162. https://doi.org/10.1139/S03-016

5. Kim DG et al (2020) Addition of biochar into activated sludge improves removal of antibiotic ciprofloxacin. J Water Process Eng 33:101019. https://doi.org/10.1016/j.jwpe.2019.101019
6. Kim D, Nguyen LN, Oh S (2020) Ecological impact of the antibiotic ciprofloxacin on microbial community of aerobic activated sludge. Environ Geochem Health 42:1531–1541. https://doi.org/10.1007/s10653-019-00392-6
7. Kraigher B et al (2008) Influence of pharmaceutical residues on the structure of activated sludge bacterial communities in wastewater treatment bioreactors. Water Res 42:4578–4588. https://doi.org/10.1016/j.watres.2008.08.006
8. Vasiliadou IA et al (2018) Toxicity assessment of pharmaceutical compounds on mixed culture from activated sludge using respirometric technique: the role of microbial community structure. Sci Total Environ 630:809–819. https://doi.org/10.1016/j.scitotenv.2018.02.095
9. Quintelas C et al (2020) Degradation of widespread pharmaceuticals by activated sludge: Kinetic study, toxicity assessment, and comparison with adsorption processes. J Water Process Eng 33:101061. https://doi.org/10.1016/j.jwpe.2019.101061
10. Jones OAH, Voulvoulis N, Lester JN (2007) The occurrence and removal of selected pharmaceutical compounds in a sewage treatment works utilising activated sludge treatment. Environ Pollut 145:738–744. https://doi.org/10.1016/j.envpol.2005.08.077
11. Drillia P et al (2005) On the occasional biodegradation of pharmaceuticals in the activated sludge process: the example of the antibiotic sulfamethoxazole. J Hazard Mater 122(3):259–265. https://doi.org/10.1016/j.jhazmat.2005.03.009
12. Fals P et al (2012) Impact of solid retention time and nitrification capacity on the ability of activated sludge to remove pharmaceuticals. Environ Technol 33:865–872. https://doi.org/10.1080/09593330.2011.601764
13. Peng J et al (2019) Characterizing the removal routes of seven pharmaceuticals in the activated sludge process. Sci Total Environ 650:2437–2445. https://doi.org/10.1016/j.scitotenv.2018.10.004
14. Hatoum R et al (2019) Elimination of micropollutants in activated sludge reactors with a special focus on the effect of biomass concentration. Water (Switzerland) 11:2217. https://doi.org/10.3390/w11112217
15. Wang K et al (2018) Interaction of ciprofloxacin with the activated sludge of the sewage treatment plant. Environ Sci Pollut Res 25:35064–35073. https://doi.org/10.1007/s11356-018-3413-0
16. Suarez S, Omil F, Lema JM (2010) Fate and removal of pharmaceuticals and personal care products (PPCPs) in a conventional activated sludge treatment process. WIT Trans Ecol Environ 135:255–265. https://doi.org/10.2495/WP100221
17. Urase T, Kikuta T (2005) Separate estimation of adsorption and degradation of pharmaceutical substances and estrogens in the activated sludge process. Water Res 39(7):1289–1300. https://doi.org/10.1016/j.watres.2005.01.015
18. Carvalho PN et al (2013) Activated sludge systems removal efficiency of veterinary pharmaceuticals from slaughterhouse wastewater. Environ Sci Pollut Res 20:8790–8800. https://doi.org/10.1007/s11356-013-1867-7
19. Polesel F et al (2015) Factors influencing sorption of ciprofloxacin onto activated sludge: experimental assessment and modelling implications. Chemosphere 119:105–111. https://doi.org/10.1016/j.chemosphere.2014.05.048
20. Zenker A et al (2014) Bioaccumulation and biomagnification potential of pharmaceuticals with a focus to the aquatic environment. J Environ Manage 133:378–387. https://doi.org/10.1016/j.jenvman.2013.12.017
21. Lagesson A et al (2016) Bioaccumulation of five pharmaceuticals at multiple trophic levels in an aquatic food web—insights from a field experiment. Sci Total Environ 568:208–215. https://doi.org/10.1016/j.scitotenv.2016.05.206
22. Mujtaba G et al (2018) Removal of nutrients and COD through co-culturing activated sludge and immobilized *Chlorella vulgaris*. Chem Eng J 343:155–162. https://doi.org/10.1016/j.cej.2018.03.007

23. Kamali M, Persson KM et al (2019) Sustainability criteria for assessing nanotechnology applicability in industrialwastewater treatment: Current status and future outlook. Environ Int 125(Jan):261–276. https://doi.org/10.1016/j.envint.2019.01.055
24. Popovic T, Avramenko Y (2013) Applicability of sustainability indicators to wastewater treatment processes. Comp Aided Chem Eng 32:931–936. https://doi.org/10.1016/B978-0-444-63234-0.50156-1
25. Kamali M, Gameiro T et al (2019) Enhanced biodegradation of phenolic wastewaters with acclimatized activated sludge—a kinetic study. Chem Eng J 378:122186. https://doi.org/10.1016/j.cej.2019.122186
26. Kamali M, Costa ME et al (2019) Sustainability of treatment technologies for industrial biowastes effluents. Chem Eng J 370:1511–1521. https://doi.org/10.1016/j.cej.2019.04.010
27. Shukla SK et al (2015) Combining activated sludge process with membrane separation to obtain recyclable quality water from paper mill effluent. Clean Technol Environ Policy 17(3):781–788. https://doi.org/10.1007/s10098-014-0836-2
28. Young MN, Marcus AK, Rittmann BE (2013) A Combined Activated Sludge Anaerobic Digestion Model (CASADM) to understand the role of anaerobic sludge recycling in wastewater treatment plant performance. Bioresour Technol 136:196–204. https://doi.org/10.1016/j.biortech.2013.02.090
29. Van den Broeck RMR, Van Impe JFM, Smets IYM (2009) Assessment of activated sludge stability in lab-scale experiments. J Biotechnol 141:147–154. https://doi.org/10.1016/j.jbiotec.2009.02.019
30. Comas J et al (2008) Risk assessment modelling of microbiology-related solids separation problems in activated sludge systems. Environ Model Softw 23:1250–1261. https://doi.org/10.1016/j.envsoft.2008.02.013
31. Liang J et al (2021) Dewaterability improvement and environmental risk mitigation of waste activated sludge using peroxymonosulfate activated by zero-valent metals: Fe_0 vs. Al_0. Chemosphere 280:130686. https://doi.org/10.1016/j.chemosphere.2021.130686
32. Cañote SJB et al (2021) Life cycle assessment of upflow anaerobic sludge blanket sludge management and activated sludge systems aiming energy use in the municipality of Itajubá, Minas Gerais, Brazil. J Mater Cycles Waste Manage 23:1810–1830. https://doi.org/10.1007/s10163-021-01253-0
33. Villamil JA et al (2020) Anaerobic co-digestion of the process water from waste activated sludge hydrothermally treated with primary sewage sludge. A new approach for sewage sludge management. Renew Energy 146:435–443. https://doi.org/10.1016/j.renene.2019.06.138
34. Xin X et al (2019) An integrated approach for waste activated sludge management towards electric energy production/resource reuse. Bioresour Technol 274:225–231. https://doi.org/10.1016/j.biortech.2018.11.092
35. Postacchini L et al (2016) Life cycle assessment comparison of activated sludge, trickling filter, and high-rate anaerobic-aerobic digestion (HRAAD). Water Sci Technol 73:2353–2360. https://doi.org/10.2166/wst.2016.087
36. Falås P et al (2012) Suspended biofilm carrier and activated sludge removal of acidic pharmaceuticals. Water Res 46:1167–1175. https://doi.org/10.1016/j.watres.2011.12.003
37. Gonzalez-Gil L et al (2021) Feeding composition and sludge retention time both affect (co-)metabolic biotransformation of pharmaceutical compounds in activated sludge systems. J Environ Chem Eng 9:105123. https://doi.org/10.1016/j.jece.2021.105123
38. Freitas RXA et al (2019) Characterization of the primary sludge from pharmaceutical industry effluents and final disposition. Processes 7:231. https://doi.org/10.3390/pr7040231
39. Yadav MK et al (2019) Understanding the removal and fate of selected drugs of abuse in sludge and biosolids from Australian wastewater treatment operations. Chinese Academy of Engineering 5:872–879. https://doi.org/10.1016/j.eng.2019.07.012
40. Boonnorat J et al (2019) Effect of hydraulic retention time on micropollutant biodegradation in activated sludge system augmented with acclimatized sludge treating low-micropollutants wastewater. Chemosphere 230:606–615. https://doi.org/10.1016/j.chemosphere.2019.05.039

41. Kim S et al (2005) Removal of antibiotics in wastewater: effect of hydraulic and solid retention times on the fate of tetracycline in the activated sludge process. Environ Sci Technol 39(15):5816–5823. https://doi.org/10.1021/es050006u
42. Verma N et al (2014) Effects of anaerobic digester sludge age on odors from dewatered biosolids. Proc Water Environ Fed 2006:1119–1141. https://doi.org/10.2175/193864706783749864
43. Schröder P et al (2007) Using phytoremediation technologies to upgrade waste water treatment in Europe. Environ Sci Pollut Res 14:490–497. https://doi.org/10.1065/espr2006.12.373
44. Hussain A, Dubey SK, Kumar V (2015) Kinetic study for aerobic treatment of phenolic wastewater. Water Resour Industry 11:81–90. https://doi.org/10.1016/j.wri.2015.05.002
45. Schnicke HFS, Märkl RMH (2001) Determination of the kinetic parameters of the phenol-degrading thermophile *Bacillus themoleovorans* sp. A2. Appl Microbiol Biotechnol 57:744–750. https://doi.org/10.1007/s002530100823
46. Stasinakis AS et al (2007) Investigation of triclosan fate and toxicity in continuous-flow activated sludge systems. Chemosphere 68:375–381. https://doi.org/10.1016/j.chemosphere.2007.01.047
47. Slater FR et al (2011) Pandemic pharmaceutical dosing effects on wastewater treatment: no adaptation of activated sludge bacteria to degrade the antiviral drug Oseltamivir (Tamiflu®) and loss of nutrient removal performance. FEMS Microbiol Lett 315:17–22. https://doi.org/10.1111/j.1574-6968.2010.02163.x
48. Nippes RP et al (2021) Since January 2020 Elsevier has created a COVID-19 resource centre with free information in English and Mandarin on the novel coronavirus COVID-19. The COVID-19 resource centre is hosted on Elsevier Connect, the company's public news and information. Process Saf Environ Prot 152:568–582
49. Kamali M et al (2022) Acclimatized activated sludge for enhanced phenolic wastewater treatment using pinewood biochar. Chem Eng J 427:131708. https://doi.org/10.1016/j.cej.2021.131708
50. Al-Khateeb LA, Almotiry S, Salam MA (2014) Adsorption of pharmaceutical pollutants onto graphene nanoplatelets. Chem Eng J 248:191–199. https://doi.org/10.1016/j.cej.2014.03.023
51. Kryuchkova M et al (2021) Pharmaceuticals removal by adsorption with montmorillonite nanoclay. Int J Mol Sci 22:9670. https://doi.org/10.3390/ijms22189670
52. Shan D et al (2018) Intercalation of rigid molecules between carbon nanotubes for adsorption enhancement of typical pharmaceuticals. Chem Eng J 332:102–108. https://doi.org/10.1016/j.cej.2017.09.054
53. Yu Z, Peldszus S, Huck PM (2009) Adsorption of selected pharmaceuticals and an endocrine disrupting compound by granular activated carbon. 2. Model prediction. Environ Sci Technol 43:1474–1479. https://doi.org/10.1021/es7032185
54. Dwandaru WSB et al (2022) Silver nanoparticle-graphene oxide mixture as anti-bacterial against *Staphylococcus aureus*. AIP Conf Proc 2022:020015. https://doi.org/10.1063/1.5141628
55. Mokkapati VRSS et al (2017) Membrane properties and anti-bacterial/anti-biofouling activity of polysulfone-graphene oxide composite membranes phase inversed in graphene oxide non-solvent. RSC Adv 7:4378–4386. https://doi.org/10.1039/C6RA25015G
56. Shao W et al (2015) Anti-bacterial performances and biocompatibility of bacterial cellulose/graphene oxide composites. RSC Adv 5:4795–4803. https://doi.org/10.1039/c4ra13057j
57. Falletti L, Conte L (2007) Upgrading of activated sludge wastewater treatment plants with hybrid moving-bed biofilm reactors. Ind Eng Chem Res 46:6656–6660. https://doi.org/10.1021/ie061635v
58. Ooi GTH et al (2018) Biological removal of pharmaceuticals from hospital wastewater in a pilot-scale staged moving bed biofilm reactor (MBBR) utilising nitrifying and denitrifying processes. Bioresour Technol 267:677–687. https://doi.org/10.1016/j.biortech.2018.07.077
59. Tang K et al (2020) Municipal wastewater treatment targeting pharmaceuticals by a pilot-scale hybrid attached biofilm and activated sludge system (HybasTM). Chemosphere 259:127397. https://doi.org/10.1016/j.chemosphere.2020.127397

60. Ma D et al (2021) Critical review of advanced oxidation processes in organic wastewater treatment. Chemosphere 275:130104. https://doi.org/10.1016/j.chemosphere.2021.130104
61. Miklos DB et al (2018) Evaluation of advanced oxidation processes for water and wastewater treatment—a critical review. Water Res 139:118–131. https://doi.org/10.1016/j.watres.2018.03.042
62. Mowla A, Mehrvar M, Dhib R (2014) Combination of sonophotolysis and aerobic activated sludge processes for treatment of synthetic pharmaceutical wastewater. Chem Eng J 255:411–423. https://doi.org/10.1016/j.cej.2014.06.064
63. da Silva JRP et al (2021) Study of effects of pharmaceuticals on the activated sludge process combining advanced oxidation using ultraviolet/hydrogen peroxide to increase their removal and mineralization of wastewater. J Environ Chem Eng 9:104576. https://doi.org/10.1016/j.jece.2020.104576
64. Mir-Tutusaus JA et al (2021) Prospects on coupling UV/H_2O_2 with activated sludge or a fungal treatment for the removal of pharmaceutically active compounds in real hospital wastewater. Sci Total Environ 773:145374. https://doi.org/10.1016/j.scitotenv.2021.145374
65. Mansour D et al (2015) Mineralization of synthetic and industrial pharmaceutical effluent containing trimethoprim by combining electro-Fenton and activated sludge treatment. J Taiwan Inst Chem Eng 53:58–67. https://doi.org/10.1016/j.jtice.2015.02.022
66. Alfonso-Muniozguren P et al (2021) A review on pharmaceuticals removal from waters by single and combined biological, membrane filtration and ultrasound systems. Ultrason Sonochem 76:105656. https://doi.org/10.1016/j.ultsonch.2021.105656

Chapter 5
Pharmaceutically Active Compounds in Anaerobic Digestion Processes—Biodegradation and Fate

5.1 Introduction

Anaerobic digestion (AD) has been widely used for the treatment of a wide range of industrial and nonindustrial effluents due to parameters such as high efficiency for the removal of chemical oxygen demand (COD) and biological oxygen demand (BOD), as well as high stability [1–3]. Production of biogas is also the main advantage of AD processes that can satisfy renewable energy strategies [4, 5]. In anaerobic environments, organic compounds undergo the fermentation process, which occurs in the absence of inorganic electron acceptors such as O_2. Under these conditions, protons and HCO_3^- receive the electrons produced in the medium by the activity of anaerobic microorganisms, leading to the transformation of complex organic compounds into low molecular weight intermediates such as volatile fatty acids (VFAs). VFAs are then converted to hydrogen and acetate by hydrogen-producing acetogens. Finally, H_2 and CO_2 are converted to CH_4 by methanogens present in the medium [6].

However, the efficiency of AD processes can be influenced by the presence of recalcitrant and nonbiodegradable compounds, such as pharmaceutically active compounds (PhACs), which are currently widely consumed and released into water bodies [7]. It can potentially affect the stability and performance of AD systems. Moreover, the toxic nature of some pharmaceutical compounds can inhibit AD microorganisms from their normal activities for the biodegradation of organic compounds, leading to failure of the system. There are various parameters that can influence the performance of AD systems to deal with such compounds, including the temperature (i.e., Psychrophilic (0–15 °C), mesophilic (20–45 °C), and thermophilic (50–80 °C)), the microbial community present in the medium, hydraulic retention time (HRT), pH, the initial COD and BOD of the influents, and the presence of AD inhibitory elements (such as sulfide compounds) [8].

This chapter has aimed to provide in-depth knowledge regarding the effects of various AD parameters and the mechanisms involved in the interactions between the AD microorganisms and PACs and their impacts on the overall performance of

M. Kamali et al., *Advanced Wastewater Treatment Technologies for the Removal of Pharmaceutically Active Compounds*, Green Energy and Technology,
https://doi.org/10.1007/978-3-031-20806-5_5

the system, including the biogas production and treatment of wastewaters to recover clean water resources.

5.2 AD for PhAC-Containing Effluents

The type and extent of the effects that PhACs can have on AD systems, as well as the mechanisms involved, are directly related to the types of pollutants and the operating conditions. There are studies in the literature stating that various PhACs can cause various effects on AD microorganisms and the overall performance of AD systems to deal with effluents (or solid wastes) containing such compounds.

The toxic effects that can be expected from the presence of PhACs to the AD microorganisms are directly related to some parameters, such as the type and concentration of the pharmaceutical and the operating conditions. Some PhACs represent dose-dependent toxic effects on AD microorganisms. For instance, 15 min half inhibitory concentrations (15 min-IC_{50}) of antibiotics, including ciprofloxacin, kanamycin, lincomycin, and amoxicillin, have been determined in a study to be 5.63 mg/L, 5.11 mg/L, 4.32 mg/L, and 3.99 mg/L, respectively [9, 10]. This value for chloromycetin, aureomycin, and polymyxin was identified as 429.90 mg/L, 12.06 mg/L, and 6.24 mg/L, respectively. There are also reports indicating the inhibitory effects of PhACs on the AD process. It has been observed for some PhACs, such as erythromycin, sulfamethoxazole, and tetracycline, even at low concentrations (1 mg/L) [11, 12]. However, some PhACs, such as cefazolin, can cause only a lag phase for methane production [13]. Severe inhibitory effects of specific pharmaceuticals such as dichloromethane have also been observed when treating effluents from the pharmaceutical industry [14].

The microbial communities present in the medium are also a determinant parameter for the removal of PhACs from effluents using AD processes. Recent studies have concluded that the presence of microorganisms from the phylum including *Thermotogae, Bacteroidetes, Proteobacteria, Spirochaetes, Firmicutes, Synergistetes, Chlorobi, Chloroflexi, Euryarchaeota, Actinobacteria, and Elusimicrobia* can considerably promote the degradation of PhACs and the methane production yield [15]. Figure 5.1 represents the microorganisms that can contribute to the degradation of PhACs under anaerobic treatment conditions.

Extracellular electron transfer (EET) is the mechanism of shuttling electrons between microbes in a biological system [16]. Direct electron transfer (DIET) is the dominant EET and occurs mainly by electrically conductive materials, conductive pili, or extracellular substances [17, 18]. In AD systems, it has been well discussed in the literature that the addition of conductive materials (CMs) can significantly improve the ETM process and hence both the methane production and the removal efficiency for complex organic compounds. In addition, CMs can aid in the enrichment of electroactive bacteria, which can further enhance the performance of AD systems for the removal of PhACs. Various types of CMs have been used to this end. In this regard, iron-based materials have high electron conductivity and low toxicity,

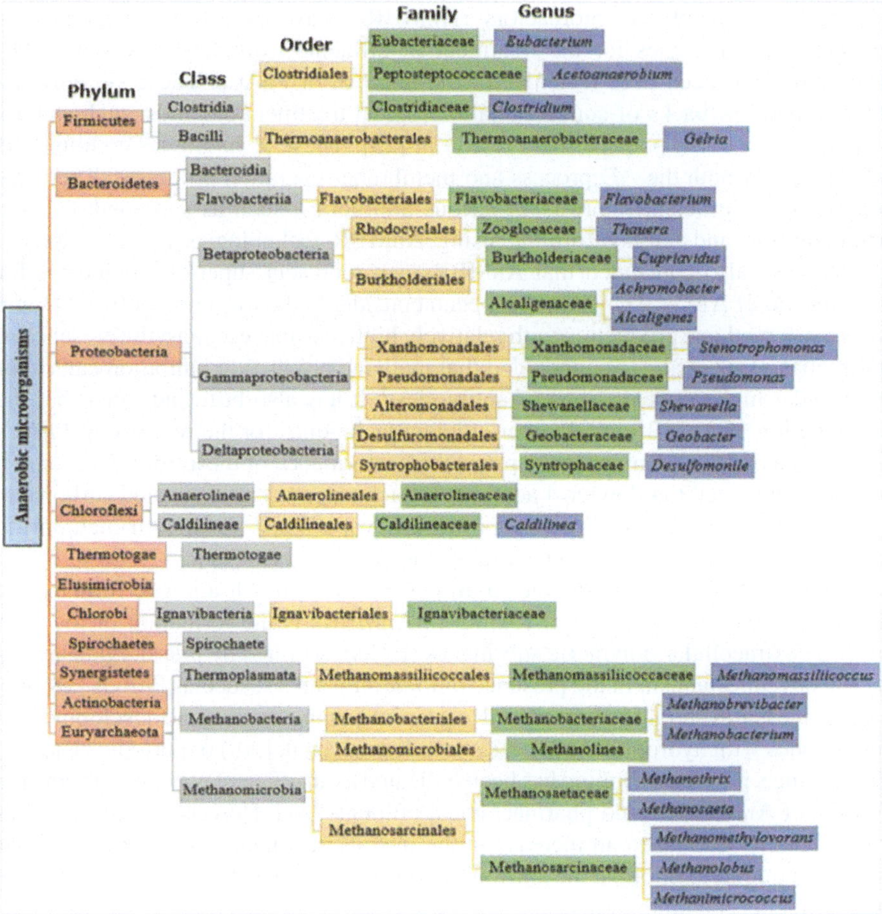

Fig. 5.1 The microbial communities that can play a role in the biodegradation of PhACs during the AD process, adapted from Aziz et al. [15]

which can make them appropriate to trigger the efficiency of AD systems. Inexpensive iron-based materials such as nZVI are good candidates and have represented acceptable performances for such applications [19]. Enrichment of acidogens, especially acetogens, by the addition of nZVI is an example of the successful application of such an attractive conductive material [20].

It has been reported that nZVI can efficiently enhance the performance of the AD process to deal with PhACs. For instance, Dai et al. [21] discussed that the addition of these iron-based materials together with granular activated carbon (GAC) has led to improvements in chemical oxygen demand (COD) removal and methane production yield. The strategy was also efficient for the removal of intermediate by-products such as dehydroepiandrosterone (DHEA) and 2,2′-methylenebis(6-tert-butyl-4-methylphenol) with possible toxic effects.

Anaerobic membrane bioreactors (AnMBRs) have also been developed as promising technologies for the treatment of effluents from various origins [22]. They have also been used recently for the removal of PhACs. These systems can address the drawbacks of conventional biological treatment methods, such as long start-up periods and poor biomass retention [23]. These technologies combine the advantages of both the AD process and membrane separation processes. There are reports of the efficient removal of various antibiotics, such as sulfamethoxazole, trimethoprim, and tetrahydrofuran, using AnMBR technologies [24–26]. Recent studies have also concluded that AnMBRs can represent superior efficiencies for the removal of ARGs. They have also been considered the most productive AD win in terms of methane production with relatively high resistance against the presence of antibiotics. Aziz et al. [15] concluded that methane production inhibition can occur only under high concentrations of antibiotics.[1] It has also been indicated that the combination of AnMBRs with other efficient techniques for the removal of PhACs (such as adsorption) can further improve the performance of these configurations. It has been indicated that low-cost adsorbents such as powdered activated carbon can promote the efficiency of AnMBRs for the removal of PhACs such as trimethoprim, sulfamethoxazole, carbamazepine, diclofenac, and triclosan [27].

However, in such configurations, fouling may affect the filtration performance of the system. The main source of fouling in AnMBRs is the formation of the cake layer by extracellular polymeric substances (EPSs), secreted by microorganisms, as a protective mechanism in the presence of toxic substances, such as PhACs [28, 29]. Hence, measures should be considered to prevent fouling, such as the development of membranes with hydrophilic properties (see Kamali et al. [30]) or periodic cleaning. For instance, physical stripping has been indicated as an efficient way to overcome the fouling of AnMBR-treated pharmaceutical effluents [31]. However, such measures can be costly, leading to an increase in the overall treatment costs. Further studies may be required to improve the antifouling properties of the membranes for dealing with effluents containing PhACs.

Implementation of combined strategies can also lead to the improvement of the removal efficiency of the PhACs. In this regard, combinations of anaerobic–aerobic processes have been considered cost-effective and efficient for effluents containing recalcitrant organic compounds. Ahmad and Eskicioglu [32] indicated that a combination of sequential anaerobic/aerobic/anoxic processes can be used for efficient methods for the removal of PhACs, including ibuprofen and fenofibric, as indicated in Fig. 5.2.

---

[1] > 25 mg/L, > 1 mg/L, > 10 mg/L, > 80 mg/L, > 90 mg/L, > 130 mg/L, and > 10 mg/L for sulfamethoxazole, tetracycline, ofloxacin, ciprofloxacin, sulfamerazine, tylosin, and ceftiofur, respectively.

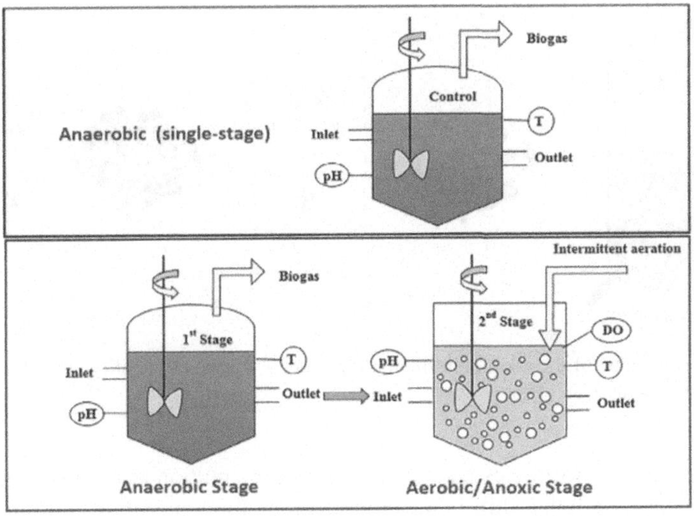

Fig. 5.2 An anaerobic/aerobic/anoxic configuration, used for the efficient removal of PhACs, adopted from Ahmad and Eskicioglu [32]

5.3 AD for PhAC-Containing Waste Sludge

PhACs are also considered an important group of micropollutants in sludge composition with possible adverse effects on the environment and living organisms [33, 34]. They can be present in various concentrations according to their origin. For instance, tonalide and galaxolide can be found in high concentrations up to 427 mg/kg in sewage sludge [35–38]. Furthermore, bactericides such as triclosan have been extensively detected in biosolids at concentrations up to 9850 μg/kg [39]. Relatively high concentrations of painkillers such as ibuprofen (up to 950 μg/kg) [36]. Various hormones and hormone-like compounds can also be frequently detected in sludge from wastewater treatment plants. Both natural (e.g., estriol, estradiol, and estrone) and synthetic (e.g., 17α-ethinylestradiol (EE2)) hormones can also be found in the composition of the sludge [40]. Such sludge wastes can create environmental issues, especially when they originate from pharmaceutical industries with high levels of various PhACs. Their disposal can also be costly, which motivates the development of sustainable and low-cost treatment methods such as anaerobic digestion [41].

As discussed in Chap. 1, there is an ongoing concern related to the increasing concentrations of these compounds in the environment, which can lead to the generation of antibiotic-resistant bacteria (ARBs) and antibiotic resistance genes (ARGs) [42, 43]. There are ways to control the release of such agents, such as pretreatment of the sludge before the AD process. For instance, thermal hydrolysis has been indicated to be efficient for the reduction of ARGs, as well as mobile genetic elements (MGEs) [44]. Such a strategy can also enhance the biogas production rate by solubilizing the organic compounds in the medium. Reduction of the ARGs in the sludge has also

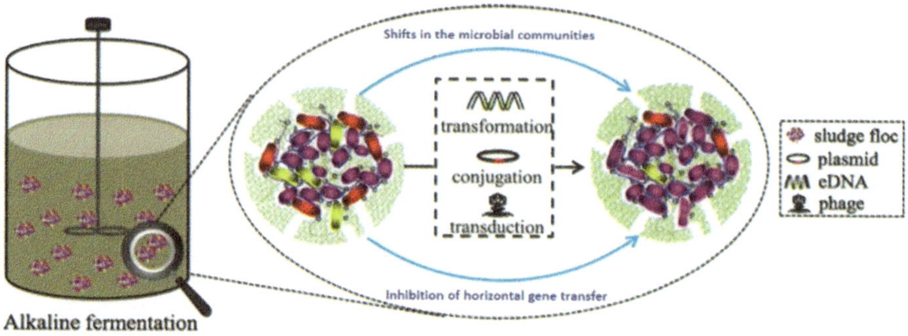

Fig. 5.3 A schematic of the alkaline fermentation process for the elimination of ARGs in sludge, adapted from Huang et al. [45]

been achieved with strategies such as alkaline fermentation at high pH values (i.e., 10), which may result in a shift in the composition of the microbial communities that could restrict the ARG hosts in the sludge (Fig. 5.3) [45].

It has been indicated that the removal of PhACs can be controlled by two main mechanisms, biotransformation and adsorption [46]. However, the capacity of AD systems for the biotransformation of such pollutants is limited, and adsorption plays the most important role in the removal of such compounds [47]. Adsorption of the PhACs can also lead to the inactivation of the microorganisms when toxic pharmaceuticals exist in the medium. This can negatively influence the efficiency of the AD systems for the biodegradation of organic compounds. For instance, Bai et al. [48] indicated a drop in methane production yield of two mesophilic AD systems for sludge treatment through the addition of antibiotics. 16S rRNA sequencing results also demonstrated a decrease in the richness while increasing the diversity of the microorganisms. They also indicated a significant change in the microbial communities at the genus level. PCR tests also revealed that the addition of PhACs could induce an increase in antibiotic resistance genes (ARGs) during the sludge AD process. Additionally, Akyol et al. [49] indicated that the addition of oxytetracycline to cattle manure digesters can reduce the abundance of Methanosarcina and syntrophic acetate-oxidizing bacteria, which are critical for methane production while leading to the occurrence of ARGs. The significant negative effects of a mixture of oxytetracycline, sulfadimethoxine, and norfloxacin on the methane production yield were also reported by Zhi et al. [50].

Lower lag phases were given by the modified Gompertz model when stimulating effects occurred. The variations in physico-chemical parameters and microbial Venn maps both showed that day 5 was a critical point for digestion time. The relative abundance of Methanosarcina was enhanced when the stimulating effect occurred, whereas Methanoculleus decreased. Different microbial characteristics were obtained for different samples from the heatmaps.

It is worth mentioning that the anaerobic digestion systems have generally represented moderate efficiencies for the removal of ARGs (approximately 50%). Hence, there is a risk of release of these agents into the environment after discharging the treated sludge. To address this issue, some techniques, such as alkaline, thermal hydrolysis, and ultrasonic pretreatment of the sludge, have been employed, resulting in a significant decrease in ARGS after the AD process [45, 51]. Carballa et al. [52] indicated that an alkaline pretreatment is superior to thermal treatment for the solubilization of organic matter (up to 80%), resulting in the removal of organic matter up to 75%. They also concluded that the implemented strategy can result in the efficient removal of PhACs such as naproxen and natural estrogens (up to 80%). However, carbamazepine degradation and removal were not observed even after pretreatments.

Physical methods such as ultrasonic and microwave have also been demonstrated to have a positive effect on the solubilization of solid and organic matter, which can lead to improvements in the removal efficiency of the AD process and the biogas yield. The basis of ultrasonic irradiation is to create hotspots as a result of the formation, rapid growth, and catastrophic collapse of bubbles in the medium, which can provide a very high local density of energy [53]. The output energy is very efficient for the solubilization of organic compounds, which can make them easier to digest by microorganisms [54–56]. Wang et al. [2, 3] indicated that the pretreatment of sludge with ultrasonic irradiation is more efficient than alkaline and thermal hydrolysis for the removal of these agents. However, there is a need for more studies to optimize pretreatment techniques for the elimination of such elements and to prevent their introduction into the environment [44]. Such pretreatments can also enhance the efficiency of the AD process for the removal of PhACs. In a relevant study, [57] indicated that the removal of 4 PhACs, including clofibric acid (73%), triclosan (76%), carbamazepine (73%), and diclofenac (64%), in sewage sludge can be enhanced by applying ultrasonic or enzymatic treatments before the thermophilic AD process. This can also lead to an increase in the methane production yield [58]. The energy input from ultrasonic irradiation can influence the microstructure of the lignocellulosic compounds and make them more appropriate for the AD process.

There are also efforts in the literature for the application of microwave irradiation[2] as a sludge pretreatment method to promote the efficiency of the AD process [59]. However, this technique has not been extensively used for PhAC-containing sludge.

It has also been well documented in the literature that the efficiency of the AD systems for the removal of PhACs from waste sludge is directly related to the operating conditions, such as pH, temperature, solid retention time (SRT), and the composition of the microbial communities used for the anaerobic digestion process [15, 48, 60].

The fate and removal of PhACs during the anaerobic digestion of sludge are highly influenced by the type and complexity of the compound. The studies have indicated the efficient biotransformation of less complex pharmaceuticals such as sulfamethoxazole, while AD represents relatively low efficiency to deal with carbamazepine. Narumiya et al. [61] studied the anaerobic digestion of sewage sludge containing various PhACs. They indicated that trimethoprim and sulfamethoxazole

[2] Described as nonionizing radiation within 0.3–300 GHz which falls between infrared light and radio waves in the electromagnetic spectrum.

could not be further detected in the digested sludge. Moderate removal of compounds such as ofloxacin, triclocarban, and triclosan (up to 50%) and almost no removal of carbamazepine were also achieved.

As demonstrated in Fig. 5.4, a decline in the removal efficiency of PhACs can be observed for most of the studied compounds by elevating the SRT from 10 to 30 days. However, there are studies indicating the positive effects of the increase in SRT on the biodegradation of some recalcitrant pharmaceuticals. For instance, Gallardo-Altamirano et al. [62] indicated that an extension in the SRT from 12 to 24 days favored the removal of PhACs, including carbamazepine, clarithromycin, codeine, gemfibrozil, ibuprofen, lorazepam, and propranolol, in a pilot-scale two-stage mesophilic AD designed to treat a sample of sewage sludge. Quantitative PCR (qPCR) and Illumina MiSeq sequencing revealed that increasing the SRT resulted in a decrease in the diversity of bacterial and archaeal communities. They discussed that the relative abundance of some bacteria under longer SRT results in better biodegradation of PhACs.

The temperature also has a determinant effect on the AD process. Although higher efficiencies have been observed in thermophilic conditions, mesophilic AD is generally preferred for the biodegradation of organic pollutants due to the need for less energy and hence satisfying sustainability considerations [8]. However, various

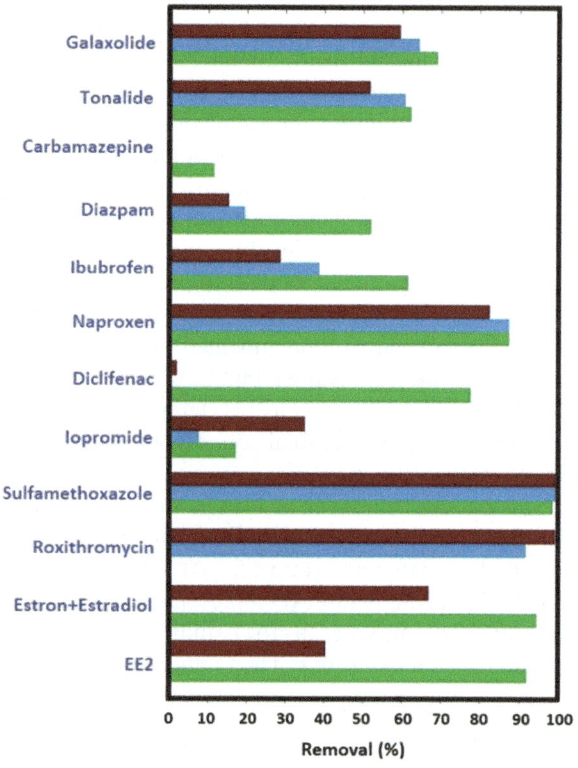

Fig. 5.4 The observed removal efficiency of various PhACs under various SRTs was adopted from Carballa et al. [63]: brown bar: 30 days, blue bar: 20 days, and green bar: 10 days

removal efficiencies have been observed for the removal of PhACs under various temperatures. For instance, a study aiming at studying the biodegradation of 20 PhACs using AD under mesophilic (37 °C) and thermophilic (55 °C) conditions revealed that AD is not efficient for some PhACs, such as galaxolide, β-estradiol, and triclosan, either in mesophilic or thermophilic compositions (removal efficiency < 20%) [36]. Very low biodegradation of ibuprofen (< 30%) was also observed in the mentioned study. However, over 80% biotransformation was observed for less recalcitrant compounds such as sulfamethoxazole. Finally, Gonzalez-Gil et al. [36] concluded that temperature had no significant impacts on the removal of PhACs from waste sludge (except for 17α-ethinylestradiol).

Microbial populations and their metabolic activities can also influence the biotransformation of PhACs from the sludge content [64]. For instance, methanogenesis bacteria have been identified with the ability to remove and biotransform 20 PhACs (above 50% in total), with higher efficiencies for antibiotics (i.e., roxithromycin and sulfamethoxazole) and neuro-drugs (i.e., citalopram and fluoxetine) [65].

Low organic loading rates (OLRs) have also been indicated to be favorable for the biodegradation of PhACs by giving more time to the microbial communities for the biodegradation of these compounds. For instance, high biodegradation rates have been observed for naproxen, sulfamethoxazole, roxithromycin and estrogens under a low OLR of 1.8 (kg VS/m^3/d) [63]. Such a condition also resulted in moderate to high removal of galaxolide, tonalide and diclofenac, diazepam, and ibuprofen. However, even under such low OLRs, very low (< 20%) removal of iopromide and no removal of carbamazepine were observed.

Any sudden change in the initial concentration of PhACs in the sludge can also affect biogas production and PhAC removal efficiency in AD processes. Some studies have explored the extent of such effects. For instance, Huang et al. [66] indicated that a shock load of macrolide antibiotics (from 0 to 2000 mg/kg) in waste-activated sludge can considerably suppress the methane production yield during the first 10 days of AD processes. However, they observed that methane production recovered slowly thereafter, but the maximum methane production rate dropped from 22 mL/(g volatile suspended solids (VSS)/d) to 15 mL/(g VSS/d).

As can be concluded from the existing studies on the biodegradation of various PhACs, AD is efficient for compounds with less- to moderate-complex PhACs. However, recalcitrant compounds such as carbamazepine can remain without any changes in the digested sludge and can be discharged to the environment. Hence, there is a need for complementary methods such as an efficient oxidation process that is able to deal with such recalcitrant PhACs [47].

Additives can also promote the efficiency of AD systems. ZVI power is a well-known magnetic material that has been used in various (waste)water treatment methods, such as Fenton oxidation [67, 68]. The addition of ZVI has also been indicated as an efficient way to improve electron transfer as an essential mechanism in the AD process (Fig. 5.5). There are also recent studies demonstrating the positive effects of the addition of ZVI on the removal of PhACs from waste sludge.

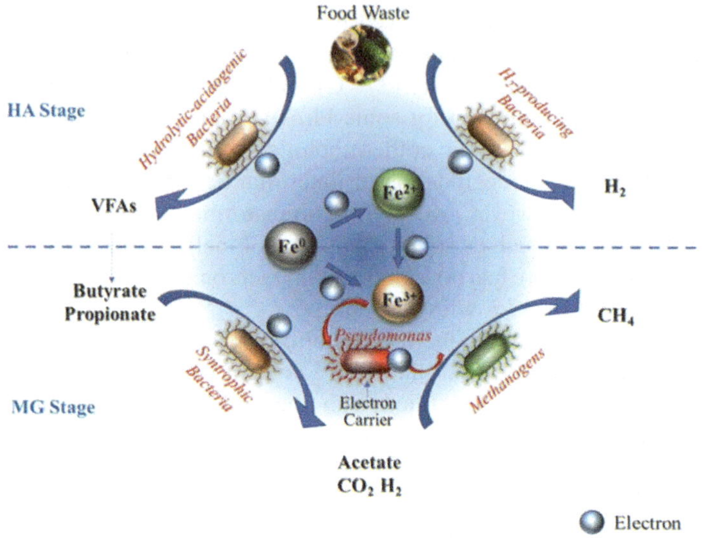

Fig. 5.5 The main mechanisms of the improvement in the removal efficiency of the AD process by the addition of ZVI, reprinted with permission from Yuan et al. [69]

Zhou et al. [70] indicated that the addition of 1 g/L of ZVI into a mesophilic AD process treating sewage sludge resulted in the improved removal of antibiotics, including sulfamerazine (97%), sulfamethoxazole (75%), tetracycline (79%), and roxithromycin (57%).[3] They also indicated that this strategy can be considered very efficient for the removal of ARGs, including AAC (6′)-IB-CR, qnrS, ermF, ermT, ermX, sul1, sul2, sul3, tetA, tetB, and tetG, from sludge during the anaerobic digestion process. The applicability of other iron-based materials has also been investigated for the improvement of methane production and the removal of PhACs. Suanon et al. [71] indicated that iron powder can considerably enhance the methane yield (by 41%) under mesophilic AD conditions. The system was also able to efficiently remove pharmaceutical compounds during the AD process.

The effects of the presence of other metallic compounds on the methane production yield in the presence of pharmaceuticals have also been investigated. A recent study by Zhao et al. [72] indicated that norfloxacin and sulfamethazine at relatively high concentrations (500 mg/kg) could not inhibit the methane production yield, but the mixture of these antibiotics with zinc oxide (ZnO) inhibited the hydrolysis, fermentation, and methanogenesis phases of the AD process. Additionally, they found that the effects of ZnO alone are less than those of the mixture on the methane production yield and the overall AD performance. It has been well indicated that ZnO can cause adverse effects to microorganisms and hence inhibit CH_4 production. For instance, growth inhabitation has been observed by introducing ZnO nanoparticles [73].

[3] The system was not efficient for the removal of ofloxacin.

Table 5.1 Further reading suggestions for more detailed coverage of the literature on the PhACs in AD processes

References	Item	Subject
Venegas et al. [75]	Table 3	Concentrations of various micropollutants in the sewage sludge
	Table 4	Effectiveness of AD process for the elimination of various micropollutants
Azizan et al. [46]	Table 1	Impacts of various PhACs on the methane yield and the VFAs accumulation
Hammer and Palmowski [76]	Table 1	The efficiency of AD processes for the removal of various PhACs
Stasinakis [47]	Table 2	Applicability of AD process under various operating conditions for the elimination of PhACs

A combination of various strategies has also been considered efficient for the enhanced biodegradation of PhACs. For instance, integration of mesophilic anaerobic treatment with the following mesophilic AD was indicated by Salgado et al. [74] for the removal of PhACs, including sulfamethoxazole (approximately 90%), caffeine (approximately 90%), oxazepam (73%), propranolol (approximately 60%), and ofloxacin (approximately 40%). However, no significant removals were observed for diclofenac and 2 hydroxy-ibuprofen, ibuprofen, and carbamazepine. The efficient biodegradation of the mentioned PhACs can be due to the combination of two redox conditions, including microaeration and anaerobic digestion. However, even such strategies cannot eliminate recalcitrant PhACs, and the risk of their release from treatment plants is considerable.

5.4 Further Reading

Table 5.1 contains items from the recent literature that the reader can be further assisted in acquiring more detailed information regarding the presence and behavior of PhACs in AD processes.

5.5 Summary

There has been an increasing pattern of the consumption of various PhACs to deal with diseases, especially under the current COVID-19 pandemic situation. Hence, there have been efforts to better understand the fate and behavior of various PhACs and the applicability of various (waste)water treatment methods for the elimination of such compounds and prevention of their release into the environment, causing secondary issues such as the generation of antibiotic-resistant bacteria (ARB) and antibacterial resistance genes (ARGs). Anaerobic digestion has been considered an interesting technique for the simultaneous removal of organic pollutants and production of methane as a sustainable energy carrier. Because of their relatively low concentrations, PhACs cannot be considered important carbon sources for the AD process. However, they can cause toxic effects or can inhibit biogas production depending on their type and concentration. AD systems can also represent various degrees of efficiency for the removal of PhACs. Adsorption is considered the main mechanism (superior to biotransformation) for the elimination of PhACs in AD systems. Hence, the addition of sustainable and low-cost adsorbents can be considered an efficient way to increase the efficiency of AD systems to remove pharmaceuticals in AD systems. There have also been efforts to compare the efficiency of various pretreatment methods and their effects on the solubilization of organic matter and volatile solids. Such a strategy has been demonstrated to have a positive effect on the removal of PhACs such as estrogens and some antibiotics but without any significant effect on the degradation of recalcitrant compounds such as carbamazepine. The addition of iron-based materials such as zero-valent iron has also been indicated to have positive effects on both PhAC removal and methane production yield. However, there are issues regarding the application of these materials for real-scale applications, such as economic considerations for the production and application of these materials. Additionally, the fate of these additives when released into the environment should be studied to prevent the possible subsequent effects from the application of such materials. Other additives, such as activated carbon and biochar, can also be considered sustainable additives that can considerably enhance the performance of AD systems for the removal of PhACs, mainly through the combination of biotransformation and adsorption of such compounds and prevention of their release into the environment. For future studies, there is a need to explore the biotransformation mechanisms of PhACs in both liquid and solid phases.

References

1. Pigoli A et al (2021) Thermophilic anaerobic digestion as suitable bioprocess producing organic and chemical renewable fertilizers: a full-scale approach. Waste Manage 124:356–367. https://doi.org/10.1016/j.wasman.2021.02.028
2. Wang C et al (2019) A short-term stimulation of ethanol enhances the effect of magnetite on anaerobic digestion. Appl Microbiol Biotechnol 103:1511–1522. https://doi.org/10.1007/s00253-018-9531-2
3. Wang M, Li R, Zhao Q (2019) Distribution and removal of antibiotic resistance genes during anaerobic sludge digestion with alkaline, thermal hydrolysis and ultrasonic pretreatments. Front Environ Sci Eng 13:43. https://doi.org/10.1007/s11783-019-1127-2
4. Curry N, Pillay P (2012) Biogas prediction and design of a food waste to energy system for the urban environment. Renew Energy 41:200–209
5. Lim M et al (2020) Removal of organic micropollutants in anaerobic membrane bioreactors in wastewater treatment: critical review. Environ Sci: Water Res Technol 6:1230–1243. https://doi.org/10.1039/c9ew01058k
6. Meyer T, Edwards EA (2014) Anaerobic digestion of pulp and paper mill wastewater and sludge. Water Res 65:321–349. https://doi.org/10.1016/j.watres.2014.07.022

7. Lv L et al (2017) Microbial community composition and function in a pilot-scale anaerobic-anoxic-aerobic combined process for the treatment of traditional Chinese medicine wastewater. Biores Technol 240:84–93. https://doi.org/10.1016/j.biortech.2017.01.053
8. Kamali M et al (2016) Anaerobic digestion of pulp and paper mill wastes—an overview of the developments and improvement opportunities. Chem Eng J 298:162–182. https://doi.org/10.1016/j.cej.2016.03.119
9. Ji JY, Xing YJ, Ma ZT, Zhang M et al (2013a) Acute toxicity of pharmaceutical wastewaters containing antibiotics to anaerobic digestion treatment. Chemosphere 91:1094–1098. https://doi.org/10.1016/j.chemosphere.2013.01.009
10. Ji JY, Xing YJ, Ma ZT, Cai J et al (2013b) Toxicity assessment of anaerobic digestion intermediates and antibiotics in pharmaceutical wastewater by luminescent bacterium. J Hazard Mater 246–247:319–323. https://doi.org/10.1016/j.jhazmat.2012.12.025
11. Aydin S, Ince B et al (2015) Combined effect of erythromycin, tetracycline and sulfamethoxazole on performance of anaerobic sequencing batch reactors. Biores Technol 186:207–214. https://doi.org/10.1016/j.biortech.2015.03.043
12. Aydin S, Cetecioglu Z et al (2015) Inhibitory effects of antibiotic combinations on syntrophic bacteria, homoacetogens and methanogens. Chemosphere 120:515–520. https://doi.org/10.1016/j.chemosphere.2014.09.045
13. Beneragama N et al (2013) The combined effect of cefazolin and oxytertracycline on biogas production from thermophilic anaerobic digestion of dairy manure. Biores Technol 133:23–30. https://doi.org/10.1016/j.biortech.2013.01.032
14. Svojitka J et al (2017) Performance of an anaerobic membrane bioreactor for pharmaceutical wastewater treatment. Biores Technol 229:180–189. https://doi.org/10.1016/j.biortech.2017.01.022
15. Aziz A et al (2022) Anaerobic digestion in the elimination of antibiotics and antibiotic-resistant genes from the environment—a comprehensive review. J Environ Chem Eng 10:106423. https://doi.org/10.1016/j.jece.2021.106423
16. Kumar A et al (2017) The ins and outs of microorganism-electrode electron transfer reactions. Nat Rev Chem 1:1–13. https://doi.org/10.1038/s41570-017-0024
17. Lovley DR (2017a) Electrically conductive pili: Biological function and potential applications in electronics. Curr Opin Electrochem 4(1):190–198. https://doi.org/10.1016/j.coelec.2017.08.015
18. Lovley DR (2017b) Syntrophy goes electric: direct interspecies electron transfer. Annu Rev Microbiol 71:643–664. https://doi.org/10.1146/annurev-micro-030117-020420
19. Zhang J et al (2020) The bio-chemical cycle of iron and the function induced by ZVI addition in anaerobic digestion: a review. Water Res 186:116405. https://doi.org/10.1016/j.watres.2020.116405
20. Liu Y et al (2012) Optimization of anaerobic acidogenesis by adding Fe_0 powder to enhance anaerobic wastewater treatment. Chem Eng J 192:179–185. https://doi.org/10.1016/j.cej.2012.03.044
21. Dai C et al (2022) Enhancing anaerobic digestion of pharmaceutical industries wastewater with the composite addition of zero valent iron (ZVI) and granular activated carbon (GAC). Biores Technol 346:126566. https://doi.org/10.1016/j.biortech.2021.126566
22. Chen WH et al (2019) Removals of pharmaceuticals in municipal wastewater using a staged anaerobic fluidized membrane bioreactor. Int Biodeterior Biodegr 140:29–36. https://doi.org/10.1016/j.ibiod.2019.03.008
23. Huang B et al (2018) Treatment of pharmaceutical wastewater containing B-lactams antibiotics by a pilot-scale anaerobic membrane bioreactor (AnMBR). Chem Eng J 341:238–247. https://doi.org/10.1016/j.cej.2018.01.149
24. Hu D et al (2017) Performance evaluation and microbial community dynamics in a novel AnMBR for treating antibiotic solvent wastewater. Biores Technol 243:218–227. https://doi.org/10.1016/j.biortech.2017.06.095
25. Monsalvo VM et al (2014) Removal of trace organics by anaerobic membrane bioreactors. Water Res 49:103–112. https://doi.org/10.1016/j.watres.2013.11.026

26. Song X et al (2016) Effects of salinity build-up on the performance of an anaerobic membrane bioreactor regarding basic water quality parameters and removal of trace organic contaminants. Biores Technol 216:399–405. https://doi.org/10.1016/j.biortech.2016.05.075
27. Xiao Y et al (2017) Removal of selected pharmaceuticals in an anaerobic membrane bioreactor (AnMBR) with/without powdered activated carbon (PAC). Chem Eng J 321:335–345. https://doi.org/10.1016/j.cej.2017.03.118
28. Li H et al (2019) Production of polyhydroxyalkanoates by activated sludge: correlation with extracellular polymeric substances and characteristics of activated sludge. Chem Eng J 361:219–226. https://doi.org/10.1016/j.cej.2018.12.066
29. Stöckl M et al (2019) Extracellular polymeric substances from *Geobacter sulfurreducens* biofilms in microbial fuel cells. ACS Appl Mater Interfaces 11:8961–8968. https://doi.org/10.1021/acsami.8b14340
30. Kamali M et al (2019) Sustainability considerations in membrane-based technologies for industrial effluents treatment. Chem Eng J 368:474–494. https://doi.org/10.1016/j.cej.2019.02.075
31. Kaya Y et al (2019) Investigation of membrane fouling in an anaerobic membrane bioreactor (AnMBR) treating pharmaceutical wastewater. J Water Process Eng 31:100822. https://doi.org/10.1016/j.jwpe.2019.100822
32. Ahmad M, Eskicioglu C (2019) Fate of sterols, polycyclic aromatic hydrocarbons, pharmaceuticals, ammonia and solids in single-stage anaerobic and sequential anaerobic/aerobic/anoxic sludge digestion. Waste Manage 93:72–82. https://doi.org/10.1016/j.wasman.2019.05.018
33. Cucina M et al (2017) Recovery of energy and plant nutrients from a pharmaceutical organic waste derived from a fermentative biomass: integration of anaerobic digestion and composting. J Environ Chem Eng 5:3051–3057. https://doi.org/10.1016/j.jece.2017.06.003
34. Ebele AJ, Abou-Elwafa Abdallah M, Harrad S (2017) Pharmaceuticals and personal care products (PPCPs) in the freshwater aquatic environment. Emerg Contam 3:1–16. https://doi.org/10.1016/j.emcon.2016.12.004
35. Farré M et al (2012) Achievements and future trends in the analysis of emerging organic contaminants in environmental samples by mass spectrometry and bioanalytical techniques. J Chromatogr A 1259:86–99. https://doi.org/10.1016/j.chroma.2012.07.024
36. Gonzalez-Gil L et al (2016) Is anaerobic digestion effective for the removal of organic micropollutants and biological activities from sewage sludge? Water Res 102:211–220. https://doi.org/10.1016/j.watres.2016.06.025
37. Kim SD et al (2007) Occurrence and removal of pharmaceuticals and endocrine disruptors in South Korean surface, drinking, and waste waters. Water Res 41:1013–1021. https://doi.org/10.1016/j.watres.2006.06.034
38. de Maria IC et al (2010) Sewage sludge application to agricultural land as soil physical conditioner. Revista Brasileira de Ciencia do Solo 34:967–974. https://doi.org/10.1590/s0100-06832010000300038
39. Bester K (2003) Triclosan in a sewage treatment process—balances and monitoring data. Water Res 37:3891–3896. https://doi.org/10.1016/S0043-1354(03)00335-X
40. Lai KM, Scrimshaw MD, Lester JN (2002) The effects of natural and synthetic steroid estrogens in relation to their environmental occurrence. Crit Rev Toxicol 32:113–132. https://doi.org/10.1080/20024091064192
41. Yin F et al (2015) Study on anaerobic digestion treatment of hazardous colistin sulphate contained pharmaceutical sludge. Biores Technol 177:188–193. https://doi.org/10.1016/j.biortech.2014.11.091
42. Ben W et al (2017) Distribution of antibiotic resistance in the effluents of ten municipal wastewater treatment plants in China and the effect of treatment processes. Chemosphere 172:392–398. https://doi.org/10.1016/j.chemosphere.2017.01.041
43. Karkman A et al (2016) High-throughput quantification of antibiotic resistance genes from an urban wastewater treatment plant. FEMS Microbiol Ecol 92:1–7. https://doi.org/10.1093/femsec/fiw014

44. Tong J et al (2017) Occurrence of antibiotic resistance genes and mobile genetic elements in enterococci and genomic DNA during anaerobic digestion of pharmaceutical waste sludge with different pretreatments. Biores Technol 235:316–324. https://doi.org/10.1016/j.biortech.2017.03.104
45. Huang H et al (2017) Alkaline fermentation of waste sludge causes a significant reduction of antibiotic resistance genes in anaerobic reactors. Sci Total Environ 580:380–387. https://doi.org/10.1016/j.scitotenv.2016.11.186
46. Azizan NAZ et al (2021) Pharmaceutical compounds in anaerobic digestion: a review on the removals and effect to the process performance. J Environ Chem Eng 9(5):105926. https://doi.org/10.1016/j.jece.2021.105926
47. Stasinakis AS (2012) Review on the fate of emerging contaminants during sludge anaerobic digestion. Biores Technol 121:432–440. https://doi.org/10.1016/j.biortech.2012.06.074
48. Bai Y et al (2019) Sludge anaerobic digestion with high concentrations of tetracyclines and sulfonamides: dynamics of microbial communities and change of antibiotic resistance genes. Biores Technol 276:51–59. https://doi.org/10.1016/j.biortech.2018.12.066
49. Akyol Ç et al (2016) A comprehensive microbial insight into single-stage and two-stage anaerobic digestion of oxytetracycline-medicated cattle manure. Chem Eng J 303:675–684. https://doi.org/10.1016/j.cej.2016.06.006
50. Zhi S et al (2019) How methane yield, crucial parameters and microbial communities respond to the stimulating effect of antibiotics during high solid anaerobic digestion. Biores Technol 283:286–296. https://doi.org/10.1016/j.biortech.2019.03.083
51. Song S et al (2020) Alkaline-thermal pretreatment of spectinomycin mycelial residues: insights on anaerobic biodegradability and the fate of antibiotic resistance genes. Chemosphere 261:127821. https://doi.org/10.1016/j.chemosphere.2020.127821
52. Carballa M et al (2006) Comparison between the conventional anaerobic digestion of sewage sludge and its combination with a chemical or thermal pre-treatment concerning the removal of pharmaceuticals and personal care products. Water Sci Technol 53:109–117. https://doi.org/10.2166/wst.2006.241
53. Jin B et al (2010) Applications of ultrasound to the synthesis of nanostructured materials. Adv Mater 22:1039–1059. https://doi.org/10.1002/adma.200904093
54. Kong X et al (2021) Effects of combined ultrasonic and grinding pre-treatments on anaerobic digestion of vinegar residue: organic solubilization, hydrolysis, and CH_4 production. Environ Technol (UK):1–11. https://doi.org/10.1080/09593330.2020.1870572
55. Li C, Champagne P, Anderson BC (2013) Effects of ultrasonic and thermo-chemical pre-treatments on methane production from fat, oil and grease (FOG) and synthetic kitchen waste (KW) in anaerobic co-digestion. Biores Technol 130:187–197. https://doi.org/10.1016/j.biortech.2012.11.053
56. Zeynali R, Khojastehpour M, Ebrahimi-Nik M (2017) Effect of ultrasonic pre-treatment on biogas yield and specific energy in anaerobic digestion of fruit and vegetable wholesale market wastes. Sustain Environ Res 27(6):259–264. https://doi.org/10.1016/j.serj.2017.07.001
57. Zhou H et al (2017) Enhancement with physicochemical and biological treatments in the removal of pharmaceutically active compounds during sewage sludge anaerobic digestion processes. Chem Eng J 316:361–369. https://doi.org/10.1016/j.cej.2017.01.104
58. Braguglia CM et al (2015) The impact of sludge pre-treatments on mesophilic and thermophilic anaerobic digestion efficiency: role of the organic load. Chem Eng J 270:362–371. https://doi.org/10.1016/j.cej.2015.02.037
59. Saifuddin N, Fazlili SA (2009) Effect of microwave and ultrasonic pretreatments on biogas production from anaerobic digestion of palm oil mill effluent. Am J Eng Appl Sci 2:139–146. https://doi.org/10.3844/ajeassp.2009.139.146
60. Alenzi A et al (2021) Pharmaceuticals effect and removal, at environmentally relevant concentrations, from sewage sludge during anaerobic digestion. Biores Technol 319:124102. https://doi.org/10.1016/j.biortech.2020.124102
61. Narumiya M et al (2013) Phase distribution and removal of pharmaceuticals and personal care products during anaerobic sludge digestion. J Hazard Mater 260:305–312. https://doi.org/10.1016/j.jhazmat.2013.05.032

62. Gallardo-Altamirano MJ et al (2021) Insights into the removal of pharmaceutically active compounds from sewage sludge by two-stage mesophilic anaerobic digestion. Sci Total Environ 789:147869. https://doi.org/10.1016/j.scitotenv.2021.147869
63. Carballa M et al (2007) Fate of pharmaceutical and personal care products (PPCPs) during anaerobic digestion of sewage sludge. Water Res 41:2139–2150. https://doi.org/10.1016/j.watres.2007.02.012
64. Rout PR et al (2021) Treatment technologies for emerging contaminants in wastewater treatment plants: a review. Sci Total Environ 753:141990. https://doi.org/10.1016/j.scitotenv.2020.141990
65. Gonzalez-Gil L et al (2018) Role of methanogenesis on the biotransformation of organic micropollutants during anaerobic digestion. Sci Total Environ 622–623:459–466. https://doi.org/10.1016/j.scitotenv.2017.12.004
66. Huang X et al (2020) Clarithromycin affect methane production from anaerobic digestion of waste activated sludge. J Clean Prod 255:120321. https://doi.org/10.1016/j.jclepro.2020.120321
67. Kallel M et al (2009) Removal of organic load and phenolic compounds from olive mill wastewater by Fenton oxidation with zero-valent iron. Chem Eng J 150:391–395. https://doi.org/10.1016/j.cej.2009.01.017
68. Pan X et al (2019) Impact of nano zero valent iron on tetracycline degradation and microbial community succession during anaerobic digestion. Chem Eng J 359:662–671. https://doi.org/10.1016/j.cej.2018.11.135
69. Yuan T et al (2020) ‘c’. Waste Manage 107:91–100. https://doi.org/10.1016/j.wasman.2020.04.004
70. Zhou H et al (2021) Zero-valent iron enhanced in-situ advanced anaerobic digestion for the removal of antibiotics and antibiotic resistance genes in sewage sludge. Sci Total Environ 754:142077. https://doi.org/10.1016/j.scitotenv.2020.142077
71. Suanon F et al (2017) Application of nanoscale zero valent iron and iron powder during sludge anaerobic digestion: impact on methane yield and pharmaceutical and personal care products degradation. J Hazard Mater 321:47–53. https://doi.org/10.1016/j.jhazmat.2016.08.076
72. Zhao L et al (2019) Effects of individual and combined zinc oxide nanoparticle, norfloxacin, and sulfamethazine contamination on sludge anaerobic digestion. Biores Technol 273:454–461. https://doi.org/10.1016/j.biortech.2018.11.049
73. Wang S et al (2021) Influence of zinc oxide nanoparticles on anaerobic digestion of waste activated sludge and microbial communities. RSC Adv 11:5580–5589. https://doi.org/10.1039/d0ra08671a
74. Gonzalez-Salgado I et al (2020) Combining thermophilic aerobic reactor (TAR) with mesophilic anaerobic digestion (MAD) improves the degradation of pharmaceutical compounds. Water Res 182:116033. https://doi.org/10.1016/j.watres.2020.116033
75. Venegas M et al (2021) Presence and fate of micropollutants during anaerobic digestion of sewage and their implications for the circular economy: a short review. J Environ Chem Eng 9:104931. https://doi.org/10.1016/j.jece.2020.104931
76. Hammer L, Palmowski L (2021) Fate of selected organic micropollutants during anaerobic sludge digestion. Water Environ Res 93:1910–1924. https://doi.org/10.1002/wer.1603

Chapter 6
Microbial Fuel Cells for the Bioelectricity Generation from Effluents Containing Pharmaceutically Active Compounds

6.1 Introduction

Overconsumption of water in industrial and nonindustrial sectors has caused the scarcity of clean water resources, as well as threats to the environment and the leaving organisms [1]. Hence, there have been efforts among the scientific community to develop efficient methods for the removal of such compounds from polluted (waste)waters.

For domestic wastewaters, methods such as activated sludge and anaerobic digestion have gained popularity due to some advantages, such as relatively low operating costs and their capability to receive large volumes of wastewaters [2, 3]. However, the presence of toxic and nonbiodegradable organic compounds in the content of most industrial effluents can make biological treatment techniques difficult to adopt. Hence, physico-chemical (waste)water treatment technologies such as adsorption [4], coagulation [5], membrane filtration [6], and advanced oxidation processes [7–9] have received attention to deal with such highly polluted effluents. However, these technologies have relatively high operating costs compared to biological treatment technologies and can bring some environmental impacts because most of them are highly dependent on an external source of energy to operate [10].

It has traditionally been an approach to control the quality of municipal (waste)waters based on overall wastewater quality indicators such as chemical oxygen demand (COD), biological oxygen demand (BOD), and total suspended solids (TSS), as well as the presence and concentrations of nutrients (i.e., nitrogen and phosphorous), pathogens, heavy metals, and compounds with known health impacts (priority pollutants), such as pesticides and halogenated organic compounds [11]. However, the current research has identified several hundred trace contaminants in the composition of municipal (waste)waters, including contaminants of emerging concern (CECs) with potential environmental and health effects [12]. The presence of such pollutants, including pharmaceuticals and personal care products (PPCP),

M. Kamali et al., *Advanced Wastewater Treatment Technologies for the Removal of Pharmaceutically Active Compounds*, Green Energy and Technology,
https://doi.org/10.1007/978-3-031-20806-5_6

even under low concentrations (ng/L to μg/L), has raised serious concerns since these pollutants are persistent, toxic, and have endocrine effects [13–15].

The possible adverse effects of pharmaceuticals in water bodies were first discussed in 1965 [16]. This has led to numerous studies on the removal of such compounds from water bodies, initiated in the 1970s, especially in China, the US, and Spain [17]. Since then, studies have considered the possibility of various biological physico-chemical techniques for the removal of such pollutants, especially those that cannot be removed using widely used wastewater municipal wastewater treatment methods such as activated sludge (see Chap. 3). In this stage, there is a need for the rapid development of efficient, cost-effective, and ecofriendly techniques with the capability of fast commercialization to deal with effluents containing pharmaceutical compounds [18].

In this chapter, the possibility of removing such compounds using microbial fuel cells (MFCs), as the technologies that benefit from the capability of specific microorganisms for the generation of electricity by degrading organic compounds, has been discussed, and recommendations for future studies have been provided.

6.2 Microbial Fuel Cells: Fundamentals and Mechanisms

Microbial fuel cells have emerged in recent years as sustainable technologies that can be used for the simultaneous degradation of organic compounds and the generation of electricity, which can be further used as clean energy [19]. Such technologies can also be implemented for the reduction of heavy metals (such as hexavalent chromium) from polluted waters and hence can be employed for wastewater polluted with both organic and inorganic compounds [20].

There are principally two types of MFCs, including single- or dual-chamber configurations, which have been frequently studied in recent years to deal with effluents originating from municipal or highly polluted industrial effluents, such as textile [21], pulp and paper [22], and the food industry [23]. In single-chamber MFCs, both anode and cathode electrodes are designed in a single reactor. Such a configuration is currently considered a low-cost candidate for real-scale wastewater applications [24]. However, the power density provided by such configurations is less compared to the dual-chamber MFCs, which normally comprise the anode, and the cathode chambers separated by a proton exchange membrane (PEM) [25]. The research is currently focused on overcoming the technical limitations of these configurations, such as the development of antifouling PEMs [26]. Figure 6.1 represents the schematics of the mentioned configurations.

In MFCs, the microorganisms responsible for the decomposition of organic compounds are concentrated around the anode, which has the role of transferring the produced electricity by an external circuit to the cathode to create the electricity current. Recent studies have identified the microbial communities with the optimum performance for bioelectricity generation. *Escherichia coli-K-12* is among the most efficient microorganisms with a high growth rate potential [29]. *Bacillus subtilis*,

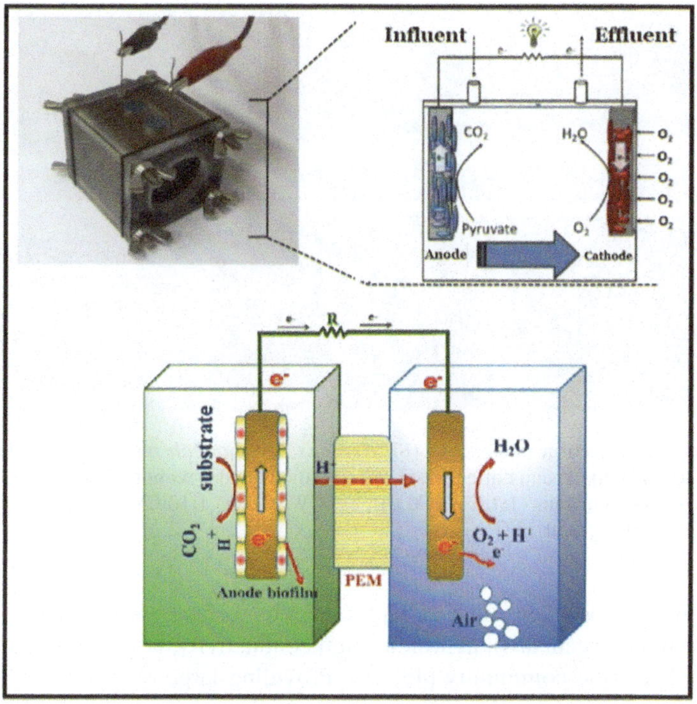

Fig. 6.1 A schematic of the single-chamber (up) and dual-chamber MFCs adopted from Abu-Reesh [27] and Rahmani et al. [28]

Lactobacillus acidophilus, and *Staphylococcus aureus* have also been considered microorganisms that have been used efficiently in MFCs but with less efficiency than *E. coli* [30–32]. Figure 6.2 illustrates the colonization of *E. coli* on various anode structures made of various materials for MFCs.

Protons are also generated in the anode area and migrate to the cathode. This occurs in dual-chamber MFCs via PEMs [34–36]. In the cathode (or cathodic chamber), the oxygen reduction reaction (ORR) occurs through the recombination of electrons and protons mediated by oxygen, which results in the generation of H_2O or H_2O_2 [37, 38]. In air–cathode microbial fuel cells (ACMFCs), natural oxygen from the atmosphere penetrates into the system to promote ORRs [39].

There are mechanisms already identified for the transfer of the electrons generated in the anodic chamber, including (a) electrically conductive pili, which play an important role in direct electron transfer between electron-producing microorganisms and the anode; (b) electron transfer by the conductive materials; and finally (c) electron transfer by the extracellular substances excreted by the microorganisms (see Rossi et al. [40]).

In addition to the trends for the development of antifouling MFCs (see Noori et al. [41], Shabani et al. [42]), fabrication of low-cost and biocompatible anodes

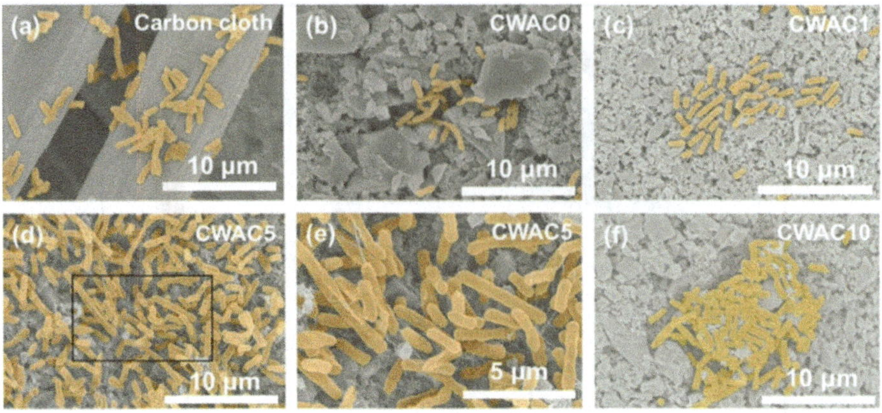

Fig. 6.2 Scanning electron microscopy (SEM) of *Escherichia coli* on various anode materials, including carbon cloth (**a**) and coffee waste carbonization anodes without KOH (CWAC0) (b) and with different KOH portions (1:1 CWAC0 (**b**), 1:5 CWAC0 (**c**) 1:10 CWAC0 (**d**)), reprinted with permission from Hung et al. [33]

with advanced properties, such as large specific surface area (to facilitate adhesion of the microorganisms), as well as high conductivity, is currently a main trend among the scientific community [43, 44]. Providing large reactive sites as well as high conductivity and mechanical stability are also currently under the spotlight to support efficient oxygen reduction reactions in the cathode side of MFCs [23, 45]. Furthermore, the importance of supplying enough oxygen concentrations in the cathode chamber has led to the development of technologies such as coating the cathode with oxygen-reducing nanomaterials (such as cerium oxide [46]) and natural air-breathing cathodes (see Wang et al. [47]) to enhance the power output of MFCs, which is critical for their application in real (waste)water treatment practices.

6.3 Microbial Fuel Cells for the Degradation of Pharmaceutically Active Compounds

The efficiency of MFCs for the degradation of a number of pharmaceuticals has been examined in recent years. The tested pharmaceuticals cover both those that can be degraded by conventional biological wastewater treatment systems, such as activated sludge, and those that are commonly considered toxic to or nonbiodegradable by the microorganisms responsible for their decomposition using such systems.

As discussed in Chap. 4, ibuprofen can be efficiently degraded by activated sludge processes (up to 90%) [48]. There are studies demonstrating the effectiveness of MFCs for bioelectricity generation from this compound. In a recent study [49, 50], it was demonstrated that a microbial fuel cell equipped with an anode fabricated from devil fish bones can exhibit appropriate properties, such as a large specific surface area and porosity, high electrical conductivity, biocompatibility, and surface roughness, and hence can be considered for bioelectricity generation from ibuprofen. The MFC resulted in the generation of 4.26 mW/m^2 during a 14-day cycle (14 days), 175% higher than that of the MFC with carbon felt, as a conventional anode material. The authors claimed that the strategy can be beneficial for both the production of low-cost MFCs and the control of the population of invasive species in countries such as Mexico, which have caused environmental and economic issues. However, the system represented a relatively low degradation efficiency for the removal of the pharmaceutical (34% mainly degradation), which calls for the further optimization of such systems. Such a strategy has also been examined for the degradation of ibuprofen [49, 50] using a single-chamber MFC. The anode fabricated from devil fish bone char induced biofilm formation, resulting in a maximum power density generation of 4.26 mW/m^2, 175% higher than that of the carbon felt electrode.

It seems that some pharmaceuticals that can inhibit methanogenesis for methane production have no significant effects on electricity-generating microorganisms. As an example, the presence of 5.6 μM mevastatin, as a hypolipidemic statin drug that inhibits hydroxymethylglutaryl~SCoA (HMG-CoA) reductase to lower cholesterol, resulted in the generation of 0.2 v electricity in a single-chamber air–cathode MFC [51]. The study also demonstrated that the presence of the pollutant caused an increase in the Faraday efficiency, from 35 to 49%. Efficient degradation of the pollutant (over 90%) can clearly demonstrate the applicability of the MFC technology for the simultaneous bioelectricity generation and methane production from such recalcitrant organic pollutants.

As mentioned earlier, reduction of the treatment costs using MFC technologies is a key point to push these methods for commercialization. Conventional materials such as carbon felt, carbon brushes, and carbon cloth are relatively expensive and cannot provide the maximum performance for the system [52]. Hence, there has been a trend for the application of low-cost materials that can provide high electrical conductivity as well as acceptable physico-chemical and mechanical properties. There are strategies for the fabrication of low-cost natural-based materials such as biochar, which have been used thus far for environmental applications such as the adsorption or degradation of various types of environmental pollutants [34–36, 53]. There are also studies demonstrating that such sustainable materials can promote the colonization of microbial communities, which can be attributed to the sheltering effects of carbon-based materials [34–36]. Various feedstocks can be used for the fabrication of carbonaceous materials to be used in MFCs as electrodes. In a relevant study [49, 50], devil fish bone was used for the preparation of biocompatible anodes, which resulted in efficient carbamazepine (50 mg/L) removal up to 90%, resulting in a maximum power density generation of 5.4 mW/m^2, which was twofold higher than that of the MFC fabricated using carbon felt. In this regard, there is a need for wider application of natural-based materials for the fabrication of various MFC components to be used for the degradation of contaminants of emerging concerns such as pharmaceutical and personal care products.

There have also been some novel designs of MFCs that have been studied for the removal of PhACs. The application of multielectrode MFCs can be considered an efficient technique that can enhance the kinetics of bioelectricity generation. This can enlarge the active area of the electrodes to facilitate adhesion of the microorganisms. Figure 6.3 shows a schematic of the multielectrode MFCs either in single-chamber or in dual-chamber modes. As indicated by [54], this strategy is highly effective for the removal of malachite green (98.15 ± 0.92% within 24 h of operation, 1.52 ± 0.08 W/m^3), compared to 88% after 5 days using a single-electrode electrode.

Other novel ideas have also been examined for the MFC configurations to deal with pharmaceutical compounds. The parabolic design of membrane-less MFCs fabricated from graphite electrodes has demonstrated an efficient strategy for the treatment of pharmaceutical effluents (2.01 W/m^3 at 168 mA/m^2) with a maximum COD removal over 80% [56] (Fig. 6.4).

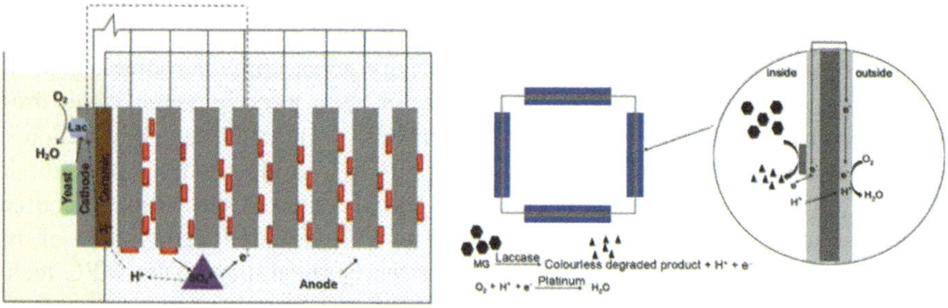

Fig. 6.3 Schematics of the dual-chamber (left) and single-chamber (right) multielectrode MFCs for bioelectricity generation from organic and inorganic pollutants, adapted from Chaijak and Sato [55] and Pol and Chaijak [54]

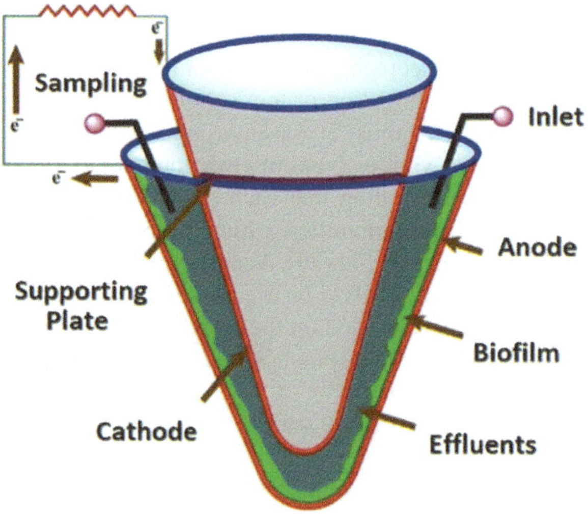

Fig. 6.4 A schematic of parabolic graphitic membrane-less MFCs for the treatment of pharmaceutical effluents, adapted from Rashid et al. [56]

The operating parameters are also of high importance for the application of MFCs for the degradation of pharmaceutical-containing effluents. pH, salinity, temperature, and hydraulic retention time (HRT) are the main factors in this regard that can influence the removal efficiency of PhACs and the involved mechanisms. According to the literature, system conductivity and, hence, the proton transfer rate are improved by increasing the salinity of the effluents in MFCs [57]. However, the high salinity of streams can inhibit the performance of the system considering that some bioelectricity generation species are not able to tolerate such extreme conditions [58, 59]. A possible solution for such conditions is the addition of specific bacterial strains to enhance the degradation of PhACs and to improve bioelectricity generation. This process is called bioaugmentation and has been explored in various biological systems for the treatment of highly polluted effluents [60, 61]. A bioaugmentation strategy has been studied for the treatment of pharmaceutical effluents using halophilic (salt-loving) bacteria (Fig. 6.5). Under a high salinity of 37 g/L, a high COD removal of 90% was achieved using such a system with an initial COD of 3 g/L [62]. In such systems, microorganisms such as *Rhodococcus*, *Ochrobactrum*, *Bacillus*, and *Marinobacter* are dominant, with known abilities to accumulate cations such as K^+ and anions such as Cl^- in their cytoplasm and organic osmolytes [63].

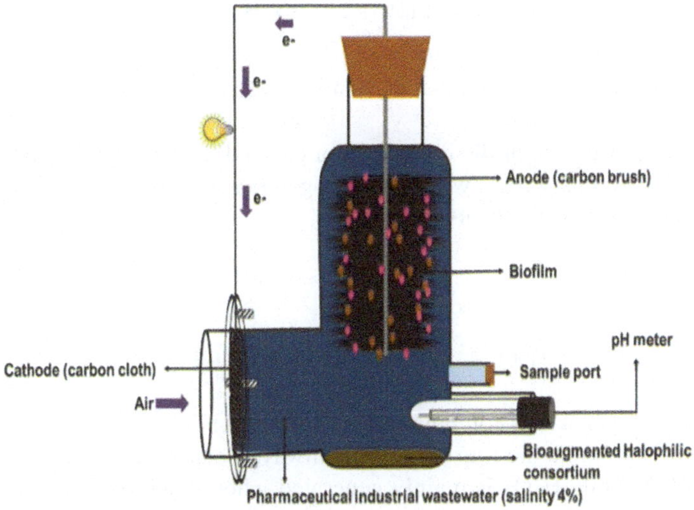

Fig. 6.5 Bioaugmentation is an effective strategy for bioelectricity generation from pharmaceutical effluents with high salinity, adapted from Pugazhendi et al. [62]

6.4 Combined Technologies

Various integration strategies can be designed based on the nature and mechanisms involved in the performance of MFCs. This section has aimed to discuss those combined technologies that have been designed and implemented thus far to deal with pharmaceutical compounds.

Combining MFCs with Fenton oxidation can be considered an attractive way to degrade recalcitrant and nonbiodegradable compounds in the cathodic chamber. In principle, Fenton oxidation, as an advanced oxidation process (AOP), functions by generating hydroxyl radicals (·OH) through Fe-mediated decomposition of hydrogen peroxide (H_2O_2). Hydroxyl radicals have strong oxidizing potential (2.8 eV) and can attack and decompose complex organic compounds efficiently [64]. Iron-based materials such as iron ions, iron oxides, and zero-valent iron have been used to conduct Fenton-based reactions [65]. There are also variations of Fenton-based reactions, including electro-Fenton, photo-Fenton, and photoelectro-Fenton, which have been designed and implemented for the treatment of a wide range of industrial and nonindustrial effluents. As discussed before, the cathode receives electrons from the anode and can undergo either two- or four-electron pathways (Eqs. 6.1 and 6.2).

$$O_2 + 2H^+ + 2e^- \rightarrow H_2O_2 \tag{6.1}$$

$$O_2 + 4H^+ + 4e^- \rightarrow H_2O \tag{6.2}$$

In addition, electrons in the cathode area can reduce Fe^{3+} to Fe^{2+} according to Eq. 6.3. Fe^{2+} can finally react with the hydrogen peroxide resulting from Eq. 6.3 to form powerful hydroxyl radicals (Eq. 6.4).

$$Fe^{3+} + e^- \rightarrow Fe^{2+} \tag{6.3}$$

$$H_2O_2 + Fe^{2+} + {}^{\cdot}OH + OH^- + Fe^{3+} \tag{6.4}$$

A typical schematic of a combination of MFCs and Fenton processes is illustrated in Fig. 6.6. To promote the four-electron pathway, metal-based cathodes have been recommended, while H_2O_2 is mainly generated when graphite-based cathodes are used in MFCs [66, 67].

There are reports for the application of such a technique for the treatment of pharmaceutical effluents. For instance, paracetamol degraded 70% within 9 h of reaction time, in which highly acidic pH values favored the degradation of this pharmaceutical by generating the highest amount of hydroxyl radicals in the medium [68]. Carbamazepine is also a recalcitrant pharmaceutical that cannot be degraded efficiently using conventional biological treatment methods such as activated sludge (see Chap. 4). More than 90% of the carbamazepine was removed using such a combination using the Fe–Mn cathode [39], with the proposed degradation pathway

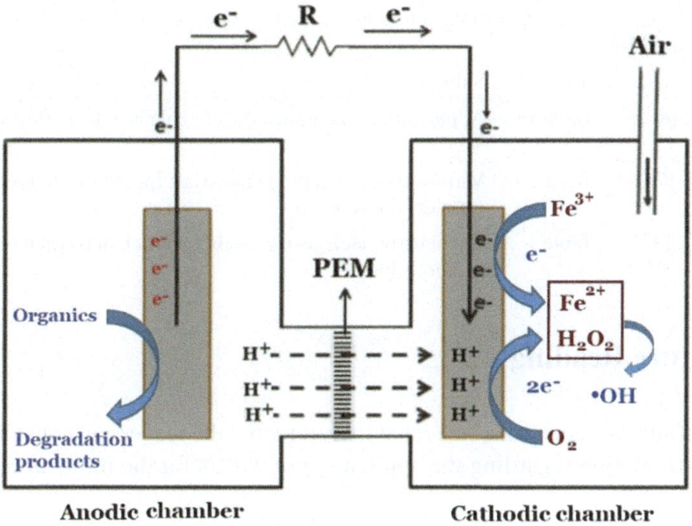

Fig. 6.6 A schematic of an MFC-Fenton combination for the generation of hydroxyl radicals to deal with a wide range of organic and nonorganic pollutants

indicated in Fig. 6.7. Degradation of sulfamethoxazole has also been reported in subsequent anodic (56%) and cathodic Fenton (95%) reactions within 24 h of reaction (anolyte was introduced to the cathodic chamber) [69].

The strategy of introducing the treated catholyte to the anodic chamber can also be considered for the degradation of PhACs. However, there is a need to explore the effectiveness of cathodic treatment processes for the degradation of pharmaceuticals and on the mineralization of pollutants. This is worth mentioning that in some cases, the degradation products can bring higher toxic effects compared to the parent compounds, which can result in the failure of the analytic treatment, which is normally based on the activity of certain microorganisms.

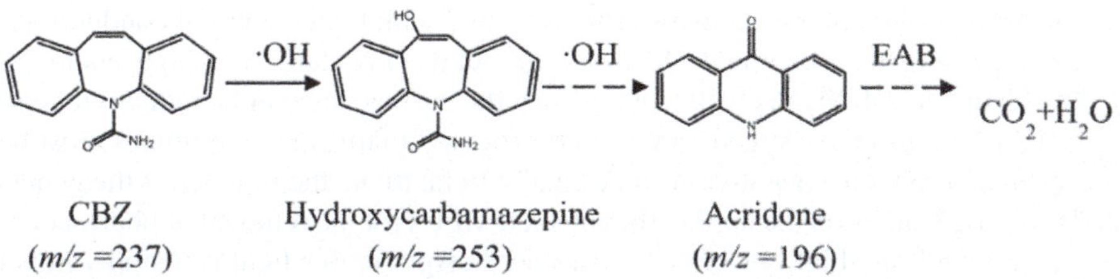

Fig. 6.7 The proposed pathway for the degradation of CBX using a combination of MFCs and Fenton reactions, reprinted with permission from Wang et al. [39]

Table 6.1 Further reading suggestions for more detailed coverage of the literature on the removal of PhACs using MFCs

References	Item	Subject
Dey et al. [70]	Table 1	Application of nanomaterials to enhance the performance of MFCs
Dange et al. [71]	Table 2	Various types of cathode materials for the oxygen reduction reactions in MFCs
Kumar et al. [72]	Table 1	Novel carbonaceous materials for efficient oxygen reduction reactions in MFCs

6.5 Future Reading

Table 6.1 contains items from the recent literature that the reader can consult for more detailed information regarding the application of MFCs for the removal of PhACs.

6.6 Summary

As discussed in this section, MFCs can be used as efficient technologies to deal with effluents contaminated by pharmaceutically active compounds. To promote the commercialization of such technologies, there is a need to (a) optimize the power generation in MFCs and (b) reduce the overall fabrication costs of the MFC components, including anodes, cathodes, and proton exchange membranes. This can be achieved by exploring inexpensive raw materials and fabrication techniques. Regarding the fabrication materials, studies have indicated that cheap and natural-based materials such as biochar can be considered ideal options for the fabrication of MFC components. Such sustainable materials need to satisfy certain criteria, such as being inexpensive and environmentally friendly, widely available, and with properties required for various MFC components. For anodes, the raw materials need to be nontoxic and have a large specific surface area and porosity to host and support colonization of the microorganisms efficiently and with high electrical conductivity to allow passing the generated electrons efficiently. For cathodes, high chemical and mechanical stability, as well as large specific surface area and conductivity, are essential to support the oxygen reduction reactions. Finally, the membranes must be porous in nature with a negative surface charge to facilitate the transfer of the generated protons. Studies on the application of such MFCs for the removal of pharmaceutical compounds are highly welcome. Another hotspot in this field is to apply novel engineering tools such as 3D printing toward the fabrication of well-designed geometries. Combining MFCs with other techniques, such as Fenton reactions, can also be considered an attractive method for the removal of pharmaceutical compounds. As discussed in this section, acidic pH values can favor this process. Hence, decreasing the pH may be required before introducing effluents to such systems, which can bring additional costs. In this regard, finding ways to optimize the performance of such systems for the generation of hydrogen peroxide is currently a topic with high interest among the scientific communities.

References

1. Tang Y et al (2020) Contaminants of emerging concern in aquatic environment: occurrence, monitoring, fate, and risk assessment. Water Environ Res 92:1811–1817. https://doi.org/10.1002/wer.1438
2. Carvalho PN et al (2013) Activated sludge systems removal efficiency of veterinary pharmaceuticals from slaughterhouse wastewater. Environ Sci Pollut Res 20:8790–8800. https://doi.org/10.1007/s11356-013-1867-7
3. Mailler R et al (2014) Biofiltration vs conventional activated sludge plants: what about priority and emerging pollutants removal? Environ Sci Pollut Res 21(8):5379–5390. https://doi.org/10.1007/s11356-013-2388-0
4. Devi P, Saroha AK (2015) Simultaneous adsorption and dechlorination of pentachlorophenol from effluent by Ni-ZVI magnetic biochar composites synthesized from paper mill sludge. Chem Eng J 271:195–203. https://doi.org/10.1016/j.cej.2015.02.087
5. Balik ÖY, Aydin S (2016) Coagulation/flocculation optimization and sludge production for pre-treatment of paint industry wastewater. Desalin Water Treat 57:12692–12699. https://doi.org/10.1080/19443994.2015.1051125
6. Kamali M et al (2019) Sustainability considerations in membrane-based technologies for industrial effluents treatment. Chem Eng J 368:474–494. https://doi.org/10.1016/j.cej.2019.02.075
7. Chuang YH et al (2017) Comparing the UV/monochloramine and UV/free chlorine advanced oxidation processes (AOPs) to the UV/hydrogen peroxide AOP under scenarios relevant to potable reuse. Environ Sci Technol:13859–13868. https://doi.org/10.1021/acs.est.7b03570
8. Soares PA et al (2016) Assessment of AOPs as a polishing step in the decolourisation of bio-treated textile wastewater: technical and economic considerations. J Photochem Photobiol, A 317:26–38. https://doi.org/10.1016/j.jphotochem.2015.10.017
9. Zhou Z et al (2019) Persulfate-based advanced oxidation processes (AOPs) for organic-contaminated soil remediation: a review. Chem Eng J 372:836–851. https://doi.org/10.1016/j.cej.2019.04.213
10. Huang YF et al (2020) Heterogeneous Fenton oxidation of trichloroethylene catalyzed by sewage sludge biochar: experimental study and life cycle assessment. Chemosphere 249:126139. https://doi.org/10.1016/j.chemosphere.2020.126139
11. Pal A et al (2014) Emerging contaminants of public health significance as water quality indicator compounds in the urban water cycle. Environ Int 71:46–62. https://doi.org/10.1016/j.envint.2014.05.025
12. Naddeo V et al (2020) Removal of contaminants of emerging concern from real wastewater by an innovative hybrid membrane process—ultraSound, adsorption, and membrane ultrafiltration (USAMe®). Ultrasonics—Sonochem 68:105237. https://doi.org/10.1016/j.ultsonch.2020.105237
13. Fent K, Weston AA, Caminada D (2006) Ecotoxicology of human pharmaceuticals. Aquat Toxicol 76:122–159. https://doi.org/10.1016/j.aquatox.2005.09.009
14. Kosma CI, Lambropoulou DA, Albanis TA (2010) Occurrence and removal of PPCPs in municipal and hospital wastewaters in Greece. J Hazard Mater 179:804–817. https://doi.org/10.1016/j.jhazmat.2010.03.075
15. Shraim A et al (2017) Analysis of some pharmaceuticals in municipal wastewater of Almadinah Almunawarah. Arab J Chem King Saud Univ 10:S719–S729. https://doi.org/10.1016/j.arabjc.2012.11.014

16. Stumm-Zollinger E, Fair GM (1965) Biodegradation of steroid hormones. J Water Poll Cont Feder 37:1506–1510
17. Davarazar M et al (2020) Treatment technologies for pharmaceutical effluents-a scientometric study. J Environ Manage 254:109800. https://doi.org/10.1016/j.jenvman.2019.109800
18. Patwardhan SB et al (2021) Recent advances in the application of biochar in microbial electrochemical cells. Fuel (October):122501. https://doi.org/10.1016/j.fuel.2021.122501
19. Lv C et al (2020) Improvement of oxygen reduction capacity by activated carbon doped with broccoli-like Co-Ni2P in microbial fuel cells. Chem Eng J 399:125601. https://doi.org/10.1016/j.cej.2020.125601
20. Uddin MJ, Jeong YK, Lee W (2021) Microbial fuel cells for bioelectricity generation through reduction of hexavalent chromium in wastewater: a review. Int J Hydrogen Energy 46:11458–11481. https://doi.org/10.1016/j.ijhydene.2020.06.134
21. Pushkar P, Mungray AK (2016) Real textile and domestic wastewater treatment by novel cross-linked microbial fuel cell (CMFC) reactor. Desalin Water Treat 57:6747–6760. https://doi.org/10.1080/19443994.2015.1013994
22. Singh P, Srivastava ASN (2017) Electricity generation by microbial fuel cell using pulp and paper mill wastewater, vermicompost and *Escherichia coli*. Indian J Biotechnol 16:211–215
23. Mohamed HO et al (2017) Graphite sheets as high-performance low-cost anodes for microbial fuel cells using real food wastewater. Chem Eng Technol 40:2243–2250. https://doi.org/10.1002/ceat.201700058
24. Obileke KC et al (2021) Microbial fuel cells, a renewable energy technology for bio-electricity generation: a mini-review. Electrochem Commun 125:107003. https://doi.org/10.1016/j.elecom.2021.107003
25. Chiu HY et al (2016) Electricity production from municipal solid waste using microbial fuel cells. Waste Manage Res 34:619–629. https://doi.org/10.1177/0734242X16649681
26. Ghasemi M et al (2013) Effect of pre-treatment and biofouling of proton exchange membrane on microbial fuel cell performance. Int J Hydrogen Energy 38:5480–5484. https://doi.org/10.1016/j.ijhydene.2012.09.148
27. Abu-Reesh IM (2020) Single-and multi-objective optimization of a dual-chamber microbial fuel cell operating in continuous-flow mode at steady state. Processes 8:839. https://doi.org/10.3390/pr8070839
28. Rahmani AR et al (2020) Effect of different concentrations of substrate in microbial fuel cells toward bioenergy recovery and simultaneous wastewater treatment. Environ Technol (UK):1–9. https://doi.org/10.1080/09593330.2020.1772374
29. Choudhury P, Bhunia B, Bandyopadhyaya TK (2021) Screening technique on the selection of potent microorganisms for operation in microbial fuel cell for generation of power. J Electrochem Sci Eng 11:129–142. https://doi.org/10.5599/jese.924
30. Hirose N et al (2021) Microbial fuel cells using α-amylase-displaying *Escherichia coli* with starch as fuel. J Biosci Bioeng 132:519–523. https://doi.org/10.1016/j.jbiosc.2021.07.008
31. Nimje VR et al (2009) Stable and high energy generation by a strain of *Bacillus subtilis* in a microbial fuel cell. J Power Sources 190:258–263. https://doi.org/10.1016/j.jpowsour.2009.01.019
32. Sun M et al (2010) Effects of a transient external voltage application on the bioanode performance of microbial fuel cells. Electrochim Acta 55:3048–3054. https://doi.org/10.1016/j.electacta.2010.01.020
33. Hung YH, Liu TY, Chen HY (2019) Renewable coffee waste-derived porous carbons as anode materials for high-performance sustainable microbial fuel cells. ACS Sustain Chem Eng 7:16991–16999. https://doi.org/10.1021/acssuschemeng.9b02405
34. Kamali M, Aminabhavi TM, Tarelho LAC et al (2022a) Acclimatized activated sludge for enhanced phenolic wastewater treatment using pinewood biochar. Chem Eng J 427:131708. https://doi.org/10.1016/j.cej.2021.131708
35. Kamali M, Sweygers N et al (2022b) Biochar for soil applications-sustainability aspects, challenges and future prospects. Chem Eng J 428(March 2021):131189. https://doi.org/10.1016/j.cej.2021.131189

36. Kamali M, Aminabhavi TM, Abbassi R et al (2022c) Engineered nanomaterials in microbial fuel cells—recent developments, sustainability aspects, and future outlook. Fuel 310:122347. https://doi.org/10.1016/j.fuel.2021.122347
37. Merino-Jimenez I et al (2017) Enhanced MFC power production and struvite recovery by the addition of sea salts to urine. Water Res 109:46–53. https://doi.org/10.1016/j.watres.2016.11.017
38. Theodosiou P, Greenman J, Ieropoulos I (2019) Towards monolithically printed MFCS: Development of a 3d-printable membrane electrode assembly (MEA). Int J Hydrogen Energy 44:4450–4462. https://doi.org/10.1016/j.ijhydene.2018.12.163
39. Wang W et al (2018) A microbial electro-Fenton cell for removing carbamazepine in wastewater with electricity output. Water Res 139:58–65. https://doi.org/10.1016/j.watres.2018.03.066
40. Rossi R, Cavina M, Setti L (2016) Characterization of electron transfer mechanism in mediated microbial fuel cell by entrapped electron mediator in saccharomyces cerevisiae. Chem Eng Trans 49:559–564. https://doi.org/10.3303/CET1649094
41. Noori MT et al (2019) Biofouling effects on the performance of microbial fuel cells and recent advances in biotechnological and chemical strategies for mitigation. Biotechnol Adv 37:107420. https://doi.org/10.1016/j.biotechadv.2019.107420
42. Shabani M et al (2021) Enhancement of microbial fuel cell efficiency by incorporation of graphene oxide and functionalized graphene oxide in sulfonated polyethersulfone membrane. Renew Energy 179:788–801. https://doi.org/10.1016/j.renene.2021.07.080
43. Fan Y, Sharbrough E, Liu H (2008) Quantification of the internal resistance distribution of microbial fuel cells. Environ Sci Technol 42:8101–8107. https://doi.org/10.1021/es801229j
44. Philamore H et al (2015) Cast and 3D printed ion exchange membranes for monolithic microbial fuel cell fabrication. J Power Sources 289:91–99. https://doi.org/10.1016/j.jpowsour.2015.04.113
45. Li M, Zhou S, Xu M (2017) Graphene oxide supported magnesium oxide as an efficient cathode catalyst for power generation and wastewater treatment in single chamber microbial fuel cells. Chem Eng J 328:106–116. https://doi.org/10.1016/j.cej.2017.07.031
46. Peng W et al (2016) Controlled growth cerium oxide nanoparticles on reduced graphene oxide for oxygen catalytic reduction. Electrochim Acta 191:669–676. https://doi.org/10.1016/j.electacta.2016.01.129
47. Wang Z et al (2017) Progress of air-breathing cathode in microbial fuel cells. J Power Sources 356:245–255. https://doi.org/10.1016/j.jpowsour.2017.02.004
48. Quintelas C et al (2020) Degradation of widespread pharmaceuticals by activated sludge: kinetic study, toxicity assessment, and comparison with adsorption processes. J Water Process Eng 33:101061. https://doi.org/10.1016/j.jwpe.2019.101061
49. Aguilera Flores MM, Ávila Vázquez V, Medellín Castillo NA, Cardona Benavides A et al (2021a) Biodegradation of carbamazepine and production of bioenergy using a microbial fuel cell with bioelectrodes fabricated from devil fish bone chars. J Environ Chem Eng 9:106692. https://doi.org/10.1016/j.jece.2021.106692
50. Aguilera Flores MM, Ávila Vázquez V, Medellín Castillo NA, Carranza Álvarez C et al (2021b) Ibuprofen degradation and energy generation in a microbial fuel cell using a bioanode fabricated from devil fish bone char. J Environ Sci Health—Part A Toxic/Hazar Substan Environ Eng 56:874–885. https://doi.org/10.1080/10934529.2021.1934357
51. Akul NB et al (2021) Effects of mevastatin on electricity generation in microbial fuel cells. Polish J Environ Stud 30:5407–5412. https://doi.org/10.15244/pjoes/133402
52. Santoro C et al (2017) Microbial fuel cells: from fundamentals to applications: a review. J Power Sources 356:225–244. https://doi.org/10.1016/j.jpowsour.2017.03.109
53. Kamali M et al (2021) Biochar in water and wastewater treatment—a sustainability assessment. Chem Eng J 420:129946. https://doi.org/10.1016/j.cej.2021.129946
54. Pol AS, Chaijak P (2021) Malachite green removal and bioelectricity generation using a novel design multi-electrode microbial fuel cell. Acta Sci Pol 20:69–76
55. Chaijak P, Sato C (2021) Power recovery and sulfate removal from rubber wastewater with the novel model multi-electrode microbial fuel cell. Pollution 7:417–424. https://doi.org/10.22059/poll.2021.313409.934

56. Rashid T et al (2021) Design and feasibility study of novel paraboloid graphite based microbial fuel cell for bioelectrogenesis and pharmaceutical wastewater treatment. J Environ Chem Eng 9:104502. https://doi.org/10.1016/j.jece.2020.104502
57. Karthikeyan R et al (2016) Influence of ionic conductivity in bioelectricity production from saline domestic sewage sludge in microbial fuel cells. Biores Technol 200:845–852. https://doi.org/10.1016/j.biortech.2015.10.101
58. Banu JR et al (2019) Biopolymer production in bio electrochemical system: literature survey. Bioresour Technol Rep 7:100283. https://doi.org/10.1016/j.biteb.2019.100283
59. Lefebvre O et al (2012) Effect of increasing anodic NaCl concentration on microbial fuel cell performance. Biores Technol 112:336–340. https://doi.org/10.1016/j.biortech.2012.02.048
60. Van Der Gast CJ et al (2003) Bioaugmentation strategies for remediating mixed chemical effluents. Biotechnol Prog 19:1156–1161. https://doi.org/10.1021/bp020131z
61. Saravanane R, Murthy DVS, Krishnaiah K (2001) Bioaugmentation and anaerobic treatment of pharmaceutical effluent in fluidized bed reactor. J Environ Sci Health—Part A Toxic/Hazardous Substan Environ Eng 36:779–791. https://doi.org/10.1081/ESE-100103760
62. Pugazhendi A et al (2022) Bioaugmentation of electrogenic halophiles in the treatment of pharmaceutical industrial wastewater and energy production in microbial fuel cell under saline condition. Chemosphere 288:132515. https://doi.org/10.1016/j.chemosphere.2021.132515
63. Hu X et al (2020) Salt tolerance mechanism of a hydrocarbon-degrading strain: salt tolerance mediated by accumulated betaine in cells. J Hazard Mater 392:122326. https://doi.org/10.1016/j.jhazmat.2020.122326
64. Amaral-Silva N et al (2016) Fenton's treatment as an effective treatment for elderberry effluents: economical evaluation. Environ Technol (UK) 37:1208–1219. https://doi.org/10.1080/09593330.2015.1107624
65. Babuponnusami A, Muthukumar K (2014) A review on Fenton and improvements to the Fenton process for wastewater treatment. J Environ Chem Eng 2:557–572. https://doi.org/10.1016/j.jece.2013.10.011
66. Chakraborty I et al (2020) Waste-derived biochar: applications and future perspective in microbial fuel cells. Biores Technol 312:123587. https://doi.org/10.1016/j.biortech.2020.123587
67. Sawant SY, Han TH, Cho MH (2017) Metal-free carbon-based materials: promising electrocatalysts for oxygen reduction reaction in microbial fuel cells. Int J Mol Sci 18:25. https://doi.org/10.3390/ijms18010025
68. Zhang L, Yin X, Li SFY (2015) Bio-electrochemical degradation of paracetamol in a microbial fuel cell-Fenton system. Chem Eng J 276:185–192. https://doi.org/10.1016/j.cej.2015.04.065
69. Li S et al (2020) Bio-electro-Fenton systems for sustainable wastewater treatment: mechanisms, novel configurations, recent advances, LCA and challenges: an updated review. J Chem Technol Biotechnol 95:2083–2097. https://doi.org/10.1002/jctb.6332
70. Dey N et al (2022) Nanomaterials as potential high performing electrode materials for microbial fuel cells. Appl Nanosci (Switzerland):0123456789. https://doi.org/10.1007/s13204-022-02371-3
71. Dange P et al (2022) A comprehensive review on oxygen reduction reaction in microbial fuel cells. J Renew Mater 10:665–697. https://doi.org/10.32604/jrm.2022.015806
72. Kumar T, Naik S, Jujjavarappu SE (2022) A critical review on early-warning electrochemical system on microbial fuel cell-based biosensor for on-site water quality monitoring. Chemosphere 291:133098. https://doi.org/10.1016/j.chemosphere.2021.133098

Chapter 7
Constructed Wetlands for the Elimination of Pharmaceutically Active Compounds; Fundamentals and Prospects

7.1 Introduction

The application of constructed wetlands (CWs) has been considered a sustainable method for (waste)water treatment in recent years. The basis of this ecofriendly technology is on the physical, chemical, and biological routes in natural wetlands but in controlled environmental conditions. CWs were first implemented in Europe in early 1960. There are three main components involved in the performance of CWs, including filtration (using a porous media), microorganisms, and plant species. From a mechanistic point of view, organic compounds in the content of polluted (waste)waters are degraded/transformed by the activity of the microorganisms present in the media, especially in the rhizosphere zone. Generally, a large surface area is provided by the materials in the media (e.g., sand, rock, and soil). Microorganisms adhere to the surface area of the materials, which can promote the biodegradation of organic compounds. The plants in the medium also play a very important role in CWs by providing the roots and rhizomes for the attachment of the bacteria and by oxygenation of the environment surrounding the microorganisms, which can facilitate the metabolic activities of the microorganisms. Plants also take up elements (such as N and P) from the decomposition of organic compounds by the epidermis and vascular bundles of the roots [1]. This can potentially prevent their release into the environment and create secondary environmental pollution (such as eutrophication of the lacks) and satisfy sustainability considerations.

CWs are principally divided into three main categories: free water surface flow CWs (FWS-CWs), horizontal subsurface flow CWs (HSF-CWs), and vertical subsurface flow CWs (VSF-CWs).

FWS-CWs, as natural wetlands, are used to treat (waste)water flows over their surfaces, as indicated in Fig. 7.1a. Such a construct can also be efficiently used to prevent floods. Various emergent (such as Typha and Scripus) and submerged plants, such as Elodea, have been used in FWS-CWs. This type of CW has been used to improve the quality of the (waste)waters in terms of approximately 50% removal of

M. Kamali et al., *Advanced Wastewater Treatment Technologies for the Removal of Pharmaceutically Active Compounds*, Green Energy and Technology,
https://doi.org/10.1007/978-3-031-20806-5_7

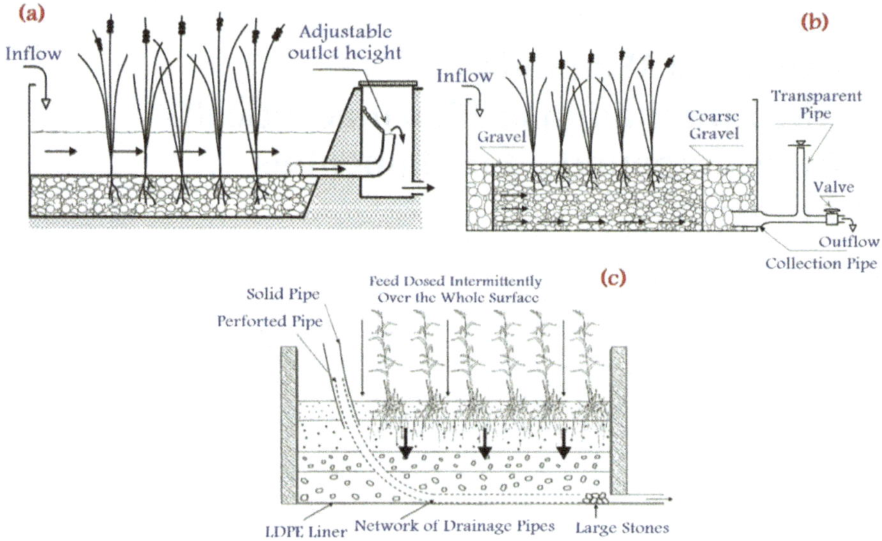

Fig. 7.1 A schematic of various CWs, including free water surface flow CWs (**a**), horizontal subsurface flow CWs (**b**), and vertical subsurface flow CWs (**c**), adapted from Wang et al. [2]

the chemical oxygen demand (COD) and biological oxygen demand. Additionally, moderate removal of metallic elements such as Fe, Cu, Pb, and Zn has been observed by the application of this kind of CW.

Figure 7.1b also illustrates HFS-CWs. In this type of CW, effluents flow horizontally through the bed of the CW. Both aerobic and anaerobic processes are involved in the degradation of organic compounds in HFS. There are also pieces of evidence for the efficient removal of COD, BOD, and total suspended solids (TSS). Hence, they have been used to treat a wide range of effluents, such as industrial and municipal (waste)waters. Regarding sustainability considerations, HFS-CWs may require more initial investments because they need more land area than other types of CWs.

Finally, VFS-CWs have been developed based on a submerged (waste)water flow (Fig. 7.1c) that leaves the CWs from the bottom. VFS-CWs principally require less land area, which can make them attractive alternatives for efficient BOD, COD, and TOC removal from polluted effluents [3].

HSF-CWs have demonstrated acceptable performances for the removal of a wide range of organic compounds, including PhACs. For instance, removal efficiencies between 85 and 99% have been observed for the removal of bisphenol A, diclofenac, naproxen, ibuprofen, and tonalide by Ávila et al. [4]. Such a configuration has also represented an acceptable efficiency for the elimination of paracetamol (over 90%) [5]. Successful removal of COD, BOD, and TSS (84%, 91%, and 93%, respectively) has also been observed using this configuration for the treatment of hospital effluents [6].

7.2 Plant Species in Constructed Wetlands

Various types of plant species have already been used in CWs. *Iris, Canna, Heliconia, Phragmites, Typhas,* and *Zantedeschia* are the most widely used genera in CWs. For instance, *Canna indica,* as an aquatic tropical plant, has been widely used in CWs for the efficient treatment of (waste)waters. Other plants, such as *Juncus effuses, Phragmites australis, Typha domingensis, Cyperus giganteus, Cyperus papyrus, Lemna valdiviana,* and *Sagitaria lancifolia,* have also represented their applicability in this regard. *Typha latifolia,* with the potential to grow at different water depths, has also been commonly used in North America (both in rural and urban areas) with the ability to grow at different water depths. This species can also tolerate a wide range of environmental conditions, such as dissolved oxygen, pH, salinity, and COD, which can make it a sustainable alternative for the treatment of highly polluted effluents [7].

There has been a trend in the application of ornamental flowering plants in CWs (Fig. 7.2). Such an approach has also resulted in the efficient treatment of (waste)waters. As we stated before [8], social acceptance is among the sustainability parameters that need to be considered for the selection of (waste)water treatment methods. Hence, it can be concluded that the application of ornamental plant species can be considered a sustainable option for the treatment of (waste)waters with moderate pollution loads, such as municipal (waste)water streams, especially in tropical and subtropical areas. There is also a need for further studies to explore the applicability of this type of CW to deal with recalcitrant compounds in the content of polluted effluents. In this regard, studying the effectiveness of a mixture of different plant species is among the research hot topics.

7.3 CWs for (Waste)Water Treatment; General Considerations

There are several reports in the literature on the application of CWs to deal with various types of effluents [9, 10]. In this regard, the treatment of combined sewers composed of sewage streams and stormwater is an example to adopt such technologies, which can provide a suitable method for urban areas. CWs for such applications can also potentially aid in preventing floods in urban areas [11]. In addition, CWs have been efficiently examined to deal with industrial effluents that are highly loaded with various organic and inorganic compounds. As an example, CWs have been employed for the treatment of mining effluents with an acidic nature. Acid mine drainage (AMD) streams can create severe surface water contamination, and it is vital to apply efficient and low-cost technologies to address them. There are reports on the effectiveness of CWs for the long-term treatment of AMDs since the 1970s and early 1980s. For such effluents, VSF-CWs have been recommended as the most appropriate type of CWs [12] due to their effectiveness in the removal of heavy metals (such as Co, Ag, Cr, Cd, Pb, Hg, Cu, Mo, Ni, and Zn), as one of the

Fig. 7.2 Some of the most widely used ornamental plant species used in CWs were adapted from Sandoval et al. [1]

most important features of AMDs. These types of CWs provide an enhanced interaction between effluents (and pollutants) and plant roots. This can activate the main mechanism by which heavy metals are taken up by plant roots. Moreover, under the existing anaerobic conditions in VSF-CWs, heavy metal removal mechanisms such as adsorption and precipitation are facilitated. It is worth mentioning that the overall efficiency of CWs to remove heavy metals is determined by the synergetic effect between the mentioned mechanisms. However, such streams can cause toxic effects on the plant species and microorganisms responsible for the removal of pollutants. This can be overcome through the careful selection of CW elements, such as planting appropriate species and inoculation of CWs with acclimatized microbial communities, with the enhanced capability to tolerate harsh environmental conditions, such as low acidity and high concentrations of heavy metals.

CWs have also been considered a sustainable option to deal with effluents with high salinity. It is worth mentioning that many types of effluents, such as brine disposal from the food industry, contain relatively high amounts of salts (cations and anions, heavy metals are not included) produced and discharged during the production processes. There are also organic compounds and heavy metals present in the content of such effluents. Hence, there is an urgent need for treatment methods that are able to address these types of pollutants simultaneously. CWs have received attention in recent years to deal with such effluents, especially in developing countries where simplicity and cost-effectiveness are critical for the selected (waste)water treatment methods. However, there are some sustainability issues with the application of CWs for this type of effluent. For instance, the efficiency of such systems can be negatively influenced by the activity of microorganisms present in CWs. To overcome this issue, introducing appropriate microorganisms that exist in natural hypersaline environments can be considered an acceptable way to promote CWs to deal with saline effluents. For instance, halophilic and halophytes and microorganisms can be used in CWs to enhance the treatment of saline effluents. Furthermore, the optimum conditions in terms of pH, temperature, and flow rate can be optimized for the efficient treatment of saline effluents using CWs [13].

7.4 CWs for the Elimination of PhACs

CWs have also been used in recent years for the elimination of PhACs from various types of (waste)water streams [14]. There are reports for the efficient treatment of hospital effluents that are normally contaminated with various PhACs using CW systems. For instance, a transportable CW designed for the treatment of hospital effluents in Belgium represented high efficiencies for COD and ammonia removal of 83% and 95%, respectively, revealing the applicability of such (waste)water treatment systems for the removal of PhACs from such effluents [15]. However, the performance of CWs for the treatment of pharmaceutical-containing effluents can be affected by variables such as the removal mechanism, operating conditions, CW configuration, and their combination with other technologies.

7.4.1 Removal Mechanism

According to the literature, mechanisms including sorption, filtration, microbial degradation, phytoremediation, etc., are involved in the efficient removal of PhACs, even under relatively high concentrations [16].

Adsorption has been identified as the main mechanism for the removal of PhACs in biological treatment systems such as anaerobic digestion and activated sludge processes [17, 18]. Liu et al. [19] indicated that the simulated red soil layer of a CW can effectively adsorb and remove pharmaceuticals such as oxytetracycline and ciprofloxacin. They also argued that physical adsorption is the main mechanism involved in such adsorptive removal processes.

The adsorption of PhACs such as ciprofloxacin by plant roots has also been reported by the recent literature as another mechanism involved in the removal of such compounds using CWs [20, 21].

The application of specific beds has also been examined for the removal of PhACs. For instance, the Leca bed with an adsorption capacity has been widely used for the removal of various organic compounds in CWs [3, 22, 23]. This configuration resulted in the efficient removal of PhACs, including carbamazepine, sulfadiazine, and ibuprofen (over 90%) [24]. Over 75% of the removal of atenolol has also been reported using this type of bed, which demonstrates its applicability for a wide range of PhACs [25].

Other substrates have also been studied for the adsorption of PhACs. For instance, light expanded clay aggregates (LECAs) have demonstrated a good capacity for the adsorption of ibuprofen, carbamazepine, and clofibric acid in a subsurface flow constructed wetland [26]. Volcanic rocks such as *Tezontle* used in CWs have also demonstrated the ability to adsorb carbamazepine (3.48 μg/g) [27].

In addition to adsorption, bioremediation has also been reported using specific types of microorganisms. Macrophytes are considered attractive candidates for the phytoremediation of PhACs. They have the ability to accumulate and degrade organic pollutants in their tissues [28, 29]. In this regard, the application of fast-growing species with very well-developed and massive root systems, such as vetiver grass (*Chrysopogon zizanioides*), can be considered a way to enhance the biodegradation of pollutants in constructed wetlands. Various reports are available indicating the tolerance of this species against highly acidic environments, as well as extreme conditions such as freezing temperatures and drought [30]. Studies are confirming the applicability of vetiver for the removal of antibiotics in CWs. Model [31] demonstrated the simultaneous removal of nutrients (nitrogen and phosphorus, 93% and 84%, respectively) and PhACs, including ciprofloxacin and tetracycline (93% and 97%, respectively).

Other macrophytes, such as *Cyperus alternifolius,* have also been reported for the biodegradation of PhACs. For instance, Liu et al. [32] indicated the efficient biodegradation of sulfamethoxazole using this plant species in a CW. *Typha latifolia L.* has also demonstrated the ability for the rapid uptake and metabolism of diclofenac at environmentally relevant concentrations (1 mg/L),[1] resulting in the formation of metabolites, including glycoside and glutathione conjugates [33]. The applicability of *Phragmites australis* for both the absorption and biodegradation of antibiotics under environmentally relevant concentrations has also been indicated in the literature (see, e.g., Liu et al. [34]). *Canna x generalis* has also recently represented its potential for the efficient removal of ciprofloxacin in the content of blackwater [35].

Notably, the uptake of toxic compounds can negatively affect the health conditions and growth of some plant species.[2] For instance, there are studies on the negative effects of ciprofloxacin and tetracycline on the Chl level in the leaves of yellow lupine [37]. Additionally, the coexistence of PhACs with other types of pollutants can have a synergistic negative effect on macrophytes with the capability of biodegradation of PhACs. For instance, the presence of tetracyclines together with Cu(II) was reported to reduce the Chl level in *Myriophyllum aquaticum* [20].

---

[1] Exposure to the diclofenac reduced the activity of glycosyltransferase by seven folds in roots, but not in shoots.

[2] Chlorophyll (Chl) content has been considered as a measure of the health of plants exposed to the toxic compounds [34, 36}.

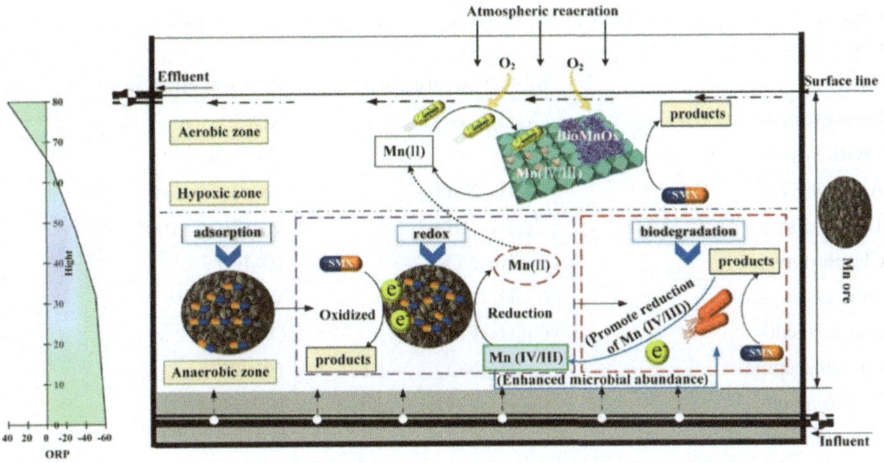

Fig. 7.3 Mechanisms involved in the removal of sulfamethoxazole with Mn ore as the additive. Both oxidation and adsorption play roles in the removal of the pharmaceutical using this system, reprinted with permission from Xu et al. [42]

The application of specific biological agents such as fungi has also been illustrated to be very efficient for the promotion of the biodegradation performance of CWs. For instance, arbuscular mycorrhizal fungi (AMF) colonized on plants such as *Glyceria maxima* improved the activities of antioxidant enzymes (i.e., superoxide dismutase and peroxidase), resulting in both increasing the efficiency of the CW for the removal of pharmaceuticals such as diclofenac and ibuprofen and reducing the decomposition metabolites of pharmaceuticals [38].

Additives have also been studied to promote the performance of constructed wetlands for the removal of PhACs. Among them, manganese (Mn) ore has been considered an efficient material due to its potential for the adsorption and oxidation of pollutants [39, 40]. The existence of Mn(III) or the free spaces in the mineral lattice causes a negative structural charge in this material, which promotes the adsorption of pollutants with a positive charge [41]. Such a strategy has been adopted for the removal of PhACs such as sulfamethoxazole (up to 50%, compared to only 8% with the gravel substrate) (Fig. 7.3) [42].

7.4.2 Operating Conditions

Operating parameters can also considerably influence the performance of CWs for the removal of PhACs. Various performances have been observed from very low to very high under various operating conditions (and CW configurations), as summarized in Table 7.1.

Table 7.1 Efficiencies observed in the literature for the removal of various PhACs using MBR technologies

PhACs	Molecular structure	Removal efficiency (%)
Ciprofloxacin	$C_{17}H_{18}FN_3O_3$	N.A.[a]
Erythromycin	$C_{37}H_{67}NO_{13}$	Very low-95[b]
Azithromycin	$C_{38}H_{72}N_2O_{12}$	Very low to over 90[c]
Trimethoprim	$C_{14}H_{18}N_4O_3$	Very low—100[b]
Clarithromycin	$C_{38}H_{69}NO_{13}$	10–100
Sulfamethoxazole	$C_{10}H_{11}N_3O_3S$	Very low-75[b]
Enrofloxacin	$C_{19}H_{22}FN_3O_3$	Over 90[d]
Carbamazepine	$C_{15}H_{12}NO$	10–90[e]
Azithromycin	$C_{38}H_{72}N_2O_{12}$	Up to 95[f]
Ibuprofen	$C_{13}H_{18}O_2$	99[g]
Diclofenac	$C_{14}H_{10}Cl_2NO_2$	Very low-75[b]
Sulfadiazine	$C_{10}H_{10}N_4O_2S$	Up to 70[g]
Metformin	$C_4H_{11}N_5$	100
Triclosan	$C_{12}H_7Cl_3O_2$	Very low-90[b]
Ranitidine	$C_{13}H_{22}N_4O_3S$	N.A.[a]
Lamotrigine	$C_9H_7Cl_2N_5$	N.A.[a]
Atenolol	$C_{14}H_{22}N_2O_3$	N.A.[a]
Propranolol	$C_{16}H_{21}NO_2$	N.A.[a]
Estrone (E1)	$C_{18}H_{22}O_2$	Very low-90[b]
17β-Estradiol (E2)	$C_{18}H_{24}O_2$	Very low-100[b]
17α-Ethynylestradiol (EE2)	$C_{20}H_{24}O_2$	Very low-100[b]

(a) Not assessed, (b) Krzeminski et al. [43], (c) Gorito et al. [44]; Bayati et al. [45], (d), Carvalho et al. [46]; Bôto et al. [47]; Santos et al. [48], (e) Özengin and Elmaci [24]; Hartl et al. [49], (f) Ávila et al. [50]; Bayati et al. [45], (g) AL Falahi et al. [51]

The hydraulic loading rate (HLR) has been demonstrated to be an important operating condition for the removal of PhACs using CWs. Liu et al. [52] indicated the indirect correlation of the HLR with the removal efficiency of sulfamethoxazole, increasing the HLR from 2 to 10 cm/d resulted in a lower adsorption efficiency of the pharmaceutical in CW soils. The organic loading rate (OLR) can also negatively affect the performance of CWs by introducing more pollutants into the system, which can impair the activity of the microorganisms [53].

Other parameters, such as depth and area, have been considered to have a positive effect on the removal of PhACs (e.g., ibuprofen, diclofenac, gemfibrozil, sulfapyridine, and ketoprofen) by promoting the aerobic and anaerobic biodegradation of PhACs [54, 55]. Furthermore, higher areas of CWs can decrease the hydraulic retention time (HRT), which is essential to deal with effluents with high inlet flow rates.

Parameters such as pH, temperature, and dissolved oxygen (DO) can also influence the performance of CWs. The effect of pH can be correlated with the types of PhACs and the mechanisms involved in their removal. For instance, higher pH values can enhance the adsorption of PhACs with positive charges [56]. Additionally, very low or very high pH values can influence the activity of some specific plant species that can be involved in the biodegradation of PhACs. Temperature can also enhance the endothermic hydrolysis reaction, leading to an improved biodegradation process [57]. The dissolved oxygen can also promote the aerobic biodegradation of PhACs diclofenac, ibuprofen, and gemfibrozil [58].

7.4.3 Combination Strategies

The combination of constructed wetlands with MFCs has also been considered an efficient and sustainable strategy for the removal of PhACs. Such systems can simultaneously remove organic compounds and generate bioelectricity, which can satisfy a part of the energy required for the performance of the combined system. Colares et al. [59] indicated that the electrode type and the operating parameters, such as hydraulic retention time, can control the performance of such combined systems. However, there are a limited number of reports on the implementation of such sustainable combinations for the removal of PhACs. Figure 7.4 illustrates a schematic of such a technology successfully implemented for the removal of sulfamethoxazole (over 96%) [52].

There are also studies indicating the effectiveness of CW-MFCs for the removal of other types of PhACs, such as sulfamethoxazole and tetracycline (over 97% for both) [60].

The mechanisms involved in the biodegradation of PhACs in CW-MFC systems have also been investigated very recently. Dai et al. [61] indicated that such combinations can influence the content of extracellular polymeric substances (EPSs)[3] and the composition of microbial communities. They observed a significant improvement in the removal of sulfamethoxazole as well as nutrients such as total nitrogen and ammonia and a lower content of EPS, which can be related to the efficient removal of the pharmaceutical in the medium. *Methylotenera* was also abundant in the combined system, leading to an efficient denitrification process.

The combination of electrolysis with constructed wetland systems has also been reported as a novel strategy for the removal of PhACs. Such a combined system without the need for any substrate has been used recently for the full degradation of sulfamethoxazole [19].

The combination of CWs with advanced oxidation processes (AOPs) has also been considered an attractive method for the efficient removal of PhACs. The basis of AOPs

[3] EPSs are consisted of proteins and polysaccharides and are normally secreted by cells as a protective mechanism to protect themselves when they are exposed to the harsh environments [42].

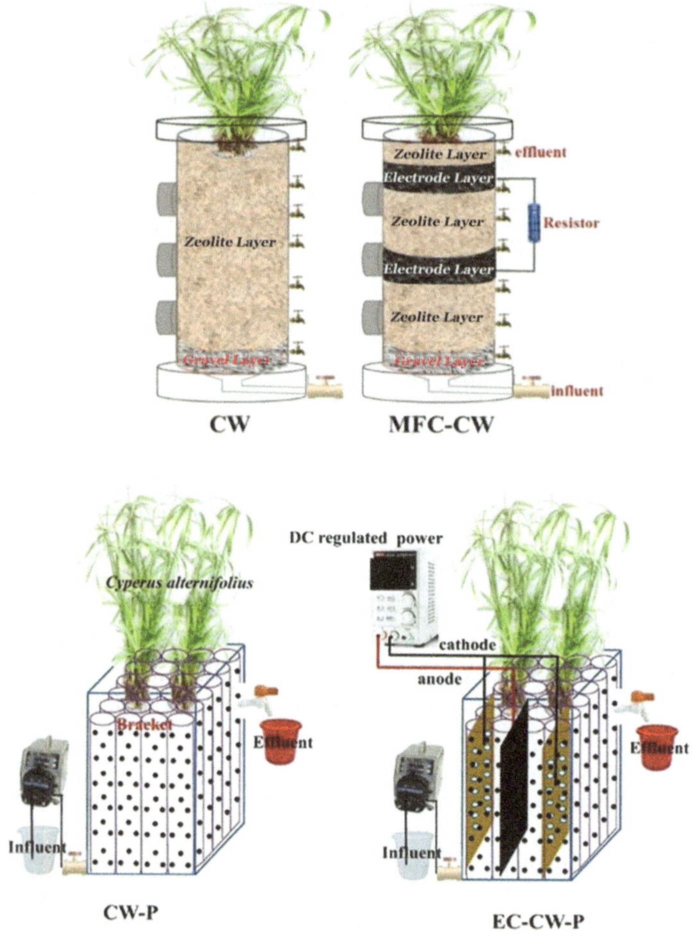

Fig. 7.4 Application of *Cyperus alternifolius* in combined systems for the biodegradation of sulfamethoxazole. Top: a schematic combination of a constructed wetland (CW) with microbial fuel cell (MFC) technology, adapted from Liu et al. [19]. Down: Electrolysis-integrated biorack CW system, adapted from Liu et al. [52]

is the generation of powerful oxidation agents in the medium for the degradation of organic pollutants. In this regard, the Fenton process, as a reaction between an iron-based compartment (Fe^{2+}) and hydrogen peroxide for the generation of hydroxyl radicals ($^{\cdot}OH$) (Eqs. 7.1 and 7.2), has received considerable attention in recent years [62, 63].

$$Fe^{2+} + H_2O_2 \rightarrow Fe^{3+} + {}^{\cdot}OH + OH^- \quad (7.1)$$

$$^{\cdot}OH + PhACs \rightarrow \text{Intermediated} + \text{Final products } (CO_2 + H_2O) \quad (7.2)$$

The utilization of an external source of light such as solar irradiation has also been demonstrated to have a positive effect on the generation of active species. Studies of the combination of CWs with photo-Fenton technology have been very recently initiated in the literature. For instance, a combination of HSF-CWs and solar photo-Fenton resulted in enhanced degradation of PhACs, including diclofenac and carbamazepine (approximately 90% for both[4]), as reported in a recent study [64].

Photocatalysis, as a branch of AOPs, is also an interesting technique for the removal of a wide range of organic pollutants. The basis of this method is to harvest photons by an appropriate photocatalyst for the generation of active species such as hydroxyl radicals or singlet oxygen, which can attack and decompose complex organic compounds[5] [65–67]. TiO_2 and ZnO have been among the most widely used catalysts, and there have been trends in the literature for the fabrication of semiconductors with appropriate energy bandgaps to utilize light in the visible range. Graphitic carbon nitride (g-C_3N_4) is an example of a new generation of visible-light active materials for the photocatalytic removal of various organic compounds, including PhACs [68, 69]. A combination of CW with photocatalytic processes can also be considered an efficient strategy to deal with effluents polluted with PhACs.

There are also examples of combinations of HFS-CW and VFS-CW configurations for the efficient treatment of pharmaceutical effluents, as reported by Shrestha et al. [70], which resulted in high TSS, COD, and BOD removals of 97%, 94%, 97%, respectively Shrestha et al. [70]. Recent literature also indicates the efficiency of this type of CW to deal with high concentrations of specific PhACs. For instance, it has been reported that a VFS-CW could remove acetaminophen (10 mg/L) under the activity of *S. validus* [71]. From a mechanistic point of view, the hydrogen peroxide generated in the shoot of this plant after exposure to acetaminophen can be activated by the peroxidase enzymes produced in the root to generate powerful oxidative species (such as hydroxyl radicals) from the breakdown of the PhACs, as indicated in Fig. 7.5.

Although CWs represent acceptable performance for the removal of numerous PhACs, there are limitations that can negatively affect the performance of these systems. Conversion of ammonia to nitrate is a common phenomenon in CWs that can result in the accumulation of nitrate in treated effluents [15, 72]. Such a drawback of CWs has been observed in other studies (e.g., in horizontal subsurface flow CWs in India [73]). The inefficiency of ammonia removal has been especially observed in HFS-CWs due to the lack of dissolved oxygen to complete the nitrification process using aerobic microorganisms [16]. This drawback of HFS-CWs can be overcome by designing hybrid HFS-VFS CWs. This is because VFS can remove ammonia efficiently [74].

In addition, there is usually a need for vast land areas for the establishment of CWs. This can be an important limiting factor for countries with limited accessible land areas [75].

[4] Less than 40% with the CW alone.

[5] The mechanisms involved in the photocatalytic processes have been described and discussed in Chap. 11.

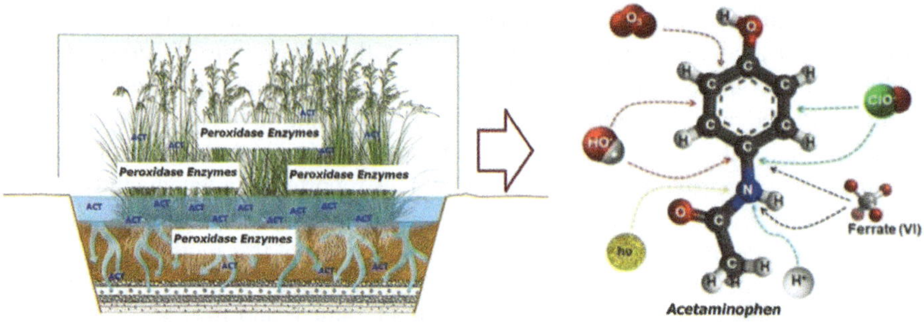

Fig. 7.5 Biodegradation of ACT using the oxidative species generated after exposure of *S. validus* to PhAC, adopted from Vo et al. [71]

Another limitation of the application of CWs is the possibility of the formation of pharmaceutical degradation metabolites, which can potentially cause risks even greater than their mother molecules [14]. For instance, metabolites such as hydroxy-IBU, 1,2-dihydroxy-IBU, carboxy-IBU, and glucopyranosyloxy-hydroxy-IBU have been detected in various parts of plants, such as roots and shoots, when ibuprofen has been introduced into CWs [33, 76]. The release of such compounds into aquatic systems can be considered a real threat, and there is a need for more in-depth studies to trace such compounds and to improve the efficiency of CWs for the complete mineralization of PhACs.

7.5 Further Reading

Table 7.2 contains items from the recent literature that the reader can consult for more detailed information regarding the efficiency of constructed wetland technologies for the elimination of pharmaceutically active compounds.

Table 7.2 Further reading suggestions for more detailed coverage of the literature on the constructed wetland technologies for the removal of pharmaceutically active compounds

References	Item	Subject
Ilyas and van Hullebusch [54]	Table 1	Main parameters for the design of CWs for the removal of PhACs
	Table 3	Mechanisms of the removal of some selected PhACs
Nguyen et al. [77]	Table 2	Plant–bacteria partnership in CWs for the removal of PhACs

7.6 Summary

Constructed wetland systems have been rapidly gaining popularity for the treatment of effluents from various industrial and nonindustrial origins. This is mainly due to their robust nature, low cost, and aesthetic aspects they can bring into (waste)water treatment processes. Several mechanisms, including sorption, phytoremediation, biodegradation, and photodegradation, can be involved in the removal of PhACs using constructed wetlands. Numerous studies have discussed the applicability of these systems for the removal of PhACs but depend on parameters such as configurations, plant and microorganism species, and operating conditions. Combinations of CWs with other sustainable technologies, such as microbial fuel cells, have also indicated very efficient simultaneous degradation of PhACs and the generation of bioelectricity as a green source of energy. However, there are limitations to the implementation of CWs, such as the need for large land areas and continuous maintenance. Furthermore, the formation and release of the metabolites of PhAC degradation in CWs can also be considered an issue for the application of such technologies. Further studies are recommended on the formation and elimination of metabolites to ensure the quality of the final treated effluents.

References

1. Sandoval L et al (2019) Role of wetland plants and use of ornamental flowering plants in constructed wetlands for wastewater treatment: a review. Appl Sci 9:1–17. https://doi.org/10.3390/app9040685
2. Wang M et al (2017) Constructed wetlands for wastewater treatment in cold climate—a review. J Environ Sci (China) 57:293–311. https://doi.org/10.1016/j.jes.2016.12.019
3. Parde D et al (2021) A review of constructed wetland on type, treatment and technology of wastewater. Environ Technol Innov 21:101261. https://doi.org/10.1016/j.eti.2020.101261
4. Ávila C et al (2010) Capacity of a horizontal subsurface flow constructed wetland system for the removal of emerging pollutants: an injection experiment. Chemosphere 81:1137–1142. https://doi.org/10.1016/j.chemosphere.2010.08.006
5. Ranieri E, Verlicchi P, Young TM (2011) Paracetamol removal in subsurface flow constructed wetlands. J Hydrol 404:130–135. https://doi.org/10.1016/j.jhydrol.2011.03.015
6. Dires S, Birhanu T, Ambelu A (2019) Use of broken brick to enhance the removal of nutrients in subsurface flow constructed wetlands receiving hospital wastewater. Water Sci Technol 79:156–164. https://doi.org/10.2166/wst.2019.037
7. Sanjrani MA et al (2020) Treatment of wastewater with constructed wetlands systems and plants used in this technology—a review. Appl Ecol Environ Res 18:107–127. https://doi.org/10.15666/aeer/1801_107127
8. Kamali M et al (2019) Sustainability considerations in membrane-based technologies for industrial effluents treatment. Chem Eng J 368:474–494. https://doi.org/10.1016/j.cej.2019.02.075
9. Ji Z, Tang W, Pei Y (2022) Constructed wetland substrates: a review on development, function mechanisms, and application in contaminants removal. Chemosphere 286:131564. https://doi.org/10.1016/j.chemosphere.2021.131564

10. Suhaib KH, Bhunia P (2022) Dynamics of clogging in subsurface flow constructed wetlands. J Hazard, Toxic, Radioactive Waste 26:03121004. https://doi.org/10.1061/(asce)hz.2153-5515.0000646
11. Rizzo A et al (2020) Constructed wetlands for combined sewer overflow treatment: a state-of-the-art review. Sci Total Enviro 727:138618. https://doi.org/10.1016/j.scitotenv.2020.138618
12. Pat-Espadas AM et al (2018) Review of constructed wetlands for acid mine drainage treatment. Water (Switzerland) 10:1–25. https://doi.org/10.3390/w10111685
13. Liang Y et al (2017) Constructed wetlands for saline wastewater treatment: a review. Ecol Eng 98:275–285. https://doi.org/10.1016/j.ecoleng.2016.11.005
14. Han EJ, Lee DS (2017) Significance of metabolites in the environmental risk assessment of pharmaceuticals consumed by human. Sci Total Environ 592:600–607. https://doi.org/10.1016/j.scitotenv.2017.03.044
15. Auvinen H et al (2017) Removal of pharmaceuticals by a pilot aerated sub-surface flow constructed wetland treating municipal and hospital wastewater. Ecol Eng 100:157–164. https://doi.org/10.1016/j.ecoleng.2016.12.031
16. Jain M et al (2020) A review on treatment of petroleum refinery and petrochemical plant wastewater: a special emphasis on constructed wetlands. J Environ Manage 272:111057. https://doi.org/10.1016/j.jenvman.2020.111057
17. Martínez-Alcalá I, Guillén-Navarro JM, Fernández-López C (2017) Pharmaceutical biological degradation, sorption and mass balance determination in a conventional activated-sludge wastewater treatment plant from Murcia, Spain. Chem Eng J 316:332–340. https://doi.org/10.1016/j.cej.2017.01.048
18. Peng J et al (2019) Characterizing the removal routes of seven pharmaceuticals in the activated sludge process. Sci Total Environ 650:2437–2445. https://doi.org/10.1016/j.scitotenv.2018.10.004
19. Liu X, Lu S et al (2021) Performance and mechanism of sulfamethoxazole removal in different bioelectrochemical technology-integrated constructed wetlands. Water Res 207:117814. https://doi.org/10.1016/j.watres.2021.117814
20. Guo X et al (2020) Responses of the growth and physiological characteristics of Myriophyllum aquaticum to coexisting tetracyclines and copper in constructed wetland microcosms. Environ Pollut 261:114204. https://doi.org/10.1016/j.envpol.2020.114204
21. Yan Y et al (2019) Ecotoxicological effects and accumulation of ciprofloxacin in Eichhornia crassipes under hydroponic conditions. Environ Sci Pollut Res 26:30348–30355. https://doi.org/10.1007/s11356-019-06232-5
22. Almeida A et al (2020) Nitrogen removal in vertical flow constructed wetlands: influence of bed depth and high nitrogen loadings. Environ Technol (United Kingdom) 41(17):2196–2209. https://doi.org/10.1080/09593330.2018.1557749
23. Amado L, Albuquerque A, Espírito Santo A (2012) Influence of stormwater infiltration on the treatment capacity of a LECA-based horizontal subsurface flow constructed wetland. Ecol Eng 39:16–23. https://doi.org/10.1016/j.ecoleng.2011.11.009
24. Özengin N, Elmaci A (2016) Removal of pharmaceutical products in a constructed wetland. Iran J Biotechnol 14:221–229. https://doi.org/10.15171/ijb.1223
25. Dordio A et al (2009) Atenolol removal in microcosm constructed wetlands. Int J Environ Anal Chem 89:835–848. https://doi.org/10.1080/03067310902962502
26. Dordio AV et al (2009) Preliminary media screening for application in the removal of clofibric acid, carbamazepine and ibuprofen by SSF-constructed wetlands. Ecol Eng 35:290–302. https://doi.org/10.1016/j.ecoleng.2008.02.014
27. Tejeda A, Barrera A, Zurita F (2017) Adsorption capacity of a volcanic rock-Used in constructed wetlands—for carbamazepine removal, and its modification with biofilm growth. Water (Switzerland) 9:721. https://doi.org/10.3390/w9090721
28. Vymazal J (2011) Plants used in constructed wetlands with horizontal subsurface flow: a review. Hydrobiologia 674:133–156. https://doi.org/10.1007/s10750-011-0738-9
29. Vymazal J (2018) Constructed wetlands for wastewater treatment. Encycl Ecol 45:14–21. https://doi.org/10.1016/B978-0-12-409548-9.11238-2

30. Truong P, Van TP, Vetiver E (2015) Vetiver system for prevention and treatment of contaminated water and land. In: The vetiver system for improving water quality the prevention and treatment of contaminated water and land, p 33
31. Model CW (2021) Removal of antibiotics and nutrients by vetiver grass. Toxics 9:84. Available at: http://www.vetiver.org/TVN_Water_quality2ed.pdf
32. Liu X, Xu J et al (2021) Stable and efficient sulfamethoxazole and phosphorus removal by an electrolysis-integrated bio-rack constructed wetland system. Chem Eng J 425:130582. https://doi.org/10.1016/j.cej.2021.130582
33. Bartha B, Huber C, Schröder P (2014) Uptake and metabolism of diclofenac in *Typha latifolia*—How plants cope with human pharmaceutical pollution. Plant Sci 227:12–20. https://doi.org/10.1016/j.plantsci.2014.06.001
34. Liu L et al (2013) Potential effect and accumulation of veterinary antibiotics in Phragmites australis under hydroponic conditions. Ecol Eng 53:138–143. https://doi.org/10.1016/j.ecoleng.2012.12.033
35. Sakurai KSI et al (2021) Hybrid constructed wetlands as post-treatment of blackwater: an assessment of the removal of antibiotics. J Environ Manage 278:111552. https://doi.org/10.1016/j.jenvman.2020.111552
36. Tambunan JAM, Effendi H, Krisanti M (2018) Phytoremediating batik wastewater using vetiver *chrysopogon zizanioides (L)*. Pol J Environ Stud 27:1281–1288. https://doi.org/10.15244/pjoes/76728
37. Rydzyński D et al (2017) Instability of chlorophyll in yellow lupin seedlings grown in soil contaminated with ciprofloxacin and tetracycline. Chemosphere 184:62–73. https://doi.org/10.1016/j.chemosphere.2017.05.147
38. Hu B et al (2021) Employ of arbuscular mycorrhizal fungi for pharmaceuticals ibuprofen and diclofenac removal in mesocosm-scale constructed wetlands. J Hazard Mater 409:124524. https://doi.org/10.1016/j.jhazmat.2020.124524
39. Liu W et al (2020) Reduction of methane emissions from manganese-rich constructed wetlands: role of manganese-dependent anaerobic methane oxidation. Chem Eng J 387:123402. https://doi.org/10.1016/j.cej.2019.123402
40. Yang Y et al (2019) Influence of application of manganese ore in constructed wetlands on the mechanisms and improvement of nitrogen and phosphorus removal. Ecotoxicol Environ Safety 170:446–452. https://doi.org/10.1016/j.ecoenv.2018.12.024
41. Yan S et al (2018) Natural Fe-bearing manganese ore facilitating bioelectro-activation of peroxymonosulfate for bisphenol A oxidation. Chem Eng J 354:1120–1131. https://doi.org/10.1016/j.cej.2018.08.066
42. Xu D et al (2022) Enhanced performance and mechanisms of sulfamethoxazole removal in vertical subsurface flow constructed wetland by filling manganese ore as the substrate. Sci Total Environ 812:152554. https://doi.org/10.1016/j.scitotenv.2021.152554
43. Krzeminski P et al (2019) Performance of secondary wastewater treatment methods for the removal of contaminants of emerging concern implicated in crop uptake and antibiotic resistance spread: a review. Sci Total Environ 648:1052–1081. https://doi.org/10.1016/j.scitotenv.2018.08.130
44. Gorito AM et al (2018) Constructed wetland microcosms for the removal of organic micropollutants from freshwater aquaculture effluents. Sci Total Environ 644:1171–1180. https://doi.org/10.1016/j.scitotenv.2018.06.371
45. Bayati M et al (2021) Assessing the efficiency of constructed wetlands in removing PPCPs from treated wastewater and mitigating the ecotoxicological impacts. Int J Hyg Environ Health 231:113664. https://doi.org/10.1016/j.ijheh.2020.113664
46. Carvalho PN et al (2013) Potential of constructed wetlands microcosms for the removal of veterinary pharmaceuticals from livestock wastewater. Bioresour Technol 134:412–416. https://doi.org/10.1016/j.biortech.2013.02.027
47. Bôto M, Almeida CMR, Mucha AP (2016) Potential of constructed wetlands for removal of antibiotics from saline aquaculture effluents. Water (Switzerland) 8:1–14. https://doi.org/10.3390/w8100465

48. Santos F et al (2019) Potential of constructed wetland for the removal of antibiotics and antibiotic resistant bacteria from livestock wastewater. Ecol Eng 129:45–53. https://doi.org/10.1016/j.ecoleng.2019.01.007
49. Hartl M et al (2021) Constructed wetlands operated as bioelectrochemical systems for the removal of organic micropollutants. Chemosphere 271:129593. https://doi.org/10.1016/j.chemosphere.2021.129593
50. Ávila C et al (2021) New insights on the combined removal of antibiotics and ARGs in urban wastewater through the use of two configurations of vertical subsurface flow constructed wetlands. Sci Total Environ 755:142554. https://doi.org/10.1016/j.scitotenv.2020.142554
51. AL Falahi OA et al (2021) Simultaneous removal of ibuprofen, organic material, and nutrients from domestic wastewater through a pilot-scale vertical sub-surface flow constructed wetland with aeration system. J Water Process Eng 43:102214. https://doi.org/10.1016/j.jwpe.2021.102214
52. Liu L et al (2021) Influence of hydraulic loading rate on antibiotics removal and antibiotic resistance expression in soil layer of constructed wetlands. Chemosphere 265:129100. https://doi.org/10.1016/j.chemosphere.2020.129100
53. Zhang DQ et al (2012) Pharmaceutical removal in tropical subsurface flow constructed wetlands at varying hydraulic loading rates. Chemosphere 87:273–277. https://doi.org/10.1016/j.chemosphere.2011.12.067
54. Ilyas H, van Hullebusch ED (2019) Role of design and operational factors in the removal of pharmaceuticals by constructed wetlands. Water (Switzerland) 11:2356. https://doi.org/10.3390/w11112356
55. Rühmland S et al (2015) Fate of pharmaceuticals in a subsurface flow constructed wetland and two ponds. Ecol Eng 80:125–139. https://doi.org/10.1016/j.ecoleng.2015.01.036
56. Lorphensri O et al (2006) Sorption of acetaminophen, 17α-ethynyl estradiol, nalidixic acid, and norfloxacin to silica, alumina, and a hydrophobic medium. Water Res 40:1481–1491. https://doi.org/10.1016/j.watres.2006.02.003
57. Białk-Bielińska A et al (2012) Hydrolysis of sulphonamides in aqueous solutions. J Hazard Mater 221–222:264–274. https://doi.org/10.1016/j.jhazmat.2012.04.044
58. Ávila C et al (2014) Emerging organic contaminants in vertical subsurface flow constructed wetlands: influence of media size, loading frequency and use of active aeration. Sci Total Environ 494–495:211–217. https://doi.org/10.1016/j.scitotenv.2014.06.128
59. Colares GS et al (2022) Hybrid constructed wetlands integrated with microbial fuel cells and reactive bed filter for wastewater treatment and bioelectricity generation. Environ Sci Pollut Res 29:22223–22236. https://doi.org/10.1007/s11356-021-17395-5
60. Wen H et al (2021) Removal of sulfamethoxazole and tetracycline in constructed wetlands integrated with microbial fuel cells influenced by influent and operational conditions. Environ Pollut 272:115988. https://doi.org/10.1016/j.envpol.2020.115988
61. Dai M et al (2021) Mechanism involved in the treatment of sulfamethoxazole in wastewater using a constructed wetland microbial fuel cell system. J Environ Chem Eng 9(5):106193. https://doi.org/10.1016/j.jece.2021.106193
62. Clarizia L et al (2017) Homogeneous photo-Fenton processes at near neutral pH: a review. Appl Catal B: Environ 209:358–371. https://doi.org/10.1016/j.apcatb.2017.03.011
63. Wei Y, Li G, Wang B (2011) Research on harbor oily wastewater treatment by Fenton oxidation. Adv Mater Res 322:164–168. https://doi.org/10.4028/www.scientific.net/AMR.322.164
64. Casierra-Martinez HA et al (2020) Diclofenac and carbamazepine removal from domestic wastewater using a constructed wetland-solar photo-Fenton coupled system. Ecol Eng 153:105699. https://doi.org/10.1016/j.ecoleng.2019.105699
65. Enesca A, Andronic L (2020) The influence of photoactive heterostructures on the photocatalytic removal of dyes and pharmaceutical active compounds: a mini-review. Nanomaterials 10(9):1–22. https://doi.org/10.3390/nano10091766
66. Lin L et al (2019) Adsorption and photocatalytic oxidation of ibuprofen using nanocomposites of TiO_2 nanofibers combined with BN nanosheets: degradation products and mechanisms. Chemosphere 220:921–929. https://doi.org/10.1016/j.chemosphere.2018.12.184

67. Rui Z et al (2010) Photocatalytic degradation of pesticide residues with RE_3^+-doped nano-TiO_2. J Rare Earths Chin Soc Rare Earths 28(October):353–356. https://doi.org/10.1016/S1002-0721(10)60329-8
68. Kumar A et al (2021) Construction of dual Z-scheme g-C_3N_4/$Bi_4Ti_3O_{12}$/$Bi_4O_5I_2$ heterojunction for visible and solar powered coupled photocatalytic antibiotic degradation and hydrogen production: boosting via I^-/I_3^- and Bi^{3+}/Bi^{5+} redox mediators. Appl Catal B: Environ 284(August 2020):119808. https://doi.org/10.1016/j.apcatb.2020.119808
69. Yang HC et al (2020) Polymeric g-C3N4 derived from the mixture of dicyandiamide and mushroom waste for photocatalytic degradation of methyl blue. Topics Catalysis 63:1182–1192. https://doi.org/10.1007/s11244-020-01237-8
70. Shrestha RR, Haberl R, Laber J (2001) Constructed wetland technology transfer to Nepal. Water Sci Technol 43:345–350. Available at: http://ovidsp.ovid.com/ovidweb.cgi?T=JS&PAGE=reference&D=emed5&NEWS=N&AN=2001387179
71. Vo HNP et al (2019) Removal and monitoring acetaminophen-contaminated hospital wastewater by vertical flow constructed wetland and peroxidase enzymes. J Environ Manage 250(September):109526. https://doi.org/10.1016/j.jenvman.2019.109526
72. Majumder A et al (2021) A review on hospital wastewater treatment: a special emphasis on occurrence and removal of pharmaceutically active compounds, resistant microorganisms, and SARS-CoV-2. J Environ Chem Eng 9:104812. https://doi.org/10.1016/j.jece.2020.104812
73. Khan NA et al (2020) Occurrence, sources and conventional treatment techniques for various antibiotics present in hospital wastewaters: a critical review. TrAC Trends Anal Chem 129:115921. https://doi.org/10.1016/j.trac.2020.115921
74. Xinshan S, Qin L, Denghua Y (2010) Nutrient removal by hybrid subsurface flow constructed wetlands for high concentration ammonia nitrogen wastewater. Procedia Environ Sci 2:1461–1468. https://doi.org/10.1016/j.proenv.2010.10.159
75. Taoufik N et al (2021) Comparative overview of advanced oxidation processes and biological approaches for the removal pharmaceuticals. J Environ Manage 288:112404. https://doi.org/10.1016/j.jenvman.2021.112404
76. He Y et al (2017) Metabolism of Ibuprofen by *Phragmites australis*: uptake and Phytodegradation. Environ Sci Technol 51:4576–4584. https://doi.org/10.1021/acs.est.7b00458
77. Nguyen PM et al (2019) Removal of pharmaceuticals and personal care products using constructed wetlands: effective plant-bacteria synergism may enhance degradation efficiency. Environ Sci Pollut Res 26:21109–21126. https://doi.org/10.1007/s11356-019-05320-w

Chapter 8
Membrane Separation Technologies for the Elimination of Pharmaceutically Active Compounds—Progress and Challenges

8.1 Introduction

Membrane-based technologies are among the widely used treatments used for the separation and purification of effluents from various industrial and nonindustrial origins. They have represented acceptable performances to recover clean water resources. Furthermore, they can be used for the recovery of various metallic and nonmetallic compounds [1, 2]. Various types of membrane-based technologies have been developed and used for (waste)water treatment applications, such as reverse osmosis (RO), forward osmosis (FO), nanofiltration, microfiltration, ultrafiltration, and their combinations with biological treatment methods (i.e., membrane bioreactors). Figure 8.1 represents the featured properties of various types of membranes used for (waste)water treatment.

The current trends in the literature for the application of membrane-based technologies are the fabrication of structures with high separation performance and low fouling properties, especially from low-cost carbonaceous structures such as biochar [4], clay-based materials such as zeolite [5, 6], kaolinite [7], and bentonite [7], as well as natural polymers [8]. The treatment of effluents containing pharmaceutically active compounds (PhACs) using various types of membrane structures has also gained popularity due to advantages such as no need for the addition of chemicals and ease of operation.

There are various parameters that can influence the removal efficiency of PhACs using membrane structures. The properties of the membranes, such as the pore size, surface charge, and surface roughness, can determine the separation mechanisms, including adsorption, size exclusion, and charge exclusion [9–11]. Operating pH is also an essential parameter that can influence the surface charge of both the membrane and pollutants, leading to attraction or repulsion between the membrane and the PhACs. Hence, there have been efforts in the literature to enhance the performance of membrane technologies through the manipulation of the membrane properties and optimization of the operating conditions [12, 13]. Such modifications in the

M. Kamali et al., *Advanced Wastewater Treatment Technologies for the Removal of Pharmaceutically Active Compounds*, Green Energy and Technology,
https://doi.org/10.1007/978-3-031-20806-5_8

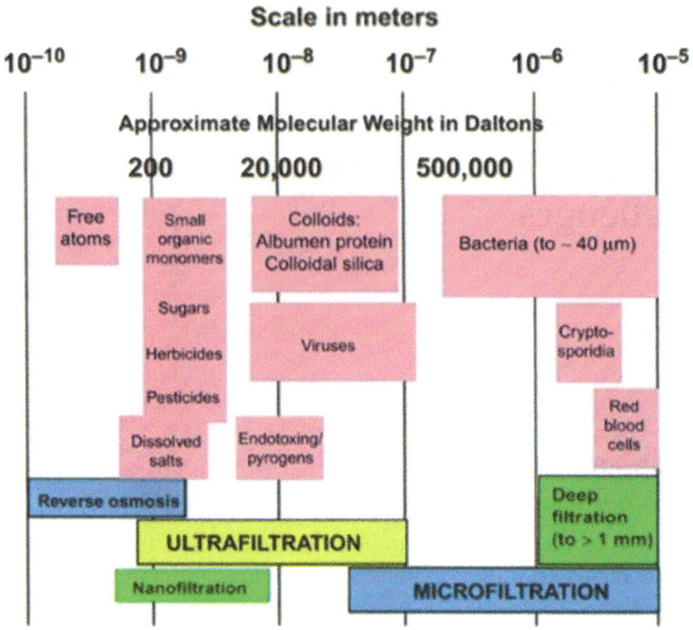

Fig. 8.1 Featured properties of various membrane separation processes, including the pore size, and their potential applications to remove various pollutants, adapted from Mallakpour and Azadi [3]

membrane structures have also been applied to mitigate the fouling process (e.g., by promoting their hydrophilicity) of the membrane structures, as one of the most important obstacles to the application of such technologies [14]. This chapter has aimed to provide an overview of various membrane-based technologies used for the removal of PhACs and the mechanisms involved in such processes and to discuss the existing challenges and the latest trends in the literature to make such technologies more sustainable and applicable for real (waste)water treatment applications.

8.2 Membrane-Based Technologies for PhAC Removal

8.2.1 Forward Osmosis and Reverse Osmosis

In the forward osmosis process, a semipermeable membrane is used to separate clean water from wastewater using a concentrated solution (e.g., salts) under the concentration difference phenomenon. The separated water will then be purified using methods such as desalination [15]. In contrast, in reverse osmosis, water is forced to pass through the membrane in response to the pressure difference, and no concentrated

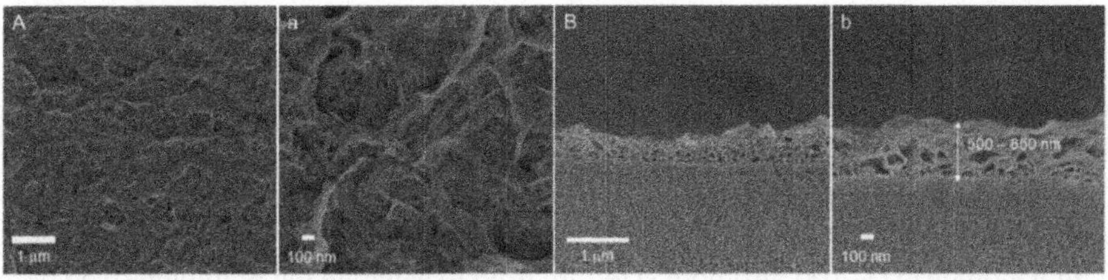

Fig. 8.2 SEM images of a TFC membrane representing the inner surface (**A**), the enlarged inner surface (**a**), the cross section (**B**), and the enlarged cross section (**b**) of the membrane used for the treatment of pharmaceutical compounds, adopted from Goh et al. [20]

solution is used to separate the water from the effluents [16]. These technologies have been previously used for the treatment of effluents from various resources and showed acceptable performance in most cases. Recently, they have been used for the removal of pharmaceutical compounds, although the removal mechanisms have been poorly studied.

In osmotic systems, processes including hydrophobicity, dipole moment, and molecular size of the pollutants are involved in the treatment of the effluents. For instance, it has been indicated that pharmaceuticals with high molecular width (such as triclosan, 0.75 nm) can be efficiently separated by typical membranes used for the FO and RO processes [17, 18]. Additionally, pH and the charge of the pharmaceuticals can determine the efficiency of the membrane structure for the rejection of PhACs. This is especially of high importance for the separation of polar pharmaceutical compounds to optimize the operating conditions to achieve acceptable performance of the system. For instance, the rejection of carbamazepine[1] with a neutral structure is normally independent of the operating pH. However, for polar compounds such as sulfamethoxazole, the rejection efficiency increases under elevated pH conditions. In fact, a negative surface charge is demonstrated at pH values above 5.83.[2] Under these conditions, repulsion can be expected between the pharmaceutical and the surface of the membrane if it is negatively charged [17, 18].

Thin-film composite membranes (TFCMs) have also received great attention in recent years for the removal of various organic compounds from polluted (waste)waters. TFCMSs normally consist of an ultra-thin selective layer combined with a highly porous substrate (Fig. 8.2) [19].

Such membranes have presented superior performance compared to conventional ones (such as cellulose triacetate) for the rejection of PhACs in FO technologies, as indicated by Jin et al. [21] for the rejection of four PhACs: carbamazepine, diclofenac, ibuprofen, and naproxen. This was mainly due to the coupling of the mechanisms, including the adsorption of the PhACs to the surface of the membrane, size exclusion, and the electrostatic repulsion between the pollutants and the membrane surface.

---

[1] Carbamazepine with a pKa of 9.73 represents a neutral nature under pH values below 9.73.

[2] Sulfamethoxazole has a pKa of 5.83.

8.2.2 Nanofiltration

Nanofiltration is a popular technique for the removal of various pollutants and has been used on a large scale since the 1980s, especially after 2000 [22]. Mery-sur-Oise treatment (340,000 m^3/d) is the largest industrial-scale nanofiltration facility worldwide built in France in 1999 [23]. In the nanofiltration process, membranes with nanometric (1–100 nm) pore sizes are used to separate the pollutants. The pore size in this type of membrane is larger than that of reverse osmosis membranes but smaller than that of ultrafiltration and microfiltration.

The nanofiltration process for the removal of PhACs relies on mechanisms including charge exclusion, size exclusion, and adsorption of the pollutants by the membrane structures. However, different mechanisms can be expected for the removal of various types of PhACs using this technology. For hydrophobic molecules, size separation can be the dominant separation mechanism by nanofiltration because of their affinity toward water molecules, which results in an increase in their volume. Hence, this technique can be especially useful for the separation of hydrophobic PhACs such as sulfaguanidine, carbamazepine, naproxentrimethoprim, hydrocortisone, and procaine [24, 25]. However, relatively low removal efficiencies can be expected using these technologies for low molecular size PhACs such as acetaminophen, which results in passing such compounds through the membrane pores [26].

The formation of complex compounds with other organic compounds present in the medium is also another way of increasing the total volume of the PhACs, which can facilitate their size exclusion using nanofiltration processes.

Electrostatic repulsion is another mechanism involved in the removal of PhACs using this technology. It is evident that neutral PhACs such as sulfamethoxazole can pass through the membrane, resulting in a lower removal efficiency of this compound, while for charged PhACs, the charge separation process plays an important role in its removal from the containing effluents [26]. It is also worth mentioning that the pKa values of PhACs can determine their surface charges under various pH conditions. For instance, carbamazepine is a base that has a pKa value of 2.3 and hence remains neutral under most water pH conditions [12]. For this compound, size exclusion seems to be the most abundant removal mechanism due to the presence of two benzene rings in its molecular structure, which results in an enlarged space volume [27]. Additionally, for compounds such as ibuprofen, a low dipole moment and a high log Know value can lead to a high removal efficiency using nanofiltration technologies [10, 28]. The effect of pH is especially of high importance for the PhACs with two pKa values, such as sulfamethoxazole[3] (pKa1 = 1.4, pKa2 = 5.8).

The membrane surface charge can also be influenced by pH [29]. Hence, finding the optimum pH at which the membrane and the PhAC represent different charges can be a key point for the removal of such compounds. For instance, Soares et al. [30] indicated that the removal of atenolol using nanofiltration can be optimized at

[3] Due to the protonation, deprotonation of the primary aromatic amine, and the sulfonamide group (as the two functional groups present in this molecule), respectively.

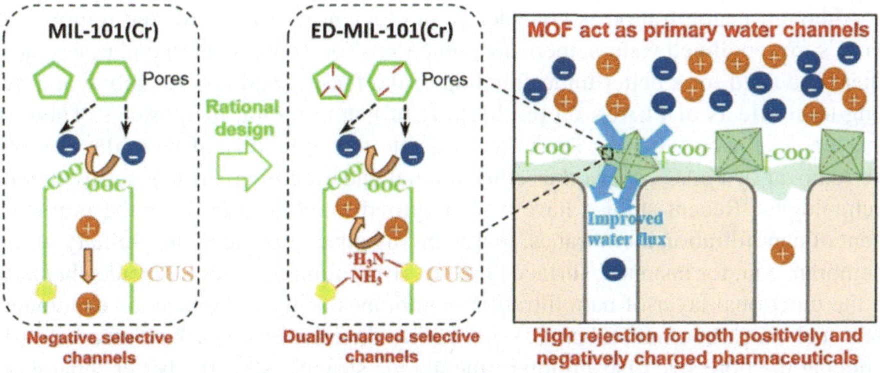

Fig. 8.3 Illustration of dually charged thin-film nanocomposites made of MOFs. The presence of –COO– groups grants a negative charge to MIL-101(Cr). ED-MIL-101(Cr) represents a dual charge property by grafting ethylenediamine (ED) onto the Cr coordinately unsaturated metal sites of MIL-101(Cr) via the presence of $–NH_3^+$ groups, adapted from Dai et al. [31]

pH = 2.5, in which both the pollutant and the membrane represent positive charges (approximately 90% removal efficiency). There has also been a recent trend in the fabrication of dually charged membranes for the efficient separation of both positively[4] and negatively[5] charged PhACs from polluted effluents. The fabrication of thin-film MOF-based structures is an example of such technologies used for the separation of PhACs, as schematically illustrated in Fig. 8.3 [31].

Adsorption can also contribute to the removal of PhACs using nanofiltration technology. In Chapt. 9, the main mechanisms involved in the adsorptive removal of PhACs have been discussed, including ion exchange, formation of hydrogen bonds, π–π interactions, hydrophobic interactions, and electrostatic interactions between the adsorbent and the adsorbate. For instance, estrone (E1) and estradiol (E2) can establish hydrogen bonding with the membrane surface [32]. However, desorption can also lead to the release of the adsorbed compounds, and after a certain time, adsorption equilibrium is reached. Adsorption of pollutants can also cause membrane fouling by blocking the pores, which can result in a decline in the flux. This can be considered a serious issue in nanofiltration technologies because of the low size of the pores [33].

The type of membrane can also determine the efficiency of the nanofiltration process for the removal of PhACs. For instance, the active and dense skin layer of polyamide nanofiltration membranes promotes the adsorption of PhACs [34, 35].

There are also a limited number of economic studies on the application of nanofiltration technologies for the removal of PhACs. For instance, it has been indicated that the removal of atenolol using commercial NF33 membranes can result in over 70% removal efficiency (pH = 9) and an estimated treatment cost of 0.53 US$/m^3 [36].

---

[4] Such as terbutaline, atenolol, and fluoxetine.

[5] Such as ketoprofen, diclofenac, and bezafibrate.

Although nanofiltration is considered an efficient technique for the removal of PhACs from polluted waters, there are some steps for its application on large scales. There is a need for a better understanding of the fouling and flux reduction and the long-term effects of PhACs on the lifetime of nanofiltration membranes. There is also a need for cost–benefit studies to evaluate the suitability of nanofiltration for the removal of PhACs over other conventional and emerging wastewater treatment technologies. Recent studies have also proposed main headlines for the improvement of nanofiltration membranes, including (a) enhancing the hydrophilicity of the membrane via, for instance, surface grafting, (b) designing water molecule channels in the functional layer of nanofiltration membranes using carbonaceous nanomaterials such as carbon nanotubes, (c) optimizing the membrane surface area, and (d) reducing the pore size distribution of membrane structures for the better separation of PhACs [37].

8.2.3 Ultrafiltration

Ultrafiltration is a low-pressure membrane-based treatment technique that has been widely used for the treatment of (waste)waters from various origins [38]. In ultrafiltration, separation through a semipermeable membrane occurs by providing a force (by pressure or concentration gradient). Hence, there is no significant difference between the fundamentals of ultrafiltration and those of nanofiltration and microfiltration processes. In fact, the mentioned technologies represent different molecular weight cutoff properties that can define their efficiencies to deal with molecules with different weights and sizes.

Polymeric membranes are the most common type used in reverse osmosis, nanofiltration, and ultrafiltration processes [39]. Modifications with inorganic materials (such as SiO_2) have also been reported as a suitable way to improve the performance of this type of membrane to enhance the surface hydrophilicity [40]. There have also been some recent reports of the successful fabrication and application of ceramic fine ultrafiltration membranes for the removal of PhACs (waste)waters. Such membrane structures can provide high thermal, mechanical, and chemical (e.g., against harsh pH conditions) stabilities. The presence of nanomaterials on the surface of ultrafiltration membranes has also been indicated as an efficient way to promote the adsorption of PhACs and their rejection by increasing the available surface area. For instance, 96% and 99% removal of atenolol and ibuprofen, respectively, was recently reported using a ZnO nanoparticle-coated ceramic membrane with a high specific surface area of 21 m^2/g [41]. However, the main drawback of ceramic membranes is their relatively high production costs [42].

Ultrafiltration membranes (0.1–0.01 μm) generally represent a larger pore size than nanofiltration membranes but less than microfiltration membranes (0.1–10 μm). Accordingly, different efficiencies for the removal of PhACs using these types of membranes can be expected. It has already been discussed in the literature that ultrafiltration exhibits less satisfactory performance than nanofiltration for the removal

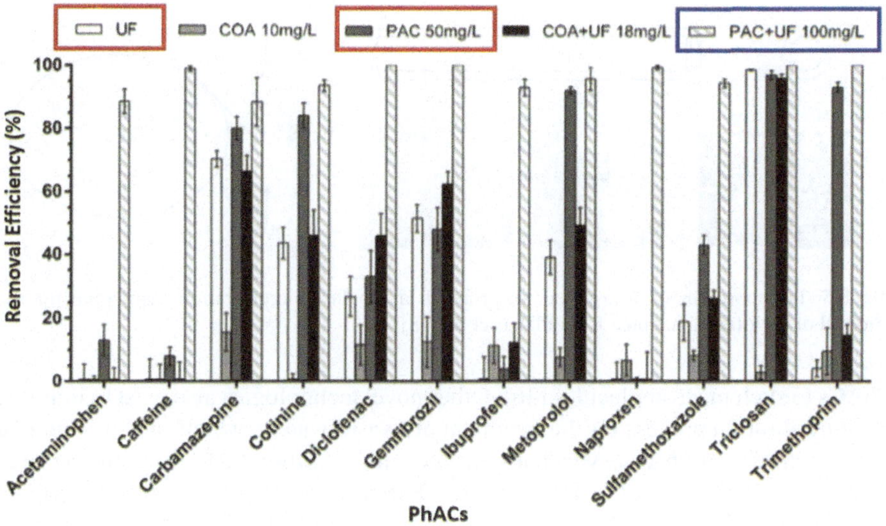

Fig. 8.4 Removal of PhACs using ultrafiltration and its combination with coagulation and adsorption using powdered activated carbon. According to the results, the combination of ultrafiltration and adsorption is the best among the studied methods for the removal of a variety of PhACs, adapted from Sheng et al. [13]

of PhACs [43]. Hence, the combination of ultrafiltration technologies with other physico-chemical techniques is currently considered a trend in the literature for the efficient removal of PhACs. For instance, it was demonstrated [13] that although ultrafiltration represented a low efficiency for the removal of a variety of PhACs, its combination with adsorption using powdered activated carbon resulted in a significant improvement in the removal of PhACs (Fig. 8.4).

Combination with novel adsorbents has also indicated an efficient way to enhance the efficiency of ultrafiltration for the removal of PhACs. For instance, the application of metal–organic frameworks (MOFs) as crystalline porous materials has been reported to be an efficient adsorbent due to the existence of coordinatively unsaturated sites. Kim et al. [44] indicated that a hybrid of ultrafiltration and MOFs is very efficient for the removal of ibuprofen and EE2.

Combination with advanced oxidation processes (AOPs) has also been explored in the literature for the improvement of the performance of ultrafiltration processes. For instance, the integration of ultrafiltration with nonthermal plasma with ultrafiltration resulted in over 90% removal of various types of PhACs, including diclofenac, carbamazepine, and sulfamethoxazole (0.80–15.15 μg/L), from conventional activated sludge process effluents [45]. The presence of other organic compounds in the effluents has also been considered a factor with the potential to affect the filtration efficiency of the ultrafiltration processes. As an example, the presence of aquatic humic substances was indicated to have a positive effect on the rejection of diclofenac and carbamazepine, probably due to the formation of large molecules by binding the

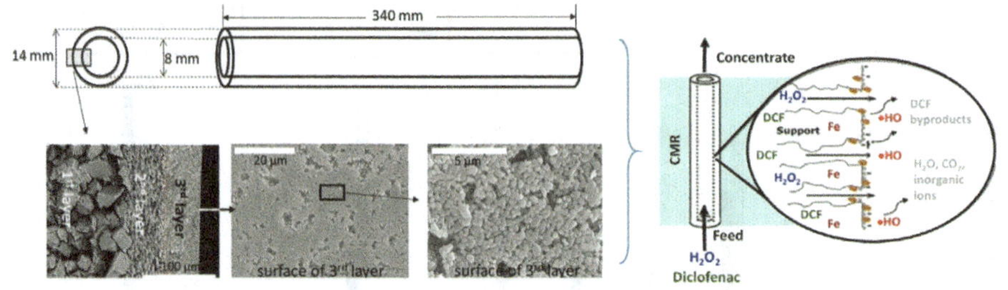

Fig. 8.5 Incorporation of iron-based materials in a tubular microfiltration membrane for the removal of diclofenac, adapted from Plakas et al. [48]

PhACs to such macromolecules [46]. Other novel technologies also exist to improve the ultrafiltration process for the treatment of (waste)waters, which can be examined for the removal of PhACs. Micellar-enhanced ultrafiltration (MEUF) is among such emerging techniques and is based on the addition of a surfactant above the critical micelle concentration (CMC), resulting in the solubilization of the compounds in micelles. The micelle will then be retained by the ultrafiltration process [38].

8.2.4 Microfiltration

Microfiltration is another membrane filtration process that has gained popularity for the treatment of (waste)waters since the 1990s. However, there is limited efficiency for the removal of contaminants of emerging concern (CECs) using microfiltration via mechanisms such as size exclusion due to the relatively high molecular weight cutoff of microfiltration membranes (approximately 300,000 g/mol)[6] [47]. Hence, a combination of microfiltration processes with other physico-chemical and biological treatment techniques or modification of the microfiltration membrane structures (e.g., to promote the adsorption or advanced oxidation of PhACs) have been developed and employed for the treatment of effluents containing PhACs.

Studies are available for the successful modification of microfiltration membranes to promote advanced oxidation processes. Figure 8.5 demonstrates the incorporation of iron-based materials for the degradation of diclofenac under the Fenton process (at $pH = 3$) [48]. A combination of ozonation with microfiltration has been demonstrated to be efficient for the removal of sulfamethoxazole, amoxicillin, bezafibrate, and ibuprofen (over 90%) [49].

Promoted adsorption of PhACs has also been reported by the fabrication of microfiltration membranes fabricated with efficient adsorbents, such as carbonaceous structures. Activated carbon is a low-cost adsorbent that can be efficiently used for the adsorption of various types of organic pollutants [50, 51], especially nonpolar and hydrophobic organic compounds [52, 53]. Carbonaceous materials with large

[6] Molecular weight of PhACs is below 1000 g/mol.

specific surface areas that can promote the adsorption process have also been used for the fabrication of microfiltration membranes. For instance, efficient adsorption of PhACs, including triclosan, acetaminophen, and ibuprofen (up to 70%), has been reported [54] using single-walled and multiwalled layers decorated on top of a PVDF membrane.

Hybrid systems comprising carbonaceous materials and microfiltration membranes have also been tested for the removal of PhACs. Such a system using an activated carbon/ceramic membrane was employed by Viegas et al. [55], reaching over 50% removal of carbamazepine, sulfamethoxazole, and atenolol, with a reasonable treatment cost of 0.21 €/m^3 for a 50,000 m^3/day wastewater treatment plant.

8.2.5 Membrane Bioreactors

MBR technologies have also received considerable attention in recent years for the removal of various types of organic pollutants, including PhACs. There are features such as low sludge production and high removal efficiency at relatively low treatment costs that have made MBRs more appealing compared to conventional (waste)water treatment methods, especially for real-scale applications. The quality of the treated water can also make it possible to discharge the treated water to the environment without any significant impact on the environment. Hence, these technologies can be considered sustainable technologies to deal with PhACs as the main class of contaminants of emerging concern. These technologies combine biological treatment processes with low-pressure membrane separation for the efficient removal of pollutants, including PhACs. The role of the membrane is to hinder the microorganisms and suspended solids and further assist the removal of PhACs. Such systems can bring advantages over conventional wastewater treatment technologies, including low sludge production as well as the high quality of the permeate in terms of the presence of pathogens and suspended solids [56]. A long sludge retention time (SRT) in MBRs allows the growth of nitrifying bacteria, which can lead to the efficient removal of PhACs [57]. The retention characteristics of MBRs for hydrophobic compounds can also result in the better treatment efficiency of these technologies compared to conventional activated sludge processes [58].

Several studies have confirmed the effectiveness of MBRs to deal with antibiotics. For instance, efficient removal of sulfamethoxazole and acetaminophen (over 99% for both) has been observed using a pilot-scale MBR [59]. The removal of estrogens has also been reported by MBRs. As an example, 17β-ethinylestradiol with an efficiency of 99% has been removed in an MBR loaded with nitrifier-enriched biomass [60].

However, the efficiency of MBRs can be highly affected by the type of PhACs, the reactor design, microbial communities, and the operating conditions. For instance, the removal of carbamazepine can vary from almost no removal to over 95% (see Table 8.1). It is evident that the type of pollutant can affect its biodegradation efficiency, as discussed in detail in previous chapters. For instance, carbamazepine has a more stable structure than PhACs, such as sulfamethoxazole, which can increase its resistance to degradation. By looking into the chemical structure of carbamazepine, it can be anticipated that the higher resistance of this compound against decomposition can be attributed to the presence of three phenolic rings present in its molecular structure. In SMX, there is a single phenolic ring that can make it easy to biodegrade (Fig. 8.6).

However, some mechanisms can lead to the low degradation efficiency of less recalcitrant PhACs. For instance, in the case of SMX, back conversion processes (such as N_4-acetylsulfamethazole to sulfamethoxazole) may occur during the biodegradation process [61]. SRT can also affect the efficiency of MBRs for the removal of PhACs. For instance, it has been demonstrated that by increasing the SRT, better efficiency of MBRs can be expected for the removal of polar PhACs such as diclofenac, sulfophenyl, and carboxylate [62]. This trend has been observed in the removal of estrogen [63].

Table 8.1 Efficiencies observed in the literature for the removal of various PhACs using MBR technologies

PhACs	Molecular structure	Removal efficiency (%)
Ciprofloxacin	$C_{17}H_{18}FN_3O_3$	15–95[a]
Erythromycin	$C_{37}H_{67}NO_{13}$	Very low to 99[a]
Azithromycin	$C_{38}H_{72}N_2O_{12}$	N.E[b]
Trimethoprim	$C_{14}H_{18}N_4O_3$	0–99[a]
Clarithromycin	$C_{38}H_{69}NO_{13}$	0–99[a]
Sulfamethoxazole	$C_{10}H_{11}N_3O_3S$	0–90[a]
Enrofloxacin	$C_{19}H_{22}FN_3O_3$	Very low to 60[a]
Carbamazepine	$C_{15}H_{12}NO$	Very low to 95[a]
Azithromycin	$C_{38}H_{72}N_2O_{12}$	Very low to 99[a]
Ibuprofen	$C_{13}H_{18}O_2$	94[c]
Diclofenac	$C_{14}H_{10}Cl_2NO_2$	0–90[a]
Metformin	$C_4H_{11}N_5$	Over 90[a]
Triclosan	$C_{12}H_7Cl_3O_2$	5–95[a]
Ranitidine	$C_{13}H_{22}N_4O_3S$	N.E[b]
Lamotrigine	$C_9H_7Cl_2N_5$	0–85[a]
Atenolol	$C_{14}H_{22}N_2O_3$	Over 90[d]
Propranolol	$C_{16}H_{21}NO_2$	63–72[e]
Estrone (E1)	$C_{18}H_{22}O_2$	60–100[a]
17β-Estradiol (E2)	$C_{18}H_{24}O_2$	40–100[a]
17α-Ethynylestradiol (EE2)	$C_{20}H_{24}O_2$	20–100[a]

(a) Krzeminski et al. [69], (b) Not Assessed, (c) Cornejo et al. [70], (d) Arya et al. [71], Díaz et al. [72], (e) Popple et al. [73]

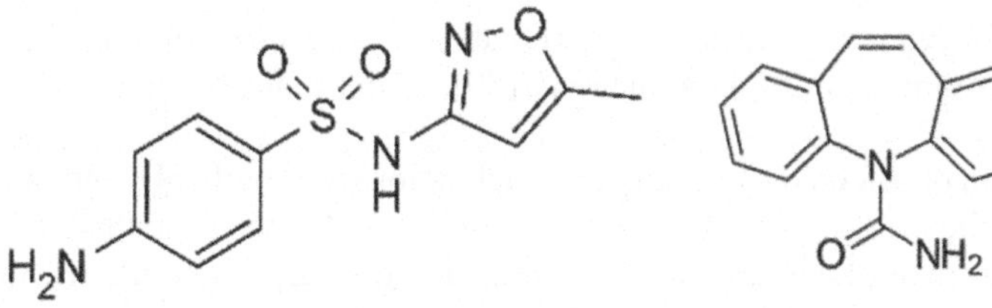

Fig. 8.6 Molecular structures of sulfamethoxazole (left) and carbamazepine (right), illustrating the presence of one and three phenolic rings in their structures, respectively. This can be anticipated as the reason for the higher resistance of carbamazepine against biodecomposition

Various configurations of MBRs have also been reported for the removal of PhACs since the beginning of the 2000s. For instance, efficient removal of COD, turbidity, and NH^{+4}-N (over 80%) has been observed for the application of submerged MBRs for the treatment of hospital effluents [64].

Some strategies can also be proposed for the effective addition of materials into MBRs for the optimum removal of PhACs. For instance, activated carbon has been examined for this purpose to support large specific surface areas for the adhesion and growth of the microbial communities responsible for the removal of PhACs. Biochar has also demonstrated its capability for supporting the colonization of microbial communities in the activated sludge process [65]. Recent studies have also indicated the potential of these materials for the adsorption of various types of pollutants, including PhACs [66, 67]. Hence, the inactivation of pharmaceutical compounds can be achieved by the addition of such biocompatible materials into the MBR process. In this regard, finding the optimum dosage of the additive materials and the operating conditions is a current trend in the relevant literature. For instance, it has been indicated that the addition of 0.5 g/L granular activated carbon (GAC) caused an increase in the biodegradation of diazepam, diclofenac, and carbamazepine by promoting the growth of nitrifiers in the system [68].

Such a strategy has also been very useful for the removal of recalcitrant PhACs. For instance, up to 90% carbamazepine removal has been observed by the addition of 1 g/L powdered activated carbon into the MBR process [9].

Table 8.1 represents the removal ranges for some important PhACs using MBR technologies.

8.3 Fouling by Pharmaceuticals

Fouling is among the most challenging issues for the application of membrane technologies in (waste)water treatment applications [74]. It has been demonstrated that mechanisms including adsorption, pore blocking, cake layer formation, and gel layer formation, or a combination of them, are responsible for the fouling of the membranes [75, 76]. Gel layer formation is generally governed by the symbiotic effects of salts

[77–79]. However, it has been concluded that cake layer formation, mediated by the concentration polarization phenomenon[7] [80], is the main fouling mechanism for PhACs [81].

Fouling can be affected by parameters such as ionic strength, pH, and the presence of cations in the medium [25, 82]. A high ionic strength can normally lead to severe fouling of the membrane. Above the critical ionic strength conditions, the energy barrier will disappear, resulting in the adhesion of foulants onto the membrane surface [83]. pH can also affect the membrane surface charge and may cause the attraction of the foulants by the membranes, resulting in the blocking of the pores [84]. The presence of cationic ions such as Ca^{2+} can also increase the fouling of membranes by the formation of complexes with organic pollutants (including PhACs) [85]. However, membrane fouling is a complex process that can be caused by a combination of various mechanisms.

The fouling process can cause issues for the long-term applicability of membrane technologies [86]. Hence, efforts have been made to overcome this issue. The combination of membrane filtration with adsorption processes is among the most popular technologies to mitigate the fouling process. For instance, it has been indicated that the combination of ultrafiltration with adsorption with MOFs for the removal of PhACs resulted in no significant fouling because the organic compounds are adsorbed by the adsorbent instead of the membrane, preventing the formation of a cake layer and resulting in a decline in the flux [44].

As discussed in Chap. 1, PhACs can have toxic effects on the microbial communities present in wastewater treatment plants, depending on the type of pollutant, operating conditions, and microbial consortium present in the medium. This can also be expected in MBRs where biological treatments are combined with the physical separation of pollutants and microorganisms from the treated water. Secretion of extracellular polymeric substances (EPSs) is a mechanism of microbial communities in the presence of potentially toxic elements [87, 88]. Various types of polymeric structures, such as humic acids, polysaccharides, proteins, nucleic acids, lipids, and uronic acids, can be present in the medium as a result of the antitoxic activities of the microorganisms. EPSs normally represent properties such as hydrophobicity, adhesion, flocculation, settling, and dewatering properties. Hence, they can significantly contribute to the fouling process [77, 78] (Fig. 8.7).

Table 8.2 represents the recent methods developed in the literature to mitigate the fouling process in various membrane-based technologies used for the removal of PhAcs.

[7] Concentration polarization is caused by the accumulation of solutes near the surface of the membrane [25].

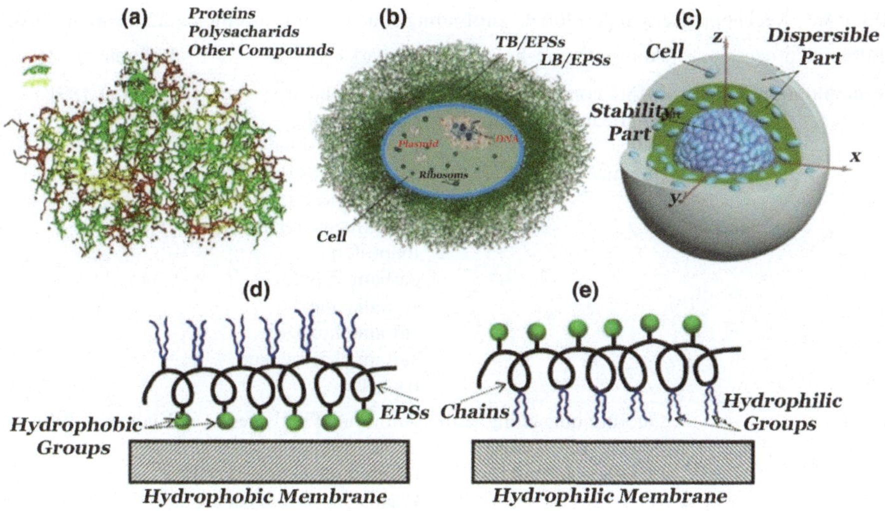

Fig. 8.7 Typical EPS structure (**a**), cell structure (**b**), and structure of the sludge flocs (**c**). **d** and **e** also represent the mechanisms of the adhesion of hydrophobic and hydrophilic EPSs onto hydrophobic membranes, adapted from Lin et al. [77, 78]

8.4 Further Reading

Table 8.3 contains items from the recent literature that the reader can consult for more detailed information regarding the application of membrane-based technologies for the treatment of PhACs.

8.5 Summary

Membrane-based technologies have gained popularity for the treatment of (waste)waters containing PhACs. Mechanisms including size exclusion, charge exclusion, and adsorption are well known and documented for the separation processes involved (i.e., forward osmosis, reverse osmosis, microfiltration, ultrafiltration, nanofiltration, and membrane bioreactors). Among them, reverse osmosis can be efficiently used for the removal of various PhACs. Nanofiltration can also represent an acceptable efficiency for the removal of PhaACs under the size exclusion mechanism. This is due to the relatively low molecular weight and size of such compounds (below 1000 g/mol). Combining other membrane filtration technologies with methods such as adsorption (e.g., with biochar) and oxidation (e.g., Fenton process or coagulation) can enhance the performance of membranes with larger pore sizes (e.g., ultrafiltration and microfiltration). However, the fouling process can bring issues for the large-scale applications of membrane technologies for the removal of

Table 8.2 Recent progress in developing antifouling strategies for the efficient removal of PhACs

Filtration process	Antifouling technique	Remarks	References
Nanofiltration (NF90)	In situ concentration polarization-enhanced radical graft polymerization	The modification resulted in the mitigation of gel layer formation, as the main fouling mechanism for the effluents containing ibuprofen, carbamazepine, sulfadiazine, sulfamethoxazole, sulfamethazine, and triclosan	Lin et al. [89]
MBRs	Quorum quenching with powdered activated carbon adsorption	Addition of activated carbon-immobilized quorum quenching strains, a 4.6-folds delay in fouling was obtained	Xiao et al. [90]
MBRs	Integration of moving-bed bioreactor process with electrooxidation, adsorptive cake, biofilm carriers, and ozonation	The integrated techniques resulted in less fouling, higher flux, and reducing the overall energy consumption for the treatment of pharmaceutical effluents	Palani et al. [91]
Ultrafiltration	Coating the acrylic ultrafiltration membranes with polyamidoamine dendrimers	The strategy resulted in a mitigation of the fouling process with a high water flux and rejection efficiency for tulathromycin	Bojaran et al. [92]
Nanofiltration	Incorporation of TiO_2, and SiO_2 into the ceramic nanofiltration membranes	Less hydrophobicity, larger molecular weight cutoffs, and lower surface roughness were achieved by the modifications applied	Zhao et al. [93]
Anaerobic membrane bioreactors	Addition of biochar	Adsorptive removal of organic compounds resulted in a significant reduction in membrane fouling	Chen et al. [94]
Microfiltration	Coupling the membrane separation with heterogeneous iron-containing microsized zeolite catalysts to perform a Fenton-like process	A high flux was maintained for the integrated filtration-oxidation process	Lan et al. [95]

(continued)

Table 8.2 (continued)

Filtration process	Antifouling technique	Remarks	References
MBRs	Coupling the MBR process with coagulation	Supplementation of the membrane with the coagulants resulted in the improvement in the permeability and declining the fouling in MBRs	Park et al. [96]

Table 8.3 Further reading suggestions for more detailed coverage of the literature on the application of membrane-based technologies for the removal of PhACs

References	Item	Subject
Lin et al. [77, 78]	Table 1	Effects of various parameters including effluent properties, operating conditions, additives, etc., on the EPS formation and properties
Wang et al. [97]	Table 6	The efficiency of commercial nanofiltration membranes such as NF90/Polyamide/102 for the removal of various types of PhACs
Du et al. [98]	Table 3	Common materials used for the fabrication of nanofiltration membranes
Mallakpour and Azadi [3]	Table 1	Summary of the reports on the nanofiltration for the treatment of pharmaceutical industry effluents

PhACs and may require periodic cleaning or replacing the membrane, which can increase the overall treatment costs. A combination of membrane filtration technologies with other physico-chemical treatment methods can also be used efficiently for mitigating the fouling of the membranes for the efficient treatment of PhACs.

References

1. Daguerre-martini S et al (2018) Nitrogen recovery from wastewater using gas-permeable membranes: impact of inorganic carbon content and natural organic matter. Water Res 137:201–210. https://doi.org/10.1016/j.watres.2018.03.013
2. Yan T et al (2018) A critical review on membrane hybrid system for nutrient recovery from wastewater. Chem Eng J 348:143–156. https://doi.org/10.1016/j.cej.2018.04.166
3. Mallakpour S, Azadi E (2021) Nanofiltration membranes for food and pharmaceutical industries. Emergent Mater (0123456789). https://doi.org/10.1007/s42247-021-00290-7
4. Kamali M, Sweygers N et al (2022) Biochar for soil applications-sustainability aspects, challenges and future prospects. Chem Eng J 428:131189. https://doi.org/10.1016/j.cej.2021.131189
5. Aftab A et al (2020) Influence of tailor-made TiO_2/API bentonite nanocomposite on drilling mud performance: towards enhanced drilling operations. Appl Clay Sci 199:105862. https://doi.org/10.1016/j.clay.2020.105862

6. Ahmadi A et al (2020) The role of bentonite clay and bentonite clay@ $MnFe_2O_4$ composite and their physico-chemical properties on the removal of Cr (III) and Cr (VI) from aqueous media. Environ Sci Pollut Res 1–14
7. Babu Valapa R et al (2017) Chapter 2—an overview of polymer–clay nanocomposites. In: Jlassi K, Chehimi MM, Thomas SBT-C-P N (eds) Elsevier, pp 29–81. https://doi.org/10.1016/B978-0-323-46153-5.00002-1
8. Mansoori S et al (2020) Membranes based on non-synthetic (natural) polymers for wastewater treatment. Polym Testing 84:106381. https://doi.org/10.1016/j.polymertesting.2020.106381
9. Li X, Hai FI, Nghiem LD (2011) Simultaneous activated carbon adsorption within a membrane bioreactor for an enhanced micropollutant removal. Biores Technol 102:5319–5324. https://doi.org/10.1016/j.biortech.2010.11.070
10. Nghiem LD, Schäfer AI, Elimelech M (2006) Role of electrostatic interactions in the retention of pharmaceutically active contaminants by a loose nanofiltration membrane. J Membr Sci 286:52–59. https://doi.org/10.1016/j.memsci.2006.09.011
11. Xiao Y et al (2017) Removal of selected pharmaceuticals in an anaerobic membrane bioreactor (AnMBR) with/without powdered activated carbon (PAC). Chem Eng J 321:335–345. https://doi.org/10.1016/j.cej.2017.03.118
12. Nghiem LD, Schäfer AI, Elimelech M (2005) Pharmaceutical retention mechanisms by nanofiltration membranes. Environ Sci Technol 39:7698–7705. https://doi.org/10.1021/es0507665
13. Sheng C et al (2016) Removal of trace pharmaceuticals from water using coagulation and powdered activated carbon as pretreatment to ultrafiltration membrane system. Sci Total Environ 550:1075–1083. https://doi.org/10.1016/j.scitotenv.2016.01.179
14. Liu X et al (2020) Fouling and cleaning protocols for forward osmosis membrane used for radioactive wastewater treatment. Nucl Eng Technol 52:581–588. https://doi.org/10.1016/j.net.2019.08.007
15. Siang K et al (2022) Organic solvent forward osmosis membranes for pharmaceutical concentration. J Membr Sci 642:119965. https://doi.org/10.1016/j.memsci.2021.119965
16. Zhao S et al (2021) Engineering antifouling reverse osmosis membranes: a review. Desalination 499:114857. https://doi.org/10.1016/j.desal.2020.114857
17. Xie M et al (2012) Comparison of the removal of hydrophobic trace organic contaminants by forward osmosis and reverse osmosis. Water Res 46:2683–2692. https://doi.org/10.1016/j.watres.2012.02.023
18. Xie M, Price WE, Nghiem LD (2012) Rejection of pharmaceutically active compounds by forward osmosis: role of solution pH and membrane orientation. Sep Purif Technol 93:107–114. https://doi.org/10.1016/j.seppur.2012.03.030
19. Dai Q et al (2020) 'Thin-film composite membrane breaking the trade-off between conductivity and selectivity for a flow battery. Nat Commun 11:1–9. https://doi.org/10.1038/s41467-019-13704-2
20. Goh KS et al (2021) Thin film composite hollow fibre membrane for pharmaceutical concentration and solvent recovery. J Membr Sci 621:119008. https://doi.org/10.1016/j.memsci.2020.119008
21. Jin X et al (2012) Rejection of pharmaceuticals by forward osmosis membranes. J Hazard Mater 227–228:55–61. https://doi.org/10.1016/j.jhazmat.2012.04.077
22. Oatley-Radcliffe DL et al (2017) Nanofiltration membranes and processes: a review of research trends over the past decade. J Water Process Eng 19:164–171. https://doi.org/10.1016/j.jwpe.2017.07.026
23. Ventresque C et al (2000) Outstanding feat of modern technology: the Mery-sur-Oise nanofiltration treatment plant (340,000 m3/d). Desalination 131:1–16. https://doi.org/10.1016/S0011-9164(00)90001-8
24. Dolar D et al (2017) Adsorption of hydrophilic and hydrophobic pharmaceuticals on RO/NF membranes: Identification of interactions using FTIR. J Appl Polym Sci 134:17–21. https://doi.org/10.1002/app.44426

25. Qu F et al (2019) 'Tertiary treatment of secondary effluent using ultrafiltration for wastewater reuse: correlating membrane fouling with rejection of effluent organic matter and hydrophobic pharmaceuticals. Environ Sci Water Res Tech Royal Soc Chem 5:672–683. https://doi.org/10.1039/c9ew00022d
26. Azaïs A et al (2014) Nanofiltration for wastewater reuse: counteractive effects of fouling and matrice on the rejection of pharmaceutical active compounds. Sep Purif Technol 133:313–327. https://doi.org/10.1016/j.seppur.2014.07.007
27. Acero JL et al (2010) Retention of emerging micropollutants from UP water and a municipal secondary effluent by ultrafiltration and nanofiltration. Chem Eng J 163:264–272. https://doi.org/10.1016/j.cej.2010.07.060
28. Vergili I (2013) Application of nanofiltration for the removal of carbamazepine, diclofenac and ibuprofen from drinking water sources. J Environ Manage 127:177–187. https://doi.org/10.1016/j.jenvman.2013.04.036
29. Oh Y et al (2017) Understanding the pH-responsive behavior of graphene oxide membrane in removing ions and organic micropollulants. J Membr Sci 541:235–243. https://doi.org/10.1016/j.memsci.2017.07.005
30. Soares EV et al (2021) The effect of ph on atenolol/nanofiltration membranes affinity. Membranes 11:1–11. https://doi.org/10.3390/membranes11090689
31. Dai R et al (2020) Dually charged MOF-based thin-film nanocomposite nanofiltration membrane for enhanced removal of charged pharmaceutically active compounds. Environ Sci Technol 54:7619–7628. https://doi.org/10.1021/acs.est.0c00832
32. Schäfer AI, Nghiem LD, Waite TD (2003) Removal of the natural hormone estrone from aqueous solutions using nanofiltration and reverse osmosis. Environ Sci Technol 37:182–188. https://doi.org/10.1021/es0102336
33. McCallum EA et al (2008) Adsorption, desorption, and steady-state removal of 17β-estradiol by nanofiltration membranes. J Membr Sci 319:38–43. https://doi.org/10.1016/j.memsci.2008.03.014
34. Gomes D et al (2020) Removal of a mixture of pharmaceuticals sulfamethoxazole and diclofenac from water streams by a polyamide nanofiltration membrane. Water Sci Technol 81:732–743. https://doi.org/10.2166/wst.2020.166
35. Liu Y et al (2018) Adsorption of pharmaceuticals onto isolated polyamide active layer of NF/RO membranes. Chemosphere 200:36–47. https://doi.org/10.1016/j.chemosphere.2018.02.088
36. Taheri E et al (2020) Retention of atenolol from single and binary aqueous solutions by thin film composite nanofiltration membrane: transport modeling and pore radius estimation. J Environ Manage 271:111005. https://doi.org/10.1016/j.jenvman.2020.111005
37. Zhang R et al (2020) How to coordinate the trade-off between water permeability and salt rejection in nanofiltration? J Mater Chem A Royal Soc Chem 8:8831–8847. https://doi.org/10.1039/d0ta02510k
38. Moreno M et al (2022) Water and wastewater treatment by micellar enhanced ultrafiltration—a critical review. J Water Process Eng 46:102574. https://doi.org/10.1016/j.jwpe.2022.102574
39. Garcia-Ivars J et al (2017) Rejection of trace pharmaceutically active compounds present in municipal wastewaters using ceramic fine ultrafiltration membranes: effect of feed solution pH and fouling phenomena. Sep Purif Technol 175:58–71. https://doi.org/10.1016/j.seppur.2016.11.027
40. Salazar H et al (2015) Poly(vinylidene fluoride-trifluoroethylene)/NAY zeolite hybrid membranes as a drug release platform applied to ibuprofen release. Colloids Surf, A 469:93–99. https://doi.org/10.1016/j.colsurfa.2014.12.064
41. Bhattacharya P et al (2020) Application of green synthesized ZnO nanoparticle coated ceramic ultrafiltration membrane for remediation of pharmaceutical components from synthetic water: reusability assay of treated water on seed germination. J Environ Chem Eng 8:103803. https://doi.org/10.1016/j.jece.2020.103803
42. Matos M et al (2016) Surfactant effect on the ultrafiltration of oil-in-water emulsions using ceramic membranes. J Membr Sci 520:749–759. https://doi.org/10.1016/j.memsci.2016.08.037

43. Kim S et al (2018) Removal of contaminants of emerging concern by membranes in water and wastewater: a review. Chem Eng J 335:896–914. https://doi.org/10.1016/j.cej.2017.11.044
44. Kim S et al (2020) A metal organic framework-ultrafiltration hybrid system for removing selected pharmaceuticals and natural organic matter. Chem Eng J 382:122920. https://doi.org/10.1016/j.cej.2019.122920
45. Back JO et al (2018) Combining ultrafiltration and non-thermal plasma for low energy degradation of pharmaceuticals from conventionally treated wastewater. J Environ Chem Eng 6:7377–7385. https://doi.org/10.1016/j.jece.2018.07.047
46. Burba P, Geltenpoth H, Nolte J (2005) Ultrafiltration behavior of selected pharmaceuticals on natural and synthetic membranes in the presence of humic-rich hydrocolloids. Anal Bioanal Chem 382:1934–1941. https://doi.org/10.1007/s00216-005-3296-z
47. Rodriguez-Mozaz S et al (2015) Pharmaceuticals and pesticides in reclaimed water: efficiency assessment of a microfiltration-reverse osmosis (MF-RO) pilot plant. J Hazard Mater 282:165–173. https://doi.org/10.1016/j.jhazmat.2014.09.015
48. Plakas KV et al (2019) Heterogeneous Fenton-like oxidation of pharmaceutical diclofenac by a catalytic iron-oxide ceramic microfiltration membrane. Chem Eng J 373:700–708. https://doi.org/10.1016/j.cej.2019.05.092
49. Oh BS et al (2007) Role of ozone for reducing fouling due to pharmaceuticals in MF (microfiltration) process. J Membr Sci 289:178–186. https://doi.org/10.1016/j.memsci.2006.11.052
50. Décima MA et al (2021) A review on the removal of carbamazepine from aqueous solution by using activated carbon and biochar. Sustainability (Switzerland) 13:11760. https://doi.org/10.3390/su132111760
51. Shan D et al (2016) Preparation of ultrafine magnetic biochar and activated carbon for pharmaceutical adsorption and subsequent degradation by ball milling. J Hazard Mater 305:156–163. https://doi.org/10.1016/j.jhazmat.2015.11.047
52. Jiang J, Pang SY, Ma J (2009) Comment on "adsorption of hydroxyl- and amino-substituted aromatics to carbon nanotubes. Environ Sci Technol 43:3398–3399. https://doi.org/10.1021/es803273b
53. Westerhoff P et al (2005) Fate of endocrine-disruptor, pharmaceutical, and personal care product chemicals during simulated drinking water treatment processes. Environ Sci Technol 39:6649–6663. https://doi.org/10.1021/es0484799
54. Wang Y et al (2015) Carbon nanotube composite membranes for microfiltration of pharmaceuticals and personal care products: capabilities and potential mechanisms. J Membr Sci 479:165–174. https://doi.org/10.1016/j.memsci.2015.01.034
55. Viegas RMC et al (2020) Pilot studies and cost analysis of hybrid powdered activated carbon/ceramic microfiltration for controlling pharmaceutical compounds and organic matter in water reclamation. Water (Switzerland) 12(1). https://doi.org/10.3390/w12010033
56. Sengar A, Vijayanandan A (2022) Effects of pharmaceuticals on membrane bioreactor: review on membrane fouling mechanisms and fouling control strategies. Sci Total Environ 808:152132. https://doi.org/10.1016/j.scitotenv.2021.152132
57. Dawas-Massalha A et al (2014) Co-metabolic oxidation of pharmaceutical compounds by a nitrifying bacterial enrichment. Bioresource 167:336–342. https://doi.org/10.1016/j.biortech.2014.06.003
58. Zandi S et al (2019) Industrial biowastes treatment using membrane bioreactors (MBRs)—a scientometric study. J Environ Manage 247:462–473. https://doi.org/10.1016/j.jenvman.2019.06.066
59. Snyder SA et al (2007) Role of membranes and activated carbon in the removal of endocrine disruptors and pharmaceuticals. Desalination 202:156–181. https://doi.org/10.1016/j.desal.2005.12.052
60. De Gusseme B et al (2009) Biological removal of 17α-ethinylestradiol by a nitrifier enrichment culture in a membrane bioreactor. Water Res 43:2493–2503. https://doi.org/10.1016/j.watres.2009.02.028

61. Huang L, Lee DJ (2015) Membrane bioreactor: a mini review on recent R&D works. Biores Technol 194:383–388. https://doi.org/10.1016/j.biortech.2015.07.013
62. Bernhard M, Müller J, Knepper TP (2006) Biodegradation of persistent polar pollutants in wastewater: comparison of an optimised lab-scale membrane bioreactor and activated sludge treatment. Water Res 40:3419–3428. https://doi.org/10.1016/j.watres.2006.07.011
63. Joss A et al (2004) Removal of estrogens in municipal wastewater treatment under aerobic and anaerobic conditions: consequences for plant optimization. Environ Sci Technol 38:3047–3055. https://doi.org/10.1021/es0351488
64. Wen X et al (2004) Treatment of hospital wastewater using a submerged membrane bioreactor. Process Biochem 39(11):1427–1431. https://doi.org/10.1016/S0032-9592(03)00277-2
65. Kamali M, Aminabhavi TM et al (2022) Acclimatized activated sludge for enhanced phenolic wastewater treatment using pinewood biochar. Chem Eng J 427:131708. https://doi.org/10.1016/j.cej.2021.131708
66. Kamali M et al (2021) Biochar in water and wastewater treatment—a sustainability assessment. Chem Eng J 420:129946. https://doi.org/10.1016/j.cej.2021.129946
67. Xiang W et al (2020) Biochar technology in wastewater treatment: a critical review. Chemosphere 252:126539. https://doi.org/10.1016/j.chemosphere.2020.126539
68. Ma C et al (2012) High concentration powdered activated carbon-membrane bioreactor (PAC-MBR) for slightly polluted surface water treatment at low temperature. Biores Technol 113:136–142. https://doi.org/10.1016/j.biortech.2012.02.007
69. Krzeminski P et al (2019) Performance of secondary wastewater treatment methods for the removal of contaminants of emerging concern implicated in crop uptake and antibiotic resistance spread: a review. Sci Total Environ 648:1052–1081. https://doi.org/10.1016/j.scitotenv.2018.08.130
70. Cornejo J et al (2020) Ibuprofen removal by a microfiltration membrane bioreactor during the startup phase. J Environ Sci Health—Part A Toxic/Hazardous Subst Environ Eng 55:374–384. https://doi.org/10.1080/10934529.2019.1697587
71. Arya V, Philip L, Murty Bhallamudi S (2016) Performance of suspended and attached growth bioreactors for the removal of cationic and anionic pharmaceuticals. Chem Eng J 284:1295–1307. https://doi.org/10.1016/j.cej.2015.09.070
72. Díaz O et al (2017) Nanofiltration/Reverse osmosis as pretreatment technique for water reuse: ultrafiltration versus tertiary membrane reactor. Clean Soil, Air, Water 45:1600014. https://doi.org/10.1002/clen.201600014
73. Popple T et al (2016) Evaluation of a sequencing batch reactor sewage treatment rig for investigating the fate of radioactively labelled pharmaceuticals: case study of propranolol. Water Res 88:83–92. https://doi.org/10.1016/j.watres.2015.09.033
74. Chen Y et al (2019) Novel insights into membrane fouling caused by gel layer in a membrane bioreactor: effects of hydrogen bonding. Biores Technol 276:219–225. https://doi.org/10.1016/j.biortech.2019.01.010
75. Le-clech P, Chen V, Fane TAG (2006) Fouling in membrane bioreactors used in wastewater treatment. J Membr Sci 284:17–53. https://doi.org/10.1016/j.memsci.2006.08.019
76. Li W et al (2018) Ceramic membrane fouling and cleaning during ultrafiltration of limed sugarcane juice. Sep Purif Technol 190:9–24. https://doi.org/10.1016/j.seppur.2017.08.046
77. Lin H et al (2014) A critical review of extracellular polymeric substances (EPSs) in membrane bioreactors: characteristics, roles in membrane fouling and control strategies. J Membr Sci 460:110–125. https://doi.org/10.1016/j.memsci.2014.02.034
78. Lin YL, Chiou JH, Lee CH (2014) Effect of silica fouling on the removal of pharmaceuticals and personal care products by nanofiltration and reverse osmosis membranes. J Hazard Mater 277:102–109. https://doi.org/10.1016/j.jhazmat.2014.01.023
79. Nyström M, Kaipia L, Luque S (1995) Fouling and retention of nanofiltration membranes. J Membr Sci 98:249–262. https://doi.org/10.1016/0376-7388(94)00196-6
80. Azaïs A et al (2016) Evidence of solute-solute interactions and cake enhanced concentration polarization during removal of pharmaceuticals from urban wastewater by nanofiltration. Water Res 104:156–167. https://doi.org/10.1016/j.watres.2016.08.014

81. Zularisam AW, Ismail AF, Salim R (2006) Behaviours of natural organic matter in membrane filtration for surface water treatment—a review. Desalination 194:211–231. https://doi.org/10.1016/j.desal.2005.10.030
82. Lin YL (2017) Effects of organic, biological and colloidal fouling on the removal of pharmaceuticals and personal care products by nanofiltration and reverse osmosis membranes. J Membr Sci 542:342–351. https://doi.org/10.1016/j.memsci.2017.08.023
83. Wang F et al (2014) Effects of ionic strength on membrane fouling in a membrane bioreactor. Biores Technol 156:35–41. https://doi.org/10.1016/j.biortech.2014.01.014
84. Lee J et al (2017) Protein fouling in carbon nanotubes enhanced ultrafiltration membrane: fouling mechanism as a function of pH and ionic strength. Sep Purif Technol 176:323–334. https://doi.org/10.1016/j.seppur.2016.10.061
85. Jarusutthirak C, Mattaraj S, Jiraratananon R (2007) Influence of inorganic scalants and natural organic matter on nanofiltration membrane fouling. J Membr Sci 287:138–145. https://doi.org/10.1016/j.memsci.2006.10.034
86. Kamali M et al (2019) Sustainability considerations in membrane-based technologies for industrial effluents treatment. Chem Eng J 368:474–494. https://doi.org/10.1016/j.cej.2019.02.075
87. Li H et al (2019) Production of polyhydroxyalkanoates by activated sludge: correlation with extracellular polymeric substances and characteristics of activated sludge. Chem Eng J 361:219–226. https://doi.org/10.1016/j.cej.2018.12.066
88. Stöckl M et al (2019) Extracellular polymeric substances from geobacter sulfurreducens biofilms in microbial fuel cells. ACS Appl Mater Interfaces 11:8961–8968. https://doi.org/10.1021/acsami.8b14340
89. Lin YL, Tsai JZ, Hung CH (2019) Using in situ modification to enhance organic fouling resistance and rejection of pharmaceutical and personal care products in a thin-film composite nanofiltration membrane. Environ Sci Pollut Res 26:34073–34084. https://doi.org/10.1007/s11356-018-3234-1
90. Xiao Y et al (2018) In tandem effects of activated carbon and quorum quenching on fouling control and simultaneous removal of pharmaceutical compounds in membrane bioreactors. Chem Eng J 341:610–617. https://doi.org/10.1016/j.cej.2018.02.073
91. Palani KN et al (2019) Development of integrated membrane bioreactor and numerical modeling to mitigate fouling and reduced energy consumption in pharmaceutical wastewater treatment. J Ind Eng Chem Korean Soc Ind Eng Chem 76:150–159. https://doi.org/10.1016/j.jiec.2019.03.028
92. Bojaran M, Akbari A, Yunessnia lehi A (2019) Novel ultrafiltration membranes with the least fouling properties for the treatment of veterinary antibiotics in the pharmaceutical wastewater. Polym Adv Technol 30:1716–1723. https://doi.org/10.1002/pat.4603
93. Zhao Y et al (2018) Effects of organic fouling and cleaning on the retention of pharmaceutically active compounds by ceramic nanofiltration membranes. J Membr Sci 563:734–742. https://doi.org/10.1016/j.memsci.2018.06.047
94. Chen L et al (2020) Biological performance and fouling mitigation in the biochar-amended anaerobic membrane bioreactor (AnMBR) treating pharmaceutical wastewater. Biores Technol 302:122805. https://doi.org/10.1016/j.biortech.2020.122805
95. Lan Y et al (2020) Feasibility of a heterogeneous Fenton membrane reactor containing a Fe-ZSM5 catalyst for pharmaceuticals degradation: membrane fouling control and long-term stability. Sep Purif Technol 231:115920. https://doi.org/10.1016/j.seppur.2019.115920
96. Park J, Yamashita N, Tanaka H (2018) Membrane fouling control and enhanced removal of pharmaceuticals and personal care products by coagulation-MBR. Chemosphere 197:467–476. https://doi.org/10.1016/j.chemosphere.2018.01.063
97. Wang S et al (2021) A review of advances in EDCs and PhACs removal by nanofiltration: mechanisms, impact factors and the influence of organic matter. Chem Eng J 406:126722. https://doi.org/10.1016/j.cej.2020.126722
98. Du Y et al (2022) Recent advances in the theory and application of nanofiltration: a review. Curr Pollut Rep 8:51–80. https://doi.org/10.1007/s40726-021-00208-1

Chapter 9
Adsorptive Techniques for the Removal of Pharmaceutically Active Compounds—Materials and Mechanisms

9.1 Introduction

Pollution of water bodies with various types of pollutants has been considered an issue of global concern [1, 2]. In this regard, a class of organic pollutants, called contaminants of emerging concern (CECs), has received particular attention since their concentrations are continuously increasing in the environment, causing ecological or health impacts [3, 4]. However, this category of pollutants[1] has not been fully covered by the current environmental laws. Pharmaceutically active compounds (PhACs) are a main class of CECs that have caused real concerns due to the increasing rate of the consumption of these products to cure humans and animals, especially during pandemic situations, such as COVID-19 [6]. Hence, there have been efforts in the literature for the development of low-cost and efficient methods to remove these compounds from polluted water bodies. These methods can be divided into biological (e.g., anaerobic digestion, microbial fuel cells, and constructed wetlands) and physico-chemical (such as membrane filtration, advanced oxidation processes, and adsorption) technologies.

The adsorption process is based on steps including the diffusion of the adsorbates on the adsorbent surface in the medium, migration of the adsorbates into the pores of the adsorbent via the intermolecular forces between the adsorbent and the adsorbate, and finally, formation of a monolayer of the adsorbate on the adsorbent [7]. This process has been widely used for the removal of various types of environmental pollutants, including heavy metals [8], organic compounds [9], and nutrients [10], from aquatic media. This process can bring advantages for large-scale (waste)water

[1] Main categories of CECs include pharmaceuticals and personal care products, endocrine-disrupting chemicals, disinfection by-products, flame retardants, microplastics, and nanomaterials [4, 5].

M. Kamali et al., *Advanced Wastewater Treatment Technologies for the Removal of Pharmaceutically Active Compounds*, Green Energy and Technology,
https://doi.org/10.1007/978-3-031-20806-5_9

treatment applications, such as simplicity of the removal processes involved, cost-effectiveness, especially when novel waste valorization products (such as biochar-based materials) are used, and efficiency in dealing with various pollutants [11, 12]. However, there are issues to be overcome for the wider application of novel adsorption processes, mainly regarding the cost-effectiveness and reusability of the spent adsorbents and the permanent elimination of the adsorbed pollutants to satisfy sustainability considerations in this regard [13, 14].

This chapter aims to discuss the main mechanisms involved in the adsorptive elimination of PhACs from polluted (waste)waters and to introduce the latest achievements of the scientific community in terms of the development of sustainable materials to be used for effective adsorption of such contaminants of emerging concern. The last part of the chapter focuses on the novel technologies developed for the reuse of used materials. Such recycled materials can be highly beneficial from technical, economic, and environmental points of view and make adsorption methods viable alternatives for real applications.

9.2 Adsorption Mechanisms

Various mechanisms, such as electrostatic interactions, ion exchange, hydrogen bonding, π–π interactions, and hydrophobic interactions, are involved in the removal of organic contaminants. This section intends to examine these mechanisms while presenting some examples reported in the literature for the removal of PhACs.

Electrostatic interactions between the adsorbate and the adsorbent are highly dependent on their positive/negative electrical surface charges, which account for attractive or repulsive forces, hence affecting the adsorption kinetics. It has been well demonstrated in the literature that pH can play an important role in the establishment of the material surface charge, thus conditioning the electrostatic forces in the medium. Zeta potential, which is defined as the electrical potential at the slipping plane, is a function of the pH. In many cases, such as in CuO nanomaterials [15], a decrease in the pH of the medium can result in an increase in the positive charge on the material surface that can facilitate the attraction of negatively charged adsorbates. The zero-point charge is the pH value at which a material presents a neutral surface charge, and hence, minimum electrostatic interactions can be expected at this point [16].

The inherent properties of a material and its surface functional groups can determine the material's surface charge under various pH conditions. For instance, the carboxylic acid group (COOH) can be readily deprotonated under neutral or alkaline pH values, giving rise to negative surface charges [17]. For instance, the deprotonation of salicylic acid COOH groups at a pH higher than 5 can cause a repulsive interaction with negatively charged (e.g., oxygenated) groups existing on the adsorbent surface, such as activated carbon [18], thus accounting for a condition unfavorable to the adsorption of salicylic acid on activated carbon. There are many other examples in the literature that illustrate the successful adsorption of pharmaceuticals, such as antibiotics (e.g., ciprofloxacin), on carbonaceous materials such as nanotubes through electrostatic interactions [19] assisted by a convenient pH.

Ion exchange covers a range of processes in which the ions are exchanged between two electrolytes, resulting in the removal of a specific pollutant from the medium [20]. Two main mechanisms are basically involved in the ion-exchange processes, including cation exchange for the adsorption of positively charged ions and anion exchange for the removal of negatively charged ions in the medium. In addition, there are specific adsorbents called amphoteric ion exchangers, such as cross-linked amphoteric starch [21], which can be used for the adsorption of both cations and anions simultaneously. Ion-exchange resins are materials that have been widely used in the ion-exchange adsorption of water pollutants. Aluminosilicate-based materials such as bentonite (montmorillonite) and zeolite have also been introduced as low-cost adsorbents for the removal of a wide range of pollutants [22, 23]. Magnetic versions of such adsorbents, such as magnetic ion-exchange resins [20] and polydopamine-modified Fe_3O_4-pillared bentonite composites [24], have already demonstrated their superior adsorption capacity for various pollutants, including pharmaceutically active compounds such as ibuprofen, diclofenac, and sulfadiazine [20].

H-bonding is another adsorption mechanism based on the primary electrostatic attraction between a hydrogen atom with covalent bonding to a more electronegative atom and another atom with high electronegativity. A schematic of H-bonding is illustrated in Fig. 9.1. H-bonding has been recently reported as an effective mechanism for the removal of a wide range of PhACs, such as naproxen, p-chloro-m-xylenol, ketoprofen, bisphenol A, and triclosan (for instance, by functionalized metal–organic frameworks [25]). π–π interactions have also been reported as one of the main mechanisms involved in the adsorption of pharmaceutically active compounds from polluted waters. π bonds are classified as covalent chemical bonds in which two lobes of an orbital from adjacent atoms overlap laterally. The presence of surface functional groups (such as surface conjugated amino acids) has been demonstrated to have determinant effects on the adsorption of pollutants on adsorbents [26].

Finally, hydrophobic interactions occur between nonpolar molecules (adsorbate and adsorbent), resulting in the aggregation and removal of pollutants using appropriate materials introduced to the medium [29]. Hydrophobic adsorption of naproxen ($Q_{e,\mathrm{exp}}$ = 14.81–18.81 μmol/g), diclofenac ($Q_{e,\mathrm{exp}}$ = 15.73–20.00 μmol/g), and ibuprofen ($Q_{e,\mathrm{exp}}$ = 16.20–20.65 μmol/g) on NaOH-activated biochars prepared from spent coffee wastes is an example of this mechanism for the removal of pharmaceutically active compounds [30].

The mechanisms involved in the adsorption of organic compounds can also be classified as physisorption, chemisorption, ion-exchange processes, and precipitation [31]. Hydrophobic interactions, hydrogen bonding, and Yoshida's interactions (i.e., the hydrogen bonding between the aromatic rings and the OH groups) are categorized among physisorption techniques. Chemosorption routes also include electrostatic interactions, complexation, chelation, and covalent bonding. Microprecipitation, surface precipitation, and proton displacement mechanisms can also be classified as precipitation mechanisms for the removal of organic compounds.

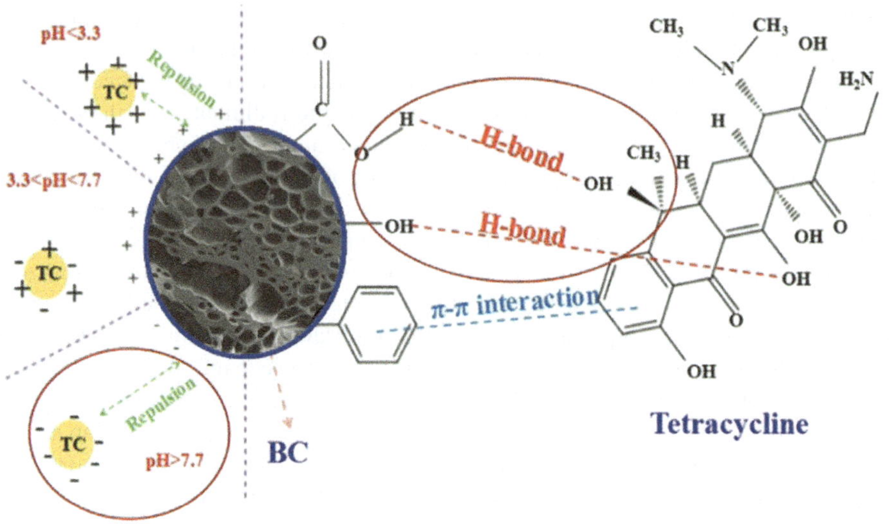

Fig. 9.1 Typical mechanism of H-bonding along with other adsorption mechanisms between biochar and tetracycline, adapted from Zhao and Dai [27]. This process is normally indicated as C–H·HAc, where the solid and the dashed lines are for the polar covalent bond, and the line denotes the hydrogen bond [28]

9.3 Sustainable Adsorbents

Sustainable adsorbents are characterized by properties such as high efficiencies for the elimination of PhACs, low fabrication costs, ease of implementation, and biocompatibility [32, 33]. This section discusses the most important types of sustainable adsorbents that have been reported in the recent literature and the opportunities for future progress in this field.

9.3.1 Carbon-Based Adsorbents

Carbonaceous materials are considered of high interest for several applications, including the treatment of polluted (waste)waters, due to their low cost and high efficiency [34]. Furthermore, they can be produced from waste resources (such as biochar from biomass wastes [35]), which may render these types of adsorbents environmentally friendly.

Various types of carbonaceous materials have been used for the adsorption of PhACs, such as activated carbon, biochar, graphene-based materials, carbon nanotubes (CNTs), and graphitic carbon nitride.

Activated carbon (AC) is a low-cost adsorbent that has been produced in various forms, including granular, powdered, and fiber forms [36]. There are reports on the efficient adsorption of various PhACs using AC, especially when activated by methods such as NH_4Cl-triggered activation (NAC) to enhance the porosity and surface functional groups, which can considerably promote the adsorption of PhACs (e.g., over 260 mg/g for amoxicillin [37]). However, the relatively high removal costs associated with activated carbon are among the most important practical issues of this type of carbonaceous material [38].

Biochar (BC) is a waste valorization product that normally results from the pyrolysis process of various carbon-rich wastes. It has been well discussed in our previous publications that the properties, and hence, the potential applications of BC, are highly influenced by parameters such as the feedstock type, pyrolysis conditions, including the peak temperature, heating rate, and biomass residence time during the pyrolysis process. Such parameters can considerably affect the specific surface area (SSA) and porosity, as well as the presence of surface functional groups, and hence can determine BC capacity for the adsorption of various organic and inorganic pollutants. As an example, highly porous BCs can be derived from lignocellulosic wastes, while sewage sludge normally results in BCs with relatively low specific surface areas [39, 40]. There are also reports for increasing the porosity of BC using techniques such as steam and sodium hydroxide activations, which may result in BCs with an extremely high specific surface area of over 700 m^2/g [41].

BC, especially the modified forms, is currently considered a cost-effective material to substitute AC for the adsorption of organic and inorganic compounds. They can be used efficiently for the removal of poorly degradable PhACs such as carbamazepine, as reviewed by Décima et al. [42]. These authors concluded that the modified BCs can offer a very high adsorption capacity of up to 3000 mg/g. Various mechanisms have also been reported for the adsorption of PhACs using BCs, including hydrogen bonding, hydrophobic partitioning, and π–π electron donor–acceptor interactions, which are highly dependent on the operating conditions, mainly the pH [43].

According to the literature, there has been a huge effort for the development of various types of BCs with the ability to remove different types of PhACs, but mainly at laboratory and pilot scales. In this regard, it would be of high importance to (i) examine the efficiency of the most efficient and economic BC for the treatment of PhACs in real wastewaters and (ii) couple such adsorption techniques with existing wastewater treatment technologies, such as anaerobic digestion and activated sludge, to prevent the rapid release of these pollutants into the environment (Fig. 9.2).

Graphene-based materials are also attractive materials for the adsorption of various organic and inorganic pollutants due mainly to the graphene-specific molecular structure in which each carbon atom is bonded to its nearest neighbor carbon atom via σ-bond. A single electron is provided by each atom to a conduction band extended over the whole sheet structure of graphene. Graphene oxide (GO) normally presents a superior adsorption capacity compared to AC and BC but a lower efficiency than carbon nanotubes. The superior performance of GO over such carbonaceous materials is mainly due to the presence of numerous oxygen-containing functional groups, such as carboxylic (–COOH), hydroxyl (–OH), epoxides (△), and carbonyl (>C=O) groups, as well as ketones and quinones, which can facilitate the adsorption

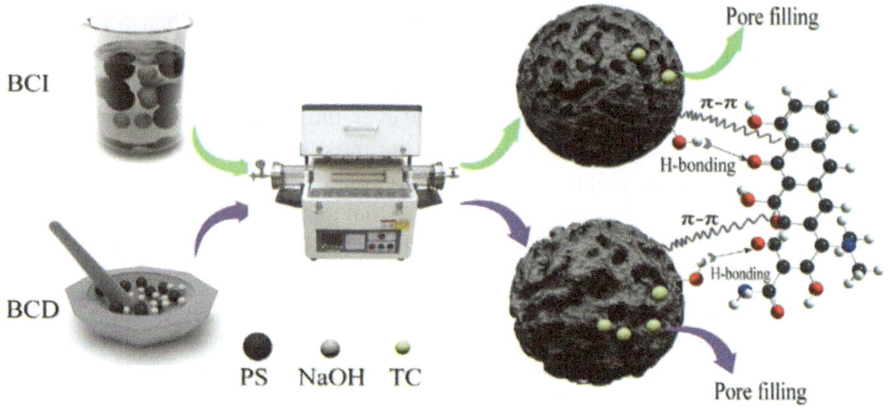

Fig. 9.2 Activation of pharmaceutical sludge biochar using NaOH for the efficient adsorption of tetracycline and the involved adsorption mechanisms, adapted from Liu et al. [44]. BCI: impregnation method and BCD: dry mixing method used for the activation of the biochar (BC)

of positively charged molecules via mechanisms involving electrostatic interactions [45, 46].

The highest adsorption capacities of 380 mg/g and 240 mg/L have been reported for ciprofloxacin and sulfamethoxazole, respectively, at a pH of 5 [47]. It has been discussed that the adsorption of PhACs on graphene-based materials implies mechanisms including H-bonding, electrostatic interactions, π–π bonding, and hydrophobic interactions. The lower production costs of graphene-based materials compared to CNTs can make these materials more attractive for real applications.

Carbon nanotubes have also presented an outstanding efficiency for the adsorption of various pharmaceutically active compounds due to their large SSA and high reactivity. It has also been reported that oxidation of CNTs can lead to the formation of surface functional groups such as epoxy groups, carbonyl groups, and hydroxyl groups [48]. Excellent adsorption of doxorubicin using oxidized CNTs (9.45×10^3 mg/g) is an example of such modifications [49]. In addition, magnetization has been considered an attractive way to provide materials with high adsorption capacity and recyclability, which are essential requirements for sustainable adsorbents [50]. Despite such advantages, it has been discussed in the literature that CNTs are not appropriate for the real-scale adsorption of pollutants, including PhACs. Therefore, there is a need to develop cost-effective methods for the production of CNTs. For instance, the conversion of low-cost carbonaceous materials into CNTs can be envisaged as an attractive way to promote CNTs for such applications (Fig. 9.3).

Graphitic carbon nitride ($g\text{-}C_3N_4$) is another type of carbonaceous material with enhanced visible-light activity that has made this material appropriate for the photocatalytic decomposition of various organic compounds [52–54]. Modifications in the structure of $g\text{-}C_3N_4$, such as creating nitrogen defects, oxygen doping, and doping with metal and nonmetal elements, can potentially tune their bandgap energy and enhance their photocatalytic activity [54, 55] (Fig. 9.4).

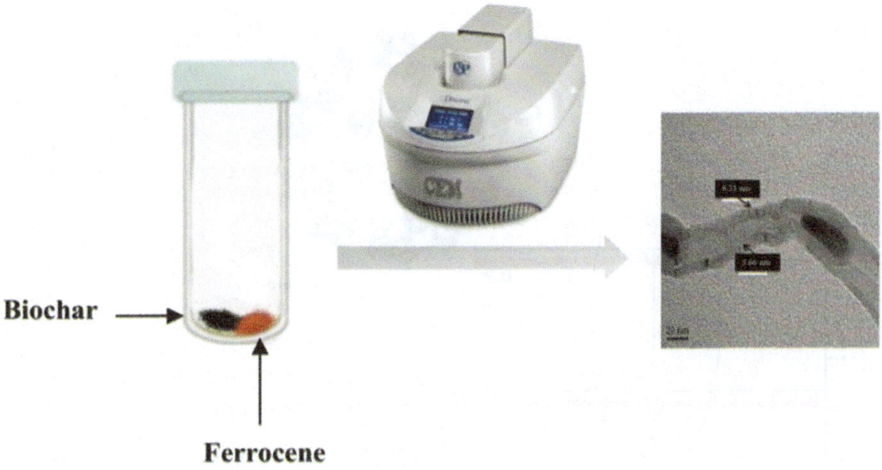

Fig. 9.3 Catalytic transformation of biochar, as a low-cost carbonaceous material, to carbon nanotubes assisted by microwave irradiation, adapted from Hildago-Oporto et al. [51]

9.3.2 *Ion-Exchange Resins*

Ion-exchange resins allow the exchange of cations and anions from the medium with those of the used resins to maintain electroneutrality. There are reports on the efficient removal of heavy metals such as copper [56] and nickel [57], as well as nutrients such as phosphate or carbonate [57, 58]. They have also been used for the recovery of valuable compounds such as boron [59] for the reduction of the overall treatment costs.

The applicability of various ion-exchange resins for the removal of PhACs from aqueous solutions has been examined: Most of them revealed a high adsorption capacity for pollutants such as diclofenac, tetracycline, ketoprofen, naproxen, sulfamethazine, ibuprofen, paracetamol, caffeine, and metformin [60–64].[2] In addition to ion exchange, other mechanisms, including H-bonding, electrostatic interactions, van der Waals forces, and π–π interactions, also play a role in PhAC removal by ion-exchange resins [65]. Such adsorbents have also been recently employed in inefficient configurations, such as fixed-bed columns,[3] to selectively remove pharmaceuticals from effluents (using specific ion-exchange resins such as anion-exchange resins (AERs), Fig. 9.5) [69, 70].

Despite the efficiency of ion-exchange resins for the removal of PhACs, there are still barriers for their large-scale applications, including high operating costs, especially compared to low-cost waste-derived materials such as biochar, and the

[2] Using conventional ion exchange resins such as MIEX®, Purolite A520E, Dowex 22, IRA938, IRA958, IRA458, IRA402, and Oasis MAX.

[3] Fixed-bed adsorption is generally performed in various steps including saturation, regeneration, and washing [66–68].

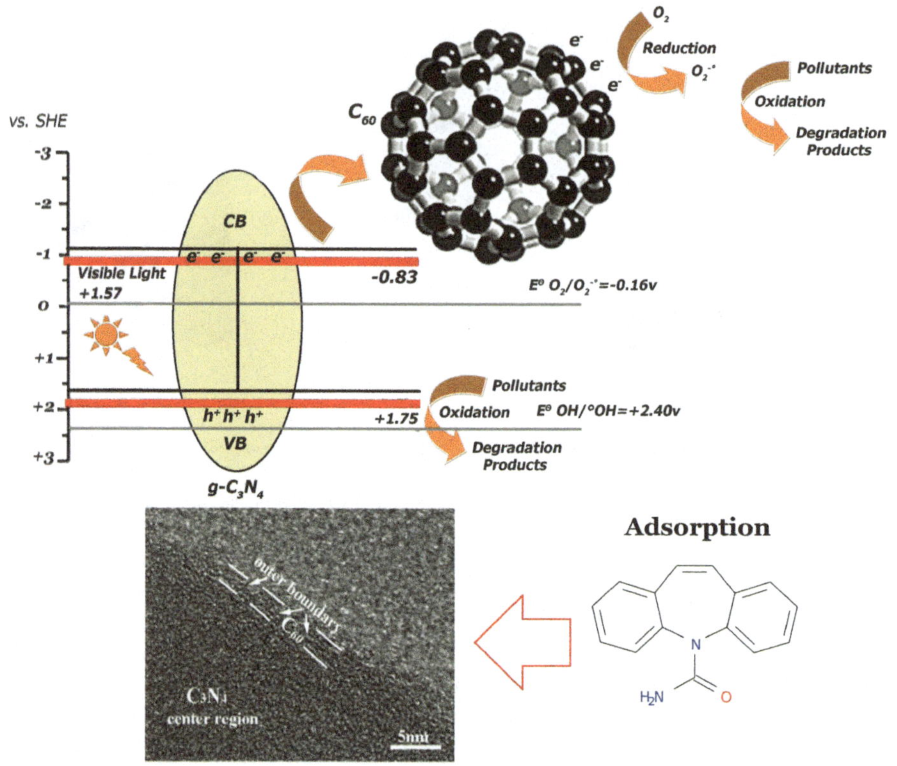

Fig. 9.4 Possibility of simultaneous adsorption and degradation of PhACs (such as carbamazepine) by graphitic carbon nitride was adopted from Zhang et al. [55]. Visible-light illumination leads to the excitation of electrons from the valence band of the adsorbent, which results in a chain of oxidative reactions

absence of practical data for the long-term applicability of such resins, especially those with laboratory optimized properties.

9.3.3 *Clay-Based Adsorbents*

Clay-based materials are generally abundant and low-cost materials with a high adsorption capacity that can make them appropriate for the adsorption and removal of PhACs from polluted (waste)waters.

Among clay-based adsorbents, bentonite is an abundant material with a very high SSA and high cation-exchange capacity [73, 74]. Its low cost has also made it an attractive alternative for real-scale (waste)water treatment applications. Typically, bentonite consists of a layered structure composed of two tetrahedral sheets of silica (SiO_4) connected to an octahedral sheet of alumina (Al_2O_3) [75]. There are reports on

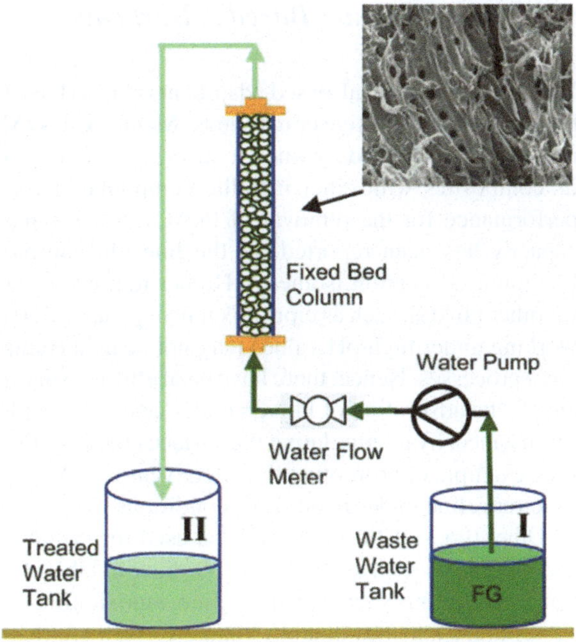

Fig. 9.5 Schematic of the possible application of efficient adsorbents for designing fixed-bed column adsorption for the removal of PhACs, adapted from Lonappan et al. [71] and Ahmed and Hossain [72]

the efficient adsorption of various types of PhACs, such as ciprofloxacin (147 mg/g) [76]. Bentonite can represent a more or less similar behavior in the adsorption of PhACs. Here, the availability and low cost of bentonite can make it an appealing alternative for the adsorption of PhACs for real (waste)water treatment applications.

Kaolinite is another clay mineral and the most abundant hydrous aluminosilicate mineral in soils and sediments. Its structure consists of a double layer of tetrahedral silica (SiO_4) and octahedral aluminum oxide/hydroxide [$AlO_2(OH)_4$]. This mineral presents a hydrophilic nature that is beneficial for a wide range of applications in ceramics, plastics, coatings, etc. [77, 78]. There are reports on the efficient adsorption of PhACs on kaolinite via cation-exchange mechanisms (and to some extent complexation).

Various configurations of the adsorption system have also been used for the adsorption of PhACs using clay-based materials to enhance the efficiency of the process. For instance, effective adsorption of ciprofloxacin has been reported using a fixed-bed reactor with calcined bentonite (12.6 mg/g) having higher stability, fewer impurities, and reduced expansion in water [79] (see Fig. 9.5).

9.3.4 Metal Oxide-Based Adsorbents

Various types of metal-based adsorbents (MAs) have been used for the adsorption of pharmaceutically active compounds. Metal oxides (MOx) and metal oxide composites are the most widely known adsorbents among MAs. Among them, MgO and its composites with other metallic compounds have revealed optimum adsorption performance for the removal of PhACs. For instance, up to 125 mg/g adsorption capacity has been reported for the linezolid antibiotic using MgO following the Langmuir adsorption isotherm. Higher removal capacities have also been reported for other PhACs, such as ciprofloxacin (e.g., up to 550 mg/g, pH = 10 [80]). However, working under high pH values may not be an advantage for real (waste)water treatment processes. Hence, there have been efforts to modify such metallic compounds to enable the adsorption of PhACs under near-neutral pH values. This can be achieved, for instance, by manipulating the surface charge of the adsorbents. S-coating of MgO is an example of promoting the effectiveness of MgO for the removal of PhACs such as tetracycline under neutral pH conditions [37].

Other types of MOXs have been used for the adsorption of PhACs. Well-known photocatalysts such as ZnO and TiO_2 have exhibited high efficiencies for the adsorption of pollutants such as tetracycline, amoxicillin, ciprofloxacin, and cefixime with adsorption capacities of approximately 100 mg/g [81, 82]. This has raised the potential for simultaneous adsorption and degradation of pollutants using such MOx. Gradual degradation of adsorbed PhACs in the presence of an appropriate light source can also enhance the performance of MOx but requires continuous regeneration of free surface active sites for the adsorption of the pollutant molecules.

In addition to MOx, zero-valent metallic particles have also been used for the adsorption of pharmaceutical compounds. For instance, zero-valent iron (nZVI) particles, especially surface-modified versions, have demonstrated high adsorption capacities for the removal of PhACs such as ciprofloxacin, diazepam, tetracycline, and chloramphenicol [83]. In this regard, nZVI particles amended with surfactants (such as polyvinylpyrrolidone) have demonstrated an excellent adsorption capacity for pollutants such as tetracycline [84]. However, there is still the main issue regarding the applicability of nZVI for real wastewater treatment: its high production costs and its rapid oxidation when introduced into water bodies. To overcome these limitations, there is a need to develop low-cost and large-scale production costs.

Transition metal carbides (TMCs) are a group of MAs that have been used for the removal of PhACs. In their special structures, carbon atoms are located in interstitial voids of the densely packed host lattice. Such hydrophilic materials normally display high mechanical stability and electrical conductivity that can make them good alternatives for the adsorption of other compounds [85]. There are reports on the application of some types of TMCs for the removal of pharmaceuticals from polluted waters. For instance, $Ti_3C_2T_x$ (T = F, O) (MXene), as a newly discovered 2D structure, has been used for the removal of PhACs, especially cationic compounds. There are studies on the application of this material for the removal of PhACs. For instance, Kim et al. [86] indicated the successful application of MXene for the removal of carbamazepine, 17 α-ethinylestradiol, amitriptyline, ibuprofen, verapamil, and diclofenac. The results showed the highest adsorption capacity for amitriptyline (58.7 mg/g, pH = 7) due to the electrostatic attraction between the adsorbent and pollutant. They also indicated a substantial increase in the adsorption capacity of the materials after a sonication process (up to 580 kHz) that separates and disperses the particles, thus providing a larger available surface area for the adsorption of pollutants. The recycling of this MXene using HCl, NaOH, and DI water indicated that sonication is a useful strategy to maintain the adsorption capacity of the material (after runs).

In metal–organic frameworks (MOFs), as a relatively new class of three-dimensional materials, the structure consists of a metal-based center connected to an organic ligand through strong coordination bonds [36]. MOFs are also of high interest for the adsorption of pharmaceutically active compounds, mainly because they present extremely high Brunauer–Emmett–Teller (BET) surface areas (up to 1400 m^2/g) [87], as well as remarkable stability. This has made these materials superior adsorbents for a wide range of pharmaceutically active compounds, such as tetracycline (e.g., 85 mg/g [88]), ibuprofen (e.g., 127 mg/g [89]), carbamazepine (e.g., 7 mg/g [90]), and triclosan (476 mg/g [91]). Various mechanisms are involved in the adsorption of pharmaceutically active compounds on MOFs. For instance, as indicated in Fig. 9.6, processes including H-bonding, electrostatic interaction, and π–π stacking underlie the adsorption of norfloxacin onto UiO-66-NH_2 [92].

There are studies indicating the facile regeneration of MOFs using organic solvents, thus pointing to them as appropriate for real applications. However, such adsorbents currently suffer from high production costs (e.g., 128 USD per gram) [93]. This can make them less competitive than low-cost (e.g., biochar) adsorbents. In this regard, there is a need to develop low-cost MOFs, for instance, from cheap feedstocks such as waste [94].

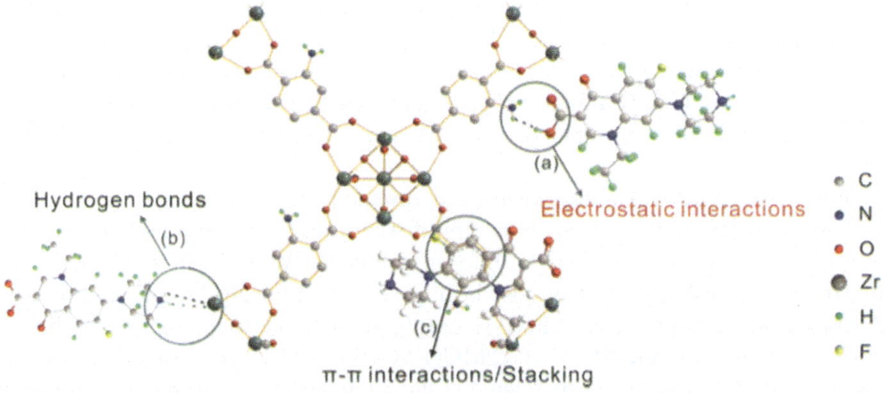

Fig. 9.6 Mechanisms involved in the adsorption of norfloxacin onto UiO-66-NH_2, reprinted with permission from Fang et al. [92]

9.3.5 Natural Biopolymers

Natural biopolymers have also been used for the removal of PhACs. Cellulose-based polymers can be considered low-cost materials for the production of efficient adsorbents for the removal of various environmental pollutants, including PhACs. Recent studies have revealed that microcrystalline cellulose can be used as a remarkable adsorbent, especially for cationic PhACs [95]. This may be explained by the presence of various hydroxyl and ester functional groups comprising both electron-poor and electron-rich phases, which create the possibility of reactions with the electron-rich and electron-poor forms of PhACs mainly through electrostatic interactions [96, 97]. Efficient adsorption of amine drugs such as tacrine under an ion-exchange mechanism is an example of the applicability of microcrystalline cellulose for the removal of PhACs [98].

Chitin is another natural biopolymer that is abundant on the exoskeletons of shellfish and crustaceans [99]. The application of adsorbents made of these types of biopolymers can be considered a sustainable option because they can be prepared from wastes of seafood crustaceans, which are widely produced worldwide. It has been demonstrated that up to 20% of prawn, lobster, and crab shells consist of chitin [100].

Chitin can also be used to produce chitosan ((poly-β-(1 → 4)-2-amino-2-deoxy-d-glucose), which is another attractive adsorbent for the removal of various types of environmental pollutants. Nontoxicity and biodegradability are among the properties that can list chitosan among the environmentally friendly adsorbents.

Alkaline N-deacetylation of chitin is a widely used method for the production of chitosan, thus satisfying environmental and economic aspects [101]. The high adsorption capacity of chitosan is related to the presence of hydroxyl and amino functional groups in the structure of this material.

Composites of chitosan have also been developed and used for the efficient adsorption of pharmaceutical compounds. An example of the mechanisms involved in the fabrication of chitosan composites with other carbonaceous materials is illustrated in Fig. 9.7. Accumulation of functional groups that can favor hydrogen bonding, π–π stacking, and electrostatic interactions between the adsorbent and the adsorbate [102, 103].

Super adsorption capacities have been revealed by such composite structures for various types of PhACs, such as ciprofloxacin (e.g., chitosan/graphene oxide, 76 mg/g [105]) and diazinon (e.g., chitosan/graphene oxide 222 mg/g [106]). Even higher adsorption capacities have also been observed with well-designed engineered chitosan-based composites (such as 473 mg/g tetracycline by genipin-cross-linked chitosan/graphene oxide-SO_3H (GC/MGO-SO_3H) [107]). Cost-effective analysis has also indicated that the costs related to the application of chitosan and chitosan-based adsorbents for the removal of PhACs are lower than those of activated carbon, ion-exchange resins, and MOFs but higher than those of fly ash, biochar, and bentonite [104, 108–111]. However, there is a need for more studies to explore the cost-effectiveness of advanced chitosan-based materials for promoting their applications

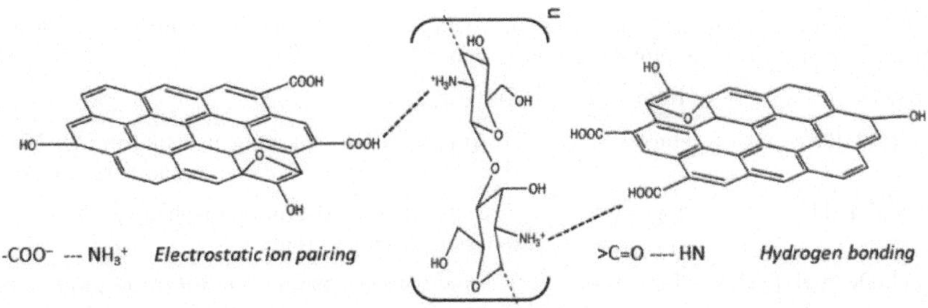

Fig. 9.7 Mechanisms of the formation of chitosan/graphene oxide including the reaction between –COOH groups of graphene oxide with –NH groups of chitosan chains, reprinted with permission from da Silva Alves et al. [104]

for real (waste)water treatment applications. Magnetic chitosan composites such as Fe_3O_4-chitosan have also shown high capabilities for the removal of various PhACs [112]. For instance, a magnetite-chitosan composite has demonstrated high efficiency for the removal of carbamazepine via a chemical adsorption mechanism[4] [113]. This can create the possibility of recovery and reuse and hence a reduction in the overall treatment costs.

It is also worth mentioning that the applicability of various adsorbents for the removal of PhACs is highly dependent on the subsequent desorption possibilities to allow the regeneration of the adsorbents and the repetition of the adsorptive removal process. For a sustainable adsorbent, it is hence essential to have an acceptable adsorption–desorption performance to minimize the respective treatment costs. It has been reported that acidic or alkaline solutions can be used for the desorption of pollutants by changing the pH of the medium [114, 115]. Nevertheless, more studies are still necessary on the desorption and regeneration of adsorbents when they are used for the removal of PhACs.

9.4 Further Reading

Table 9.1 contains items from the recent literature that the reader can consult for more detailed information regarding the adsorptive removal of pharmaceutically active compounds from polluted (waste)waters.

[4] Revealed by the kinetic data which were better fitted to the pseudosecond-order equation than to the pseudofirst-order equation, as an indication of the chemical adsorption.

Table 9.1 Further reading suggestions for more detailed coverage of the literature on the adsorption of active pharmaceutical compounds in (waste)waters

References	Item	Subject
Zango et al. [93]	Tables 1, 2	Adsorption of pharmaceuticals using various UiO (Universitetet of Oslo) metal–organic frameworks
Crini et al. [31]	Figure 5	Classification of adsorption mechanisms according to Crini, Crini, and Bado
De Andrade et al. [38]	Table 8	Thermodynamics parameters for the adsorption of PhACs using various adsorbents
Balarak et al. [116]	Table 4	The adsorption capacity of various adsorbents for the removal of antibiotics
Décima et al. [42]	Table 6	Reports on the adsorption of carbamazepine in a different medium
Singh et al. [117]	Table 1	Various types of nanomaterials developed to adsorb antibiotics
Liakos et al. [118]	Table 1	Applicability of various chitosan-based materials for the adsorption of PhACs

9.5 Summary

As discussed in previous chapters, conventional wastewater treatment technologies such as activated sludge have not been designed for the removal of PhACs from polluted (waste)waters. For instance, less than 5% carbamazepine removal has been reported in sewage wastewater treatment plants. Hence, there is a need to develop complementary technologies to enhance the capability of current wastewater treatment techniques to enable them to deal with various PhACs, especially those with severe environmental and health impacts. The recent literature has discussed the applicability of various types of adsorbents, including carbonaceous, clay-based, metal-based, metal–organic frameworks, and natural-based polymer adsorbents, for the efficient removal of PhACs. However, this process can be highly influenced by operating conditions such as temperature and pH. Carbonaceous materials such as activated carbon and carbon nanotubes have evidenced excellent efficiencies for the removal of PhACs, but their relatively high production costs have restricted their wider applications. However, biochar (as a waste-derived carbonaceous material) and low-cost clay-based materials, such as bentonite, have attracted much attention for their practical applications. Natural polymers such as chitosan can also be considered attractive options for such applications due to the low fabrication costs and the abundance of raw materials. Despite the effectiveness of the mentioned sustainable adsorbents, there is no evidence for their rapid commercialization in real applications. The combination of such technology with conventional biological treatment systems such as activated sludge can be proposed as an efficient and applicable way to prevent the release of PhACs into the environment. More studies are also welcome about the cost-effectiveness of various adsorbents, including the possibility of their regeneration and reuse.

References

1. Kumar A et al (2022) Environmental pollution remediation via photocatalytic degradation of sulfamethoxazole from waste water using sustainable $Ag_2S/Bi_2S_3/g\text{-}C_3N_4$ nano-hybrids. Earth Syst Environ 6(1):141–156. https://doi.org/10.1007/s41748-021-00223-8
2. Sirés I, Brillas E (2012) Remediation of water pollution caused by pharmaceutical residues based on electrochemical separation and degradation technologies: a review. Environ Int 40:212–229. https://doi.org/10.1016/j.envint.2011.07.012
3. Kim S et al (2018) Removal of contaminants of emerging concern by membranes in water and wastewater: a review. Chem Eng J 335(November 2017):896–914. https://doi.org/10.1016/j.cej.2017.11.044
4. Shi Q et al (2022) Contaminants of emerging concerns in recycled water: fate and risks in agroecosystems. Sci Total Environ 814:152527. https://doi.org/10.1016/j.scitotenv.2021.152527
5. Lee BCY et al (2021) Emerging contaminants: an overview of recent trends for their treatment and management using light-driven processes. Water (Switzerland) 13:2340. https://doi.org/10.3390/w13172340
6. Kuroda K et al (2021) Predicted occurrence, ecotoxicological risk and environmentally acquired resistance of antiviral drugs associated with COVID-19 in environmental waters. Sci Total Environ 776:145740. https://doi.org/10.1016/j.scitotenv.2021.145740
7. Benjelloun M et al (2021) Recent advances in adsorption kinetic models: their application to dye types. Arab J Chem 14(4):103031. https://doi.org/10.1016/j.arabjc.2021.103031
8. Da'na E (2017) Adsorption of heavy metals on functionalized-mesoporous silica: a review. Microporous Mesoporous Mater 247:145–157. https://doi.org/10.1016/j.micromeso.2017.03.050
9. Ersan G et al (2017) Adsorption of organic contaminants by graphene nanosheets: a review. Water Res 126:385–398. https://doi.org/10.1016/j.watres.2017.08.010
10. Carey DE, McNamara PJ, Zitomer DH (2015) Biochar from pyrolysis of biosolids for nutrient adsorption and turfgrass cultivation. Water Environ Res 87:2098–2106. https://doi.org/10.2175/106143015x14362865227391
11. Awad AM et al (2019) Adsorption of organic pollutants by natural and modified clays: a comprehensive review. Sep Purif Technol 228:115719. https://doi.org/10.1016/j.seppur.2019.115719
12. Tabana L et al (2020) Adsorption of phenol from waste water using calcined magnesium-zinc-aluminium layered double hydroxide clay. Sustainability 12:4273. https://doi.org/10.3390/su12104273
13. Jain SN et al (2020) Batch and continuous studies for adsorption of anionic dye onto waste tea residue: kinetic, equilibrium, breakthrough and reusability studies. J Clean Prod 252:119778. https://doi.org/10.1016/j.jclepro.2019.119778
14. Naushad M et al (2019) Adsorption kinetics, isotherm and reusability studies for the removal of cationic dye from aqueous medium using arginine modified activated carbon. J Mol Liq 293:111442. https://doi.org/10.1016/j.molliq.2019.111442
15. Kamali M et al (2020) Optimization of kraft black liquor treatment using ultrasonically synthesized mesoporous tenorite nanomaterials assisted by Taguchi design. Chem Eng J 401:126040. https://doi.org/10.1016/j.cej.2020.126040
16. Luxbacher T (2012) Electrokinetic properties of natural fibres, handbook of natural fibres. Woodhead Publishing Limited. https://doi.org/10.1533/9780857095510.1.185
17. Feng J et al (2015) Effect of hydroxyl group of carboxylic acids on the adsorption of acid red G and methylene blue on TiO_2. Chem Eng J 269:316–322. https://doi.org/10.1016/j.cej.2015.01.109
18. Bernal V, Giraldo L, Moreno-Piraján JC (2020) Adsorption of pharmaceutical aromatic pollutants on heat-treated activated carbons: effect of carbonaceous structure and the adsorbent-adsorbate interactions. ACS Omega 5:15247–15256. https://doi.org/10.1021/acsomega.0c01288

19. Cong Q, Yuan X, Qu J (2013) A review on the removal of antibiotics by carbon nanotubes. Water Sci Technol 68:1679–1687. https://doi.org/10.2166/wst.2013.420
20. Jiang M et al (2015) Adsorption of three pharmaceuticals on two magnetic ion-exchange resins. J Environ Sci (China) 31:226–234. https://doi.org/10.1016/j.jes.2014.09.035
21. Cao LQ et al (2004) Adsorption of Zn (II) ion onto crosslinked amphoteric starch in aqueous solutions. J Polym Res 11:105–108. https://doi.org/10.1023/B:JPOL.0000031066.67853.28
22. Laske C et al (2008) Decreased plasma and cerebrospinal fluid levels of stem cell factor in patients with early Alzheimer's disease. J Alzheimer's Disease 15:451–460. https://doi.org/10.3233/JAD-2008-15311
23. Toor M, Jin B (2012) Adsorption characteristics, isotherm, kinetics, and diffusion of modified natural bentonite for removing diazo dye. Chem Eng J 187:79–88. https://doi.org/10.1016/j.cej.2012.01.089
24. Ain QU et al (2020) Superior dye degradation and adsorption capability of polydopamine modified Fe3O4-pillared bentonite composite. J Hazard Mater 397:122758. https://doi.org/10.1016/j.jhazmat.2020.122758
25. Song JY, Jhung SH (2017) Adsorption of pharmaceuticals and personal care products over metal-organic frameworks functionalized with hydroxyl groups: quantitative analyses of H-bonding in adsorption. Chem Eng J 322:366–374. https://doi.org/10.1016/j.cej.2017.04.036
26. Shieh FK et al (2013) A bioconjugated design for amino acid-modified mesoporous silicas as effective adsorbents for toxic chemicals. J Hazard Mater 260:1083–1091. https://doi.org/10.1016/j.jhazmat.2013.06.066
27. Zhao J, Dai Y (2022) Tetracycline adsorption mechanisms by NaOH-modified biochar derived from waste Auricularia auricula dregs. Environ Sci Pollut Res 29:9142–9152. https://doi.org/10.1007/s11356-021-16329-5
28. Arunan E et al (2011) Definition of the hydrogen bond (IUPAC Recommendations 2011). Pure Appl Chem 83:1637–1641. https://doi.org/10.1351/PAC-REC-10-01-02
29. Chandler D (2005) Interfaces and the driving force of hydrophobic assembly. Nature 437:640–647. https://doi.org/10.1038/nature04162
30. Shin J et al (2021) Competitive adsorption of pharmaceuticals in lake water and wastewater effluent by pristine and NaOH-activated biochars from spent coffee wastes: contribution of hydrophobic and π-π interactions. Environ Pollut 270:116244. https://doi.org/10.1016/j.envpol.2020.116244
31. Crini G et al (2019) Conventional and non-conventional adsorbents for wastewater treatment. Environ Chem Lett 17:195–213. https://doi.org/10.1007/s10311-018-0786-8
32. Tan TH et al (2020) Current development of geopolymer as alternative adsorbent for heavy metal removal. Environ Technol Innov 18:100684. https://doi.org/10.1016/j.eti.2020.100684
33. Yang P et al (2020) Porous carbons derived from sustainable biomass via a facile one-step synthesis strategy as efficient CO_2 adsorbents. Ind Eng Chem Res 59:6194–6201. https://doi.org/10.1021/acs.iecr.0c00073
34. Veclani D, Tolazzi M, Melchior A (2020) Molecular interpretation of pharmaceuticals' adsorption on carbon nanomaterials: theory meets experiments. Processes 8:1–39. https://doi.org/10.3390/PR8060642
35. Kamali M et al (2021) Biochar in water and wastewater treatment—a sustainability assessment. Chem Eng J 420:129946. https://doi.org/10.1016/j.cej.2021.129946
36. Pauletto PS, Bandosz TJ (2022) Activated carbon versus metal-organic frameworks: a review of their PFAS adsorption performance. J Hazard Mater 425:127810. https://doi.org/10.1016/j.jhazmat.2021.127810
37. Moussavi G et al (2018) The catalytic destruction of antibiotic tetracycline by sulfur-doped manganese oxide (S–MgO) nanoparticles. J Environ Manage 210:131–138. https://doi.org/10.1016/j.jenvman.2018.01.004
38. De Andrade JR et al (2018) Adsorption of pharmaceuticals from water and wastewater using nonconventional low-cost materials: a review. Ind Eng Chem Res 57:3103–3127. https://doi.org/10.1021/acs.iecr.7b05137

39. Leng L et al (2021) An overview on engineering the surface area and porosity of biochar. Sci Total Environ 763:144204. https://doi.org/10.1016/j.scitotenv.2020.144204
40. Lu H et al (2013) Characterization of sewage sludge-derived biochars from different feedstocks and pyrolysis temperatures. J Anal Appl Pyrol 102:137–143. https://doi.org/10.1016/j.jaap.2013.03.004
41. Azargohar R, Dalai AK (2008) Steam and KOH activation of biochar: experimental and modeling studies. Microporous Mesoporous Mater 110:413–421. https://doi.org/10.1016/j.micromeso.2007.06.047
42. Décima MA et al (2021) A review on the removal of carbamazepine from aqueous solution by using activated carbon and biochar. Sustainability (Switzerland) 13:11760. https://doi.org/10.3390/su132111760
43. Ndoun MC et al (2021) Adsorption of pharmaceuticals from aqueous solutions using biochar derived from cotton gin waste and guayule bagasse. Biochar 3(1):89–104. https://doi.org/10.1007/s42773-020-00070-2
44. Liu H, Xu G, Li G (2021) Preparation of porous biochar based on pharmaceutical sludge activated by NaOH and its application in the adsorption of tetracycline. J Colloid Interface Sci 587:271–278. https://doi.org/10.1016/j.jcis.2020.12.014
45. Liu Z et al (2019) Nanoscale chemical mapping of oxygen functional groups on graphene oxide using atomic force microscopy-coupled infrared spectroscopy. J Colloid Interface Sci 556:458–465. https://doi.org/10.1016/j.jcis.2019.08.089
46. Şinoforoğlu M et al (2013) Graphene oxide sheets as a template for dye assembly: graphene oxide sheets induce H-aggregates of pyronin (Y) dye. RSC Adv 3:11832–11838. https://doi.org/10.1039/c3ra40531a
47. Chen H, Gao B, Li H (2015) Removal of sulfamethoxazole and ciprofloxacin from aqueous solutions by graphene oxide. J Hazard Mater 282:201–207. https://doi.org/10.1016/j.jhazmat.2014.03.063
48. Yang ST et al (2011) Removal of carbon nanotubes from aqueous environment with filter paper. Chemosphere 82:621–626. https://doi.org/10.1016/j.chemosphere.2010.10.048
49. Wang Y et al (2012) Adsorption and desorption of doxorubicin on oxidized carbon nanotubes. Colloids Surf, B 97:62–69. https://doi.org/10.1016/j.colsurfb.2012.04.013
50. Ateia M et al (2017) Green and facile approach for enhancing the inherent magnetic properties of carbon nanotubes for water treatment applications. PLoS ONE 12:1–21. https://doi.org/10.1371/journal.pone.0180636
51. Hildago-Oporto P et al (2019) Synthesis of carbon nanotubes using biochar as precursor material under microwave irradiation. J Environ Manage 244:83–91. https://doi.org/10.1016/j.jenvman.2019.03.082
52. Hu K et al (2020) Facile synthesis of Z-scheme composite of TiO_2 nanorod/g-C3N4 nanosheet efficient for photocatalytic degradation of ciprofloxacin. J Clean Prod 253:120055. https://doi.org/10.1016/j.jclepro.2020.120055
53. Shen M et al (2021) gCN-P: a coupled g-C3N4/persulfate system for photocatalytic degradation of organic pollutants under simulated sunlight. Environ Sci Pollut Res (0123456789). https://doi.org/10.1007/s11356-021-17540-0
54. Yin S et al (2015) 'Recent progress in g-C3N4 based low cost photocatalytic system: activity enhancement and emerging applications. Catal Sci Technol Royal Soc Chem 5:5048–5061. https://doi.org/10.1039/c5cy00938c
55. Zhang S et al (2019) Recent developments in fabrication and structure regulation of visible-light-driven g-C3N4-based photocatalysts towards water purification: a critical review. Catal Today 335:65–77. https://doi.org/10.1016/j.cattod.2018.09.013
56. Abd El-Halim EH, El-Gayar DA, Farag HA (2020) Treatment of wastewater by ion exchange resin using a pulsating disc. Desalin Water Treat 193:133–141. https://doi.org/10.5004/dwt.2020.25689
57. Sica M et al (2011) Kinetic study of nitrite removal from municipal wastewater using ion exchange resins. Adv Mater Res 287–290:1513–1516. https://doi.org/10.4028/www.scientific.net/AMR.287-290.1513

58. Zhu C et al (2017) Mechanisms of phosphorus removal from wastewater by ion exchange resin. Desalin Water Treat 79:347–355. https://doi.org/10.5004/dwt.2017.20890
59. Kabay N et al (2004) Removal and recovery of boron from geothermal wastewater by selective ion-exchange resins—II field test. Desalination 167:427–438. https://doi.org/10.1016/j.desal.2004.06.158
60. Choi KJ, Son HJ, Kim SH (2007) Ionic treatment for removal of sulfonamide and tetracycline classes of antibiotic. Sci Total Environ 387:247–256. https://doi.org/10.1016/j.scitotenv.2007.07.024
61. Fernández AML, Rendueles M, Díaz M (2014) Competitive retention of sulfamethoxazole (SMX) and sulfamethazine (SMZ) from synthetic solutions in a strong anionic ion exchange resin. Solvent Extr Ion Exch 32:763–781. https://doi.org/10.1080/07366299.2014.940235
62. Landry KA et al (2015) Ion-exchange selectivity of diclofenac, ibuprofen, ketoprofen, and naproxen in ureolyzed human urine. Water Res 68:510–521. https://doi.org/10.1016/j.watres.2014.09.056
63. Landry KA, Boyer TH (2013) Diclofenac removal in urine using strong-base anion exchange polymer resins. Water Res 47:6432–6444. https://doi.org/10.1016/j.watres.2013.08.015
64. Robberson KA et al (2006) Adsorption of the quinolone antibiotic nalidixic acid onto anion-exchange and neutral polymers. Chemosphere 63:934–941. https://doi.org/10.1016/j.chemosphere.2005.09.047
65. Sun J et al (2015) Effect of long-term organic removal on ion exchange properties and performance during sewage tertiary treatment by conventional anion exchange resins. Chemosphere 136:181–189. https://doi.org/10.1016/j.chemosphere.2015.05.002
66. Bashiri H, Hassani Javanmardi A (2017) A new rate equation for desorption at the solid/solution interface. Chem Phys Lett 671:1–6. https://doi.org/10.1016/j.cplett.2017.01.007
67. Costa C, Rodrigues A (1985) Design of cyclic fixed-bed adsorption processes Part II: Regeneration and cyclic operation. AIChE J 31:1655–1665. https://doi.org/10.1002/aic.690311009
68. Nur T et al (2015) Nitrate removal using Purolite A520E ion exchange resin: batch and fixed-bed column adsorption modelling. Int J Environ Sci Technol 12:1311–1320. https://doi.org/10.1007/s13762-014-0510-6
69. Staudt J et al (2020) Ciprofloxacin desorption from gel type ion exchange resin: desorption modeling in batch system and fixed bed column. Sep Purif Technol 230:115857. https://doi.org/10.1016/j.seppur.2019.115857
70. Wang W et al (2016) Effect of resin charged functional group, porosity, and chemical matrix on the long-term pharmaceutical removal mechanism by conventional ion exchange resins. Chemosphere 160:71–79. https://doi.org/10.1016/j.chemosphere.2016.06.073
71. Lonappan L et al (2018) Adsorption of diclofenac onto different biochar microparticles: dataset—characterization and dosage of biochar. Data Brief 16:460–465. https://doi.org/10.1016/j.dib.2017.10.041
72. Ahmed R, Hossain MA (2022) Optimization of a fixed bed column adsorption of fast green dye on used black tea leaves from aqueous solution. J Iran Chem Soc 19:381–391. https://doi.org/10.1007/s13738-021-02310-z
73. El-Zahhar AA (2013) Sorption of cesium from aqueous solutions using polymer supported bentonite. J Radioanal Nucl Chem 295:1693–1701. https://doi.org/10.1007/s10967-012-2246-4
74. Zhou L et al (2018) Adsorption of U(VI) onto the carboxymethylated chitosan/Na-bentonite membranes: kinetic, isothermic and thermodynamic studies. J Radioanal Nucl Chem 317:1377–1385. https://doi.org/10.1007/s10967-018-6009-8
75. Maged A et al (2020) Tuning tetracycline removal from aqueous solution onto activated 2:1 layered clay mineral: characterization, sorption and mechanistic studies. J Hazard Mater 384:121320. https://doi.org/10.1016/j.jhazmat.2019.121320
76. Genç N, Dogan EC, Yurtsever M (2013) Bentonite for ciprofloxacin removal from aqueous solution. Water Sci Technol 68:848–855. https://doi.org/10.2166/wst.2013.313

77. Hubadillah SK et al (2018) Fabrications and applications of low cost ceramic membrane from kaolin: a comprehensive review. Ceram Int 44(5):4538–4560. https://doi.org/10.1016/j.ceramint.2017.12.215
78. Zhang B et al (2019) Effect of kaolin content on the performances of kaolin-hybridized soybean meal-based adhesives for wood composites. Compos B Eng 173:106919. https://doi.org/10.1016/j.compositesb.2019.106919
79. Antonelli R et al (2021) Fixed-bed adsorption of ciprofloxacin onto bentonite clay: characterization, mathematical modeling, and DFT-based calculations. Ind Eng Chem Res 60:4030–4040. https://doi.org/10.1021/acs.iecr.0c05700
80. Sivaselvam S et al (2020) Enhanced removal of emerging pharmaceutical contaminant ciprofloxacin and pathogen inactivation using morphologically tuned MgO nanostructures. J Environ Chem Eng 8:104256. https://doi.org/10.1016/j.jece.2020.104256
81. Dao TH et al (2018) Removal of antibiotic from aqueous solution using synthesized TiO2 nanoparticles: characteristics and mechanisms. Environ Earth Sci 77:1–14. https://doi.org/10.1007/s12665-018-7550-z
82. Malakootian M, Nasiri A, Amiri Gharaghani M (2020) Photocatalytic degradation of ciprofloxacin antibiotic by TiO_2 nanoparticles immobilized on a glass plate. Chem Eng Commun 207:56–72. https://doi.org/10.1080/00986445.2019.1573168
83. Bautitz IR, Velosa AC, Nogueira RFP (2012) Zero valent iron mediated degradation of the pharmaceutical diazepam. Chemosphere 88:688–692. https://doi.org/10.1016/j.chemosphere.2012.03.077
84. Chen J et al (2012) Removal mechanism of antibiotic metronidazole from aquatic solutions by using nanoscale zero-valent iron particles. Chem Eng J 181–182:113–119. https://doi.org/10.1016/j.cej.2011.11.037
85. Xiao Y, Hwang JY, Sun YK (2016) 'Transition metal carbide-based materials: synthesis and applications in electrochemical energy. J Mater Chem A Royal Soc Chem 4:10379–10393. https://doi.org/10.1039/c6ta03832h
86. Kim S et al (2021) Enhanced adsorption performance for selected pharmaceutical compounds by sonicated Ti3C2TX MXene. Chem Eng J 406:126789. https://doi.org/10.1016/j.cej.2020.126789
87. Thi Dang Y et al (2020) Microwave-assisted synthesis of nano Hf- and Zr-based metal-organic frameworks for enhancement of curcumin adsorption. Microporous Mesoporous Mater 298:110064. https://doi.org/10.1016/j.micromeso.2020.110064
88. Cao J et al (2018) One-step synthesis of Co-doped UiO-66 nanoparticle with enhanced removal efficiency of tetracycline: simultaneous adsorption and photocatalysis. Chem Eng J 353:126–137. https://doi.org/10.1016/j.cej.2018.07.060
89. Cui J et al (2020) Soft foam-like UiO-66/Polydopamine/Bacterial cellulose composite for the removal of aspirin and tetracycline hydrochloride. Chem Eng J 395:125174. https://doi.org/10.1016/j.cej.2020.125174
90. Elhussein EAA, Şahin S, Bayazit ŞS (2020) Removal of carbamazepine using UiO-66 and UiO-66/graphene nanoplatelet composite. J Environ Chem Eng 8:2–9. https://doi.org/10.1016/j.jece.2020.103898
91. Ma J et al (2020) Preparation of magnetic metal-organic frameworks with high binding capacity for removal of two fungicides from aqueous environments. J Ind Eng Chem Korean Soc Ind Eng Chem 90:178–189. https://doi.org/10.1016/j.jiec.2020.07.010
92. Fang X et al (2020) High-efficiency adsorption of norfloxacin using octahedral UIO-66-NH2 nanomaterials: dynamics, thermodynamics, and mechanisms. Appl Surf Sci 518:146226. https://doi.org/10.1016/j.apsusc.2020.146226
93. Zango ZU et al (2021) Selective adsorption of dyes and pharmaceuticals from water by UiO metal–organic frameworks: a comprehensive review. Polyhedron 210:115515. https://doi.org/10.1016/j.poly.2021.115515
94. El-Sayed ESM, Yuan D (2020) Waste to MOFs: Sustainable linker, metal, and solvent sources for value-added MOF synthesis and applications. Green Chem 22:4082–4104. https://doi.org/10.1039/d0gc00353k

95. Cho BG et al (2022) Adsorption modeling of microcrystalline cellulose for pharmaceutical-based micropollutants. J Hazard Mater 426:128087. https://doi.org/10.1016/j.jhazmat.2021.128087
96. Hammer L, Palmowski L (2021) Fate of selected organic micropollutants during anaerobic sludge digestion. Water Environ Res 93:1910–1924. https://doi.org/10.1002/wer.1603
97. Pei A et al (2013) Surface quaternized cellulose nanofibrils with high water absorbency and adsorption capacity for anionic dyes. Soft Matter 9:2047–2055. https://doi.org/10.1039/c2sm27344f
98. Steele DF et al (2003) Adsorption of an amine drug onto microcrystalline cellulose and silicified microcrystalline cellulose samples. Drug Dev Ind Pharm 29:475–487. https://doi.org/10.1081/DDC-120018382
99. Zhang N et al (2020) Recent investigations and progress in environmental remediation by using covalent organic framework-based adsorption method: a review. J Clean Prod 277:123360. https://doi.org/10.1016/j.jclepro.2020.123360
100. Foroughi M, Azqhandi MHA (2020) A biological-based adsorbent for a non-biodegradable pollutant: modeling and optimization of Pb (II) remediation using GO-CS-Fe_3O_4-EDTA nanocomposite. J Mol Liq 318:114077. https://doi.org/10.1016/j.molliq.2020.114077
101. Shah J, Jan MR (2020) Microextraction of selected endocrine disrupting phenolic compounds using magnetic chitosan biopolymer graphene oxide nanocomposite. J Polym Environ 28:1673–1683. https://doi.org/10.1007/s10924-020-01714-x
102. Chowdhury S, Balasubramanian R (2014) Recent advances in the use of graphene-family nanoadsorbents for removal of toxic pollutants from wastewater. Adv Coll Interface Sci 204:35–56. https://doi.org/10.1016/j.cis.2013.12.005
103. Ramesha GK et al (2011) Graphene and graphene oxide as effective adsorbents toward anionic and cationic dyes. J Colloid Interface Sci 361:270–277. https://doi.org/10.1016/j.jcis.2011.05.050
104. da Silva Alves DC et al (2021) Recent developments in Chitosan-based adsorbents for the removal of pollutants from aqueous environments. Molecules 26:594. https://doi.org/10.3390/molecules26030594
105. Afzal MZ et al (2018) Enhancement of ciprofloxacin sorption on chitosan/biochar hydrogel beads. Sci Total Environ 639:560–569. https://doi.org/10.1016/j.scitotenv.2018.05.129
106. Firozjaee TT et al (2017) The removal of diazinon from aqueous solution by chitosan/carbon nanotube adsorbent. Desalin Water Treat 79:291–300. https://doi.org/10.5004/dwt.2017.20794
107. Liu Y et al (2019) Removal of pharmaceuticals by novel magnetic genipin-crosslinked chitosan/graphene oxide-SO_3H composite. Carbohyd Polym 220:141–148. https://doi.org/10.1016/j.carbpol.2019.05.060
108. Ahmed MB et al (2016) Insight into biochar properties and its cost analysis. Biomass Bioenerg 84:76–86. https://doi.org/10.1016/j.biombioe.2015.11.002
109. Babel S, Kurniawan TA (2003) Low-cost adsorbents for heavy metals uptake from contaminated water: a review. J Hazard Mater 97:219–243. https://doi.org/10.1016/S0304-3894(02)00263-7
110. Gupta VK, Ali I (2004) Removal of lead and chromium from wastewater using bagasse fly ash—a sugar industry waste. J Colloid Interface Sci 271:321–328. https://doi.org/10.1016/j.jcis.2003.11.007
111. Lin SH, Juang RS (2009) Adsorption of phenol and its derivatives from water using synthetic resins and low-cost natural adsorbents: a review. J Environ Manage 90:1336–1349. https://doi.org/10.1016/j.jenvman.2008.09.003
112. Zhang S et al (2016) Adsorption of pharmaceuticals on chitosan-based magnetic composite particles with core-brush topology. Chem Eng J 304:325–334. https://doi.org/10.1016/j.cej.2016.06.087
113. Zhang YL et al (2013) Sorption of carbamazepine from water by magnetic molecularly imprinted polymers based on chitosan-Fe_3O_4. Carbohyd Polym 97:809–816. https://doi.org/10.1016/j.carbpol.2013.05.072

114. Costa C, Rodrigues AE (1985) Experimental diffusion of phenol in and study of batch and CSTR. Chem Eng Sci 40:983–993
115. Costa C, Rodriques A (1985) Regeneration of polymeric adsorbents in a CSTR. Chem Eng Sci 40:707–713. https://doi.org/10.1016/0009-2509(85)85023-5
116. Balarak D, Khatibi AD, Chandrika K (2020) Antibiotics removal from aqueous solution and pharmaceutical wastewater by adsorption process: a review. Int J Pharm Invest 10:106–111. https://doi.org/10.5530/ijpi.2020.2.19
117. Singh S et al (2021) Adsorption and detoxification of pharmaceutical compounds from wastewater using nanomaterials: a review on mechanism, kinetics, valorization and circular economy. J Environ Manage 300:113569. https://doi.org/10.1016/j.jenvman.2021.113569
118. Liakos EV et al (2021) Chitosan adsorbent derivatives for pharmaceuticals removal from effluents: a review. Macromol 1:130–154. https://doi.org/10.3390/macromol1020011

Chapter 10
Homogeneous Advanced Oxidation Processes for the Removal of Pharmaceutically Active Compounds—Current Status and Research Gaps

10.1 Introduction

Concerns regarding the presence of pharmaceutically active compounds (PhACs) in water bodies have led to numerous studies aiming at developing efficient, economic, and environmentally friendly (waste)water treatment techniques [1]. As previously discussed in the literature, conventionally used biological wastewater treatment techniques, such as activated sludge, are not efficient enough to remove such compounds from water bodies [2]; hence, an increasing concentration of such compounds in the environment can be expected, which may lead to the generation of antibiotic-resistant bacteria (ARBs) or antibiotic resistance genes (ARGs) [3, 4]. Direct toxic effects, such as immobilization, lethality, and reproductive, behavioral, physiological, and biochemical changes in aquatic microorganisms, have also been observed for some types of PhACs, such as atenolol, propranolol, and ketoprofen [5–8]. Hence, the development of efficient physico-chemical techniques has been considered a solution to address these issues. Adsorption techniques using low-cost carbonaceous (e.g., biochar) [9, 10] and clay-based materials (such as bentonite) [11], membrane separation techniques [12, 13], and advanced oxidation processes (AOPs) [14, 15] have been rapidly developed in recent years to remove PhACs from polluted waters.

AOPs are based on the generation and release of active agents (including radical and nonradical species), which can efficiently decompose a wide range of recalcitrant organic compounds [16, 17]. As indicated in Fig. 10.1, such techniques can be divided into heterogeneous (HE-AOPs) and homogeneous (HO-AOPs) techniques based on different phases involved in the oxidation processes. Each of them can be further classified into energy-free and energy-intensive methods based on the need for an external source of energy to perform the degradation reactions.

This chapter has aimed to discuss the fundamentals involved in the performance of various types of H-AOPs for the removal of PhACs. Recommendations have also been provided to direct future studies for the development of efficient, cost-effective, and

M. Kamali et al., *Advanced Wastewater Treatment Technologies for the Removal of Pharmaceutically Active Compounds*, Green Energy and Technology,
https://doi.org/10.1007/978-3-031-20806-5_10

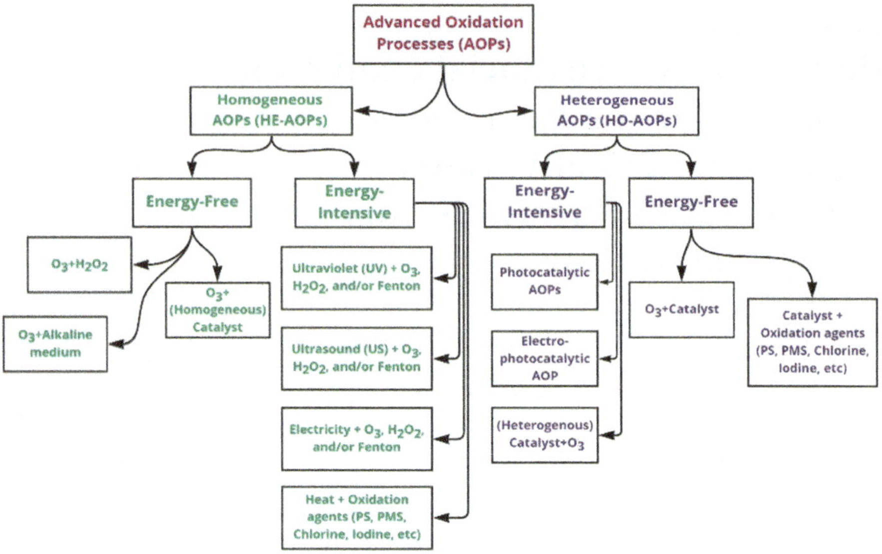

Fig. 10.1 Most widely studied and implemented AOPs for the removal of organic pollutants from (waste)waters. Blue box: Homogeneous AOPs (HO-AOPs) divided into energy-free and energy-intensive HO-AOPs

environmentally friendly techniques with the potential of being used for real-scale (waste)water treatment applications.

10.2 Energy-Free HO-AOPs

10.2.1 Ozonation

Ozone, as a powerful oxidation agent, has been used for over 100 years as an efficient drinking water treatment method all over the world for disinfection as well as the removal of odor, color, chemical oxygen demand (COD), etc., from polluted waters [18–20]. Direct reactions with ozone or the generation of oxidative radicals (ORs) (mainly hydroxyl radicals (HRs, $^{\cdot}OH$)) are considered the main pathways for the decomposition of complex organic compounds in polluted (waste)waters, as schematically illustrated in Fig. 10.2. HRs with a high standard redox potential (+2.8 V) have the capability for the efficient and nonselective degradation of various types of organic pollutants with relatively high reaction kinetics [21–23].

Various configurations of ozone-based systems have been used for the treatment of effluents from various origins. Such technologies can be used as a pretreatment step for highly polluted industrial effluents with low biodegradability (BOD_5/COD) to reduce the toxicity load of these effluents and make them appropriate for the

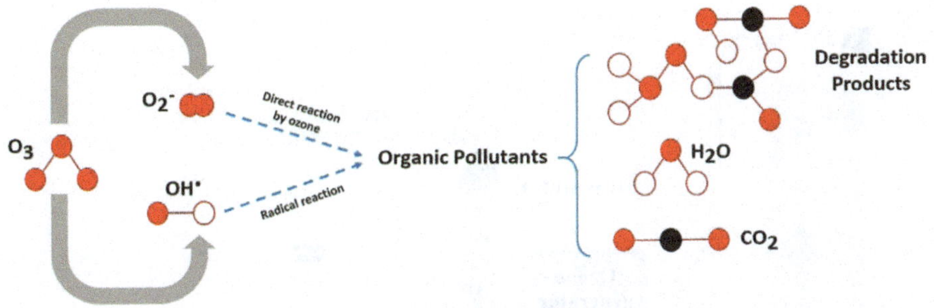

Fig. 10.2 Reaction pathways of the organic pollutants with ozonation oxidation systems, adopted from Taoufik et al. [28]

subsequent biological treatment process. Additionally, there are reports on the efficient application of such technologies as tertiary treatment steps to polish effluents already treated with other physico-chemical or biological treatment technologies [24–27].

In this section, the applicability of various ozone-based technologies, such as O_3, $O_3 + UV$, $O_3 + H_2O_2$, and catalytic ozonation, and the possible combinations of these methods for the degradation of pharmaceutically active compounds are explored, and suggestions are provided for future studies to promote ozone-based technologies for real-scale (waste)water treatment applications.

In direct ozonation processes, ozone gas is generated in situ and introduced into polluted (waste)waters to remove organic pollutants (Fig. 10.3). During the ozonation process, PhACs are degraded through direct attack either by ozone molecules or by the hydroxyl radicals formed by the decomposition of ozone molecules under alkaline reaction conditions through a chain of reactions (Eqs. 10.1–10.7) [29].

$$O_3 + OH^- \rightarrow HO_2^- + O_2^{\cdot -} \tag{10.1}$$

$$HO_2^- + O_3 \rightarrow HO_2^{\cdot -} + O_3^{\cdot -} \tag{10.2}$$

$$O_3 + O_2^{\cdot -} \rightarrow O_3^{\cdot -} + O_2 \tag{10.3}$$

$$O_3 + H^+ \leftrightarrow HO_3^{\cdot} \tag{10.4}$$

$$HO_3^{\cdot} \rightarrow {}^{\cdot}OH + O_2 \tag{10.5}$$

$$O_3 + {}^{\cdot}OH \rightarrow HO_4^{\cdot} \tag{10.6}$$

$$HO_4^{\cdot} \rightarrow HO_2^{\cdot} + O_2 \tag{10.7}$$

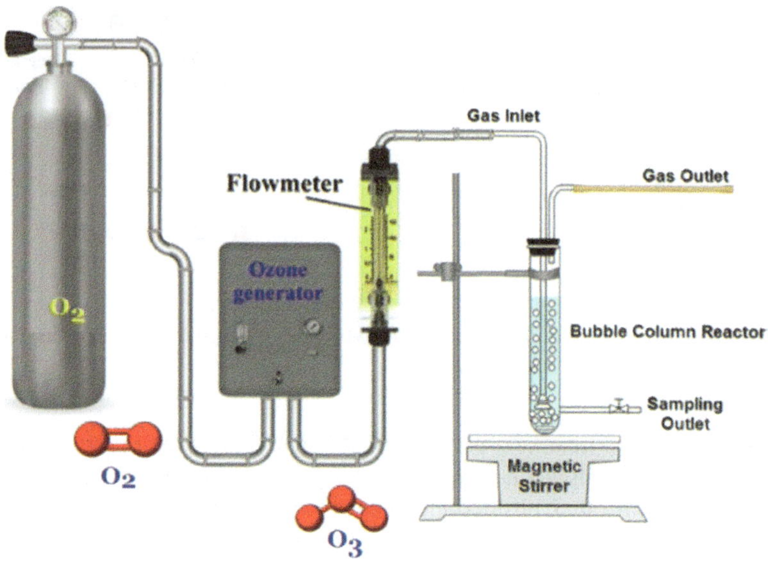

Fig. 10.3 A typical apparatus for the conversion of molecular oxygen to ozone and its application for the oxidation of organic pollutants, adapted from Aghaeinejad-Meybodi et al. [32]

Ozonation has been reported in the literature as a relatively simple and efficient method for the decomposition of some pharmaceutical micropollutants, such as estrogen 17β-estradiol (E2), some antibiotics and antiinflammatory drugs, and anticonvulsant compounds (such as carbamazepine). These pharmaceuticals are normally characterized by the presence of one or more functional groups and moieties, such as amines, sulfurs, nonaromatic carbon–carbon double bonds, and activated aromatic rings [30]. However, various degradation kinetics have been reported for different types of PhACs. For instance, very fast degradation of acetaminophen and tetracycline was observed in a recent study (5 min, ozone dosage of 5 g. Nm^{-3}), while complete removal of carbamazepine and sulfamethoxazole occurred within 15–20 min of reaction and even longer (30 min) for terbutryn [31].

There are studies to identify the mechanisms involved in the decomposition of PhACs using ozone-based treatment. For instance, it has been suggested that ozone molecules can first cause ring-opening in carbamazepine via the Criegee mechanism, in which the nonaromatic carbon–carbon double bond in carbamazepine is attacked by ozone molecules, followed by the closure of the ring to form the quinazoline moiety of 1-(2-benzaldehyde)-4-hydro(1H,3H)-quinazoline-2-one (BQM) [33]. Hence, it can be expected to detect quinazoline derivatives when ozone is used for the removal of such pharmaceuticals from polluted (waste)waters. However, different degradation pathways can be expected when ozonation is applied for the removal of other pharmaceutically active compounds. As an example, the preferred sites to be attacked by ozone in sulfamethoxazole are the functional groups $-NH_2$ and $-CH_3$ at the benzene or isoxazole ring (Fig. 10.4).

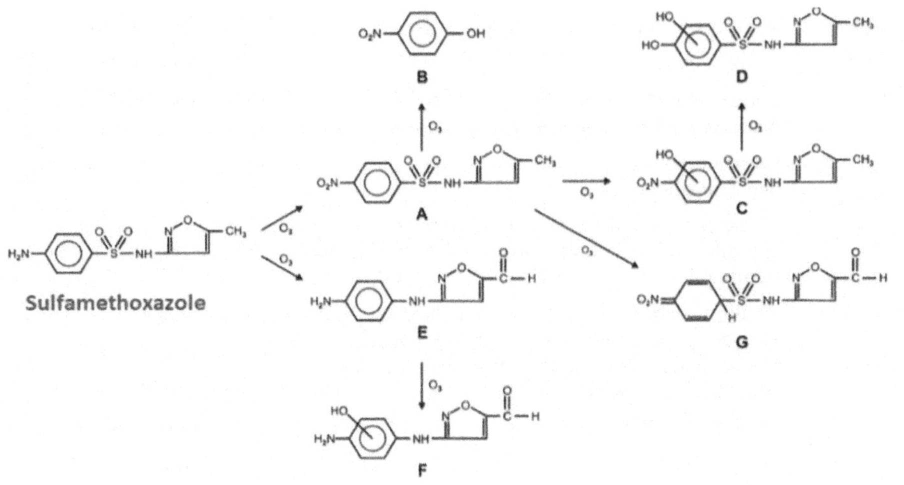

Fig. 10.4 Proposed pathway of sulfamethoxazole degradation under the ozonation process. Analysis was performed using liquid chromatography-mass spectrometry (LC–MS) analysis, adopted from Abellán et al. [39]

In addition to the efficiency of the system for the removal of PhACs, the degree of mineralization is another important parameter that can determine the real applicability of the applied AOP processes. In this regard, various efficiencies have been reported in the literature for the removal of total organic carbon (TOC) from PhAC-containing effluents. For instance, full removal of indomethacin was reported by Zhao et al. [34], while achieving a maximum of 50% of the total organic carbon removal under various ozone dosages (2–35 mg/L) within 30 min of reaction. Similar reports stating the incomplete mineralization of PhACs such as carbamazepine, diclofenac, trimethoprim, and sulfamethoxazole are also available in the literature [35]. The presence of ozone-resistant PhACs (Table 10.1) or the degradation products generated during the ozonation of the parent compounds are considered the main reason for the incomplete mineralization of the effluents containing PhACs. In such cases, applying an appropriate post-treatment is of high importance to remove the residual compounds from the treated (waste)waters. For instance, it has been recently reported that applying a post-treatment with a moving-bed biofilm reactor (MBBR) to effluents treated under a moderate ozone dosage of 0.5 g O_3/g DOC can result in the elimination of ozone-resistant pharmaceuticals such as iomeprol, iohexol, iopamidol, and gabapentin [36]. However, the applied subsequent biological method was not able to remove N-oxides formed during the ozonation of erythromycin, clarithromycin, and venlafaxine. Post-adsorption has also been demonstrated to be an effective method for the removal of persistent PhACs or their degradation products. As an example, complete removal of the degradation products of acetaminophen and amoxicillin from the ozonation process has been recently reported using chitosan/bentonite as cheap and widely available adsorbents [37]. Coupling ozonation with other advanced oxidation methods, such as photocatalytic AOPs, has also been demonstrated as an

Table 10.1 Stability of some pharmaceutically active compounds against ozonation [40]

Category	Pharmaceuticals	Sensitivity to O_3
A	Furosemide, Indomethacin, Diclofenac, Isopropylantipyrine, Naproxen, Mefenamic acid, Propranolol, and Dipyridamole	Unstable
B	Carbamazepine, Diltiazem, Lincomycin, Sulfadimethoxine, and Trimethoprim	Relatively unstable
C	Metoprolol, Roxithromycin, Erythromycin, Pirenzepine, Ciprofloxacin, Crotamiton, Atenolol, Sulpiride, and Ifenprodil	Relatively stable
D	Azithromycin, Bezafibrate, Sulfamethoxazole, Griseofulvin, Levofloxacin, Ethenzamide, Clofibric, Clarithromycin, Disopyramide, Theophylline, Nalidixic acid, and Chloramphenicol acid	Stable
E	Primidone, DEET,[1] and ketoprofen	Very stable

applicable method for the mineralization of PhACs. For instance, integrating the ozonation process with solar TiO_2-photocatalytic oxidation resulted in high mineralization rates (up to 70% TOC removal) for pharmaceuticals including atenolol, ofloxacin, and trimethoprim [38]. This is mainly due to the generation of additional oxidation agents, such as hydroxyl radicals, by the applied photocatalytic process.

However, the presence of other types of organic compounds and salts in the composition of real effluents can lead to a drop in the efficiency of ozonation-alone systems because such compounds can compete with PhACs for the consumption of the oxidative species present in the medium. To address this issue, catalytic ozonation processes (i.e., heterogeneous[2] or homogeneous) have been developed and employed for the removal of PhACs.

In homogenous catalysis, transition metal ions such as Cu^{2+}, Mn^{2+}, Fe^{3+}, Zn^{2+} are used in combination with the ozonation process for the effective generation of oxidative species in the medium [41, 42]. The decomposition of ozone molecules and generation of hydroxyl radicals are normally facilitated in the presence of heterogeneous metal catalysts (Eq. 10.8), especially under acidic conditions [43].

$$M^{n+} + O_3 + H^+ \rightarrow M^{(n+1)+} + {}^{\cdot}OH + O_2 \quad (10.8)$$

Hydroxyl radicals can also react with ozone molecules to generate $HO_2^{\cdot -}$, which can be involved in the regeneration process of metal ions by the ozonation of $HO_2^{\cdot -}$ (Eqs. 10.9, 10.10) [44].

$$O_3 + {}^{\cdot}OH \rightarrow O_2 + HO_2^{\cdot -} \quad (10.9)$$

$$M^{(n+1)+} + HO_2^{\cdot -} + OH^- \rightarrow M^{n+} + H_2O + O_2 \quad (10.10)$$

[1] N,N-Diethyl-meta-toluamide.

[2] Heterogeneous catalytic ozonation processes have been discussed in Chap. 11.

The formation of metal–organic complexes can also facilitate the decomposition of organic compounds. In such a process, ozone can react with the metallic elements of such complexes and produce reactive species close to organic pollutants [45].

Such techniques can be efficiently used for the decomposition of PhACs. For instance, ferrous or manganese catalytic ozonation processes have been demonstrated to be efficient methods for the mineralization of salicylic acid under optimum initial pH and ozone concentration conditions [46]. However, removal of the metallic elements from the treated effluent may be required to prevent the release of such compounds into the environment. Recovery of such compounds can also reduce the overall treatment costs of the system.

10.2.2 Activation of Oxidation Agents

The applicability of various oxidation agents, including hydrogen peroxide (H_2O_2), persulfate (PS), chlorine, and iodine, has been widely studied for the removal of PhACs in recent years. This can be achieved through the generation of oxidative species as a result of the activation of such oxidation agents.

The Fenton process is a conventional oxidation system that has been applied for the degradation of a wide range of organic pollutants. The reaction between Fe^{2+} and hydrogen peroxide results in the generation of hydroxyl radicals ($^{\cdot}OH$). This powerful oxidation agent can attack organic pollutants and degrade them, as indicated in Eqs. 10.11 and 10.12 [47, 48].

$$Fe^{2+} + H_2O_2 \rightarrow Fe^{3+} + {}^{\cdot}OH + OH^- \quad (10.11)$$

$$^{\cdot}OH + R \rightarrow \text{Intermediated} + \text{Final products}(CO_2 + H_2O) \quad (10.12)$$

There is also a possibility to generate $Fe(IV)O^{2+}$ during the Fenton process, according to Eq. 10.13.

$$Fe^{2+} + H_2O_2 \rightarrow H_2O + Fe(IV)O^{2+} + H_2O \quad (10.13)$$

The OH^- present in the medium can also react with H_2O_2 to produce $HO_2^{\cdot}$ (Eq. 10.14), which can further react with hydrogen peroxide, hydroxyl radicals, or Fe^{3+} (Eqs. 10.15–10.17).

$$H_2O_2 + OH^- + \rightarrow H_2O + HO_2^{\cdot} \quad (10.14)$$

$$HO_2^{\cdot} + H_2O_2 \rightarrow H_2O + {}^{\cdot}OH + O_2 \quad (10.15)$$

$$HO_2^{\cdot} + {}^{\cdot}OH \rightarrow H_2O + O_2 \quad (10.16)$$

$$HO_2^{\cdot} + Fe^{3+} \rightarrow Fe^{2+} + H^{+} + O_2 \quad (10.17)$$

The generated radicals can also scavenge the available Fe^{2+} as rate-limiting side reactions [49, 50] (Eqs. 10.18–10.19).

$$HO_2^{\cdot} + Fe^{2+} \rightarrow Fe^{3+} + HO_2^{-} \quad (10.18)$$

$$^{\cdot}OH + Fe^{2+} \rightarrow Fe^{3+} + HO^{-} \quad (10.19)$$

Regarding sustainability considerations, Fenton processes can be highlighted with high efficiency for the removal of a wide range of pharmaceutically active compounds. Furthermore, such technologies are normally easy to operate, which can make them attractive for real applications. However, the efficiency of Fenton processes is highly dependent on the operating conditions, namely, pH. The system can represent its optimum performance under acidic pH conditions, and a significant drop in its performance is normally expected by elevating the pH of the medium to the alkaline range. Such low pH values can be expected in specific industrial effluents, such as those from the bleaching steps of the pulp and paper industry, where Fenton processes can be applied as a viable technology for the removal of highly recalcitrant and toxic pollutants, such as adsorbable organic halides (AOXs) [51].

Oxidation agents such as persulfate, peroxymonosulfate, chlorine, and iodine have also been used in recent years for the generation of various oxidative agents (e.g., radicals and superoxides) [52]. Among these, sulfate radicals with a high redox potential (2.5–3.1 V) and a relatively long half-life (30–40 μs) have been used efficiently to deal with a wide range of organic pollutants. Such systems can also operate efficiently over a wide pH range from 3 to 8. This can make them appropriate to deal with effluents from various origins.

There are reports of the efficient activation of persulfate (PS) or peroxymonosulfate (PMS) for the degradation of PhACs using UV light, heat, ultrasound, and appropriate (heterogeneous and homogeneous) catalysts [53]. For instance, Milh et al. [54] reported the efficient activation of PS using heat for the rapid removal of sulfamethoxazole with an activation energy of 103 kJ/mol.

The cleavage of PS by an external source of energy results in the generation of two sulfate radicals, as indicated in Eq. 10.20 [55].

$$S_2O_8^{2-} + \text{Energy Input} \rightarrow 2SO_4^{-\cdot} \quad (10.20)$$

The generation of sulfate radicals can also be achieved through the oxidation of PS in the medium according to Eq. 10.21.

$$S_2O_8^{2-} + e^{-} \rightarrow S_2O_4^{2-} + SO_4^{-\cdot} \quad (10.21)$$

In addition to sulfate radicals, other reactive species, such as hydroxyl radicals and $H^{+,}$ can also be generated in the medium by the reaction between sulfate radicals and water molecules (Eq. 10.22).

$$SO_4^{-\cdot} + H_2O \rightarrow SO_4^{2-} + OH^{\cdot} + H^+ \quad (10.22)$$

According to the literature, the pH of the medium plays an important role in the PS activation process and the degradation pathways of the pollutants. The sulfate radicals are dominant under neutral pH values. By elevating the pH to alkaline conditions, sulfate radicals react with the available OH^-, leading to the generation of hydroxyl radicals (Eq. 10.23) [56]. The reaction between sulfate radicals and the generated hydroxyl radicals is also a possible scenario resulting in the formation of $HSO_5{}^-$ (Eq. 10.24).

$$SO_4^{-\cdot} + OH^- \rightarrow SO_4^{2-} + OH^{\cdot} \quad (10.23)$$

$$SO_4^{-\cdot} + OH^{\cdot} \rightarrow HSO_5^- \quad (10.24)$$

Decomposition of organic molecules by sulfate radicals occurs under mechanisms such as hydrogen abstraction, electron transfer, and addition to the double bonds. Furthermore, electron-donating groups can react faster with electrophilic persulfate radicals [57].

Activation of PS using metallic cations has also received attention in recent years. In this regard, elements such as copper, silver, zinc, and iron have been examined as homogenous catalysts for such processes. According to Eq. 10.25, the one-electron transfer mechanism results in the generation of sulfate radicals in the medium using metallic elements.

$$S_2O_8^{2-} + Mn^+ \rightarrow Mn^{+1} + SO_4^{2-} + SO_4^{-\cdot} \quad (10.25)$$

Among the metallic compounds, iron is the most widely used due to its non- or low-toxicity, low price, and availability [58]. There have also been some studies on the applicability of Fe/PS systems coupled with the Fenton process to further promote the treatment process for the efficient removal of PhACs such as tetracycline, sulfamethazine, bisphenol A, indomethacin, norfloxacin, sulfamethoxazole, carbamazepine, phenacetin, and paracetamol, as reported by Wu et al. [59].

Waste-derived low-cost materials such as fly ash, as a solid waste of manufacturing combustion processes, have also been proposed in recent years for the activation of PS. This is mainly due to the presence of metallic compartments such as Fe in the composition of such materials. Such a strategy can considerably reduce the overall costs of the treatment process [60]. However, this should be taken into consideration because the heavy metals present in the composition of such materials can be leached into the treated effluents, causing subsequent environmental issues. Hence, it would be of high importance to conduct further research regarding the possible leaching of various elements from the spent catalysts and the toxic nature of the effluents treated using such materials.

The combination of PS or PMS with ozone can also be considered an efficient and cost-effective method for the degradation of PhACs, with the potential for simultaneous generation of hydroxyl and sulfite radicals. The mechanisms of the reaction between reaction O_3 and $SO_5{}^{2-}$ are presented in Eqs. 10.26 to 10.33.

$$^{-}O_3SOO^{-} + O_3 \rightarrow ^{-O_3} SO_5^{-} \left(k = 21200\ M^{-1}s^{-1}\right) \quad (10.26)$$

$$^{-}O3SO_5^{-} \rightarrow SO_5^{\cdot -} + O_3^{\cdot -} \quad (10.27)$$

$$^{-}O_3SO_5^{-} \rightarrow SO_4^{2-} + 2O_2 \quad (10.28)$$

$$SO_5^{\cdot -} + O_3 \rightarrow SO_4^{\cdot -} + 2O_2\left(k = 1.6\times^{105\ M-1\ s-1}\right) \quad (10.29)$$

$$2SO_5^{\cdot -} \rightarrow 2SO_4^{\cdot -} + O_2\left(k = 2.1 \times 108\ M^{-1\ s^{-1}}\right) \quad (10.30)$$

$$2SO_5^{\cdot -} \rightarrow S_2O_8^{2-} + O_2\left(k = 2.2x^{108\ M-1\ s-1}\right) \quad (10.31)$$

$$O_3^{\cdot -} \rightleftarrows O^{\cdot -} + O_2\left(k = 2.1 \times 10^3 M^{-1}s^{-1}\right) \quad (10.32)$$

$$O^{\cdot -} + H_2O \rightarrow {}^{\cdot}OH + OH^{-}\left(k = 10^8 s^{-1}\right) \quad (10.33)$$

According to these reactions, $SO_5{}^{\cdot -}$, $O_3{}^{\cdot -}$, $SO_4{}^{\cdot -}$, $O^{\cdot -}$, and ${}^{\cdot}OH$ are generated for the decomposition of organic compounds [61, 62]. It has been indicated in a recent study that the degradation of pharmaceuticals, including metronidazole, venlafaxine, ketoprofen, atrazine, and carbamazepine, can be enhanced up to 5 times when PMS is added to the ozonation system, especially under higher pH conditions (8–10) [63]. The study also achieved a 3 times higher reaction rate compared to the UV/H_2O_2 system. Efficient degradation of acetaminophen (91% in 30 min) using an O_3/PS system has also been reported by Khashij et al. [64], compared to 63% and 22% for ozonation alone and PS alone, respectively.

Although such combinations seem very efficient for the degradation of pharmaceutically active compounds, there are concerns regarding the toxicity of the treated effluents using such methods due to the existence of sulfur-based materials resulting from the activation of PS. Additionally, the overall costs associated with the generation of ozone and purchasing oxidation agents need to be evaluated to push such technologies for commercialization.

10.3 Energy-Intensive HO-AOPs

10.3.1 Light-Assisted HO-AOPs

Various light-assisted HO-AOPs, including UV photolysis and its combination with oxidation agents (such as hydrogen peroxide), have also been examined recently for the degradation of PhACs.

UV photolysis has been considered a facile method for the removal of less recalcitrant PhACs from polluted waters. In a relevant study, 46% and 98% of the ciprofloxacin removal was achieved after 4 min and 128 min of illumination with UV and Xenon lamps, respectively. However, several degradation products were identified in the treated effluents, as illustrated in Fig. 10.5 [65]. This indicates that although photolysis is an efficient technique for the degradation of such PhACs, there is a need for more powerful methods to reach high degrees of mineralization. In this regard, techniques such as γ- or electron beam irradiation, which result in the generation of powerful hydroxyl radicals ($^{\cdot}OH$), hydrogen atoms ($H^{\cdot}$), and hydrated electrons (e_{aq}^{-}), have been shown to be effective for the mineralization of effluents containing such pollutants [66].

PhACs such as diclofenac and sulfamethoxazole can also be degraded under UV photolysis [67]. However, such techniques are not efficient for the removal of recalcitrant PhACs such as carbamazepine and trimethoprim. In such a case, the addition of oxidation agents such as H_2O_2 or ozone to the system can potentially lead to decomposing such persistent organic pollutants.

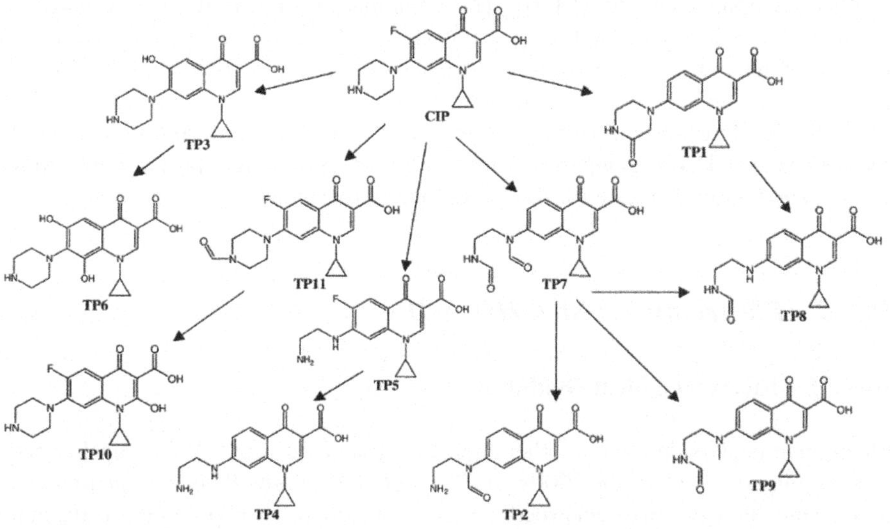

Fig. 10.5 Ciprofloxacin pathways and products of the degradation of ciprofloxacin under UV and xenon illumination, reprinted with permission from Haddad and Kümmerer [65]

The combination of UV with ozone is another way of generating highly reactive oxidative species in the medium for the removal of PhACs. It is evident that by increasing the dosage of ozone, the amount of the generated hydroxyl radicals increases, leading to the higher efficiency of the O_3/UV system. This can also result in a higher mineralization rate of the pollutant, which is essential to ensure the quality of the treated water. An example is an enhancement in the TOC removal of the sildenafil citrate (50 mg/L), reaching 75% under an inlet ozone concentration of 125 g Nm^{-3}, up to 9 times higher than that of the UV alone treatment system [68]. This can potentially lead to the elimination of the toxic byproducts formed during the ozonation of this PhAC.

The combination of UV/O_3 with hydrogen peroxide can also enhance the kinetics of PhAC removal, especially for compounds that are resistant to ozonation alone, such as clofibric acid, diazepam, and ibuprofen. In this process, UV photons are used to activate both hydrogen peroxide and ozone molecules for the generation of active species, according to Eqs. 10.34–10.36 [69].

$$O_3 + H_2O + hv \rightarrow H_2O_2 + O_2 \tag{10.34}$$

$$H_2O_2 + +hv \rightarrow 2^{\cdot}OH \tag{10.35}$$

$$2O_3 + H_2O_2 \rightarrow 2^{\cdot}OH + 3O_2 \tag{10.36}$$

An example is the application of an ozone/H_2O_2 system with an ozone dosage of 1.5 mg/L and ozone:H_2O_2 ratio of 1:0.25 for the efficient degradation of gemfibrozil (100%) and ibuprofen (80%) [70]. The combination of UV and ozone can also result in higher degradation efficiency for other PhACs, such as carbamazepine and 17-α-ethinylestradiol [71]. However, the excess hydrogen peroxide applied in such combined systems can scavenge hydroxyl radicals and may negatively affect the performance of the system. Hence, optimization of the operating conditions, mainly ozone dosage and hydrogen peroxide concentration, is of high importance to achieve an efficient degradation rate for the PhACs under reasonable operating costs.

10.3.2 Electricity-Assisted HO-AOPs

10.3.2.1 Electrochemical Oxidation

Electrochemical oxidation is another type of AOP (known as EO-AOPs) that has been progressing rapidly since the 2000s [16, 72]. EO-AOPs normally occur via direct and indirect mechanisms. In direct processes, electron transfer from pollutants is the main mechanism involved in their degradation, while in indirect processes, degradation reactions are mediated by electrogenerated active species such as hydroxyl radicals [73].

The anodes used in EO-AOPs can also be divided into active and nonactive anodes based on the processes involved in the degradation of organic pollutants. In nonactive anodes (such as boron-doped diamond (BDD) and lead oxide electrodes), physisorbed hydroxyl radicals are generated on the anode surface through the water oxidation process, as indicated in Eq. 10.37 [74, 75].

$$A_{EO} + H_2O \rightarrow A_{EO}(HO^{\cdot}) + H^+ + e^- \tag{10.37}$$

In this equation, A_{EO} represents the anode materials used in the EO-AOP process, and $A_{EO}(HO^{\cdot})$ indicates the hydroxyl radical adsorbed on the anode surface. On the other hand, active anodes (such as Pt and mixed metal oxides) support the generation of higher state oxides. This happens through the interaction between the electrode and the generated hydroxyl radicals (Eqs. 10.38, and 10.39). However, decomposition of the superoxide generated in this process leads to the oxygen evolution reaction at a lower potential, resulting in a narrow work potential window [76–78].

$$MO(HO^{\cdot}) \rightarrow MO_{x+1} + H^+ + e^- \tag{10.38}$$

$$MO_{x+1} \rightarrow MO_x + 1/2O_2 \tag{10.39}$$

In these reactions, MO_x and MO_{x+1} are the metal oxide surface and the higher state oxides that are generated in the medium.

Other types of oxidative species can also be formed under EO-AOP processes. For instance, sulfate radicals can be generated at BDD when sulfate ions are present in wastewater. While hydroxyl radicals are nonselective and can efficiently degrade various PhACs, sulfate radicals selectively react with some PhACs. For instance, the promoted degradation efficiency of ciprofloxacin and sulfamethoxazole can be achieved using such a technique and in the presence of sulfate anions [79]. It has also been indicated in the literature that parameters such as the applied current potential, the pH of the medium, the amount of boron doped to the electrode and the diamond-SP^3/SP^2 carbon ratio can greatly influence the indirect oxidation processes by E-AOPs [79–82].

Although BDD is a widely used anode for EO-AOPs, metal-doped metal oxides (such as Ni-doped ZnO [83]) and metal oxide-loaded carbonaceous materials (such as P-doped TiO_2 nanotubes [84]) have also been developed and employed for the efficient removal of PhACs. Recent studies have also indicated that cathode materials can also influence the generation of active species in the medium. Carbon felt, as a low-cost and highly conductive material, has been used in the majority of recent studies concerning EO-AOPs for the removal of PhACs [85–87]. In addition to the electrode materials, other parameters, such as pH, current density, supporting electrolyte, conductivity, and the reactor configuration, can all influence the performance of such oxidation processes [88]. Table 10.2 represents the remarks regarding the effects of the operating parameters on the degradation of the PhACs using various EO-AOPs.

Table 10.2 Most important parameters influencing the performance of EO-AOP processes for the removal of PhACs

Parameter	Remarks	Examples	References
Initial concentration	Generally, higher degradation rates can be achieved under a lower initial concentration of the PhACs	Ibuprofen	Wang et al. [89]
		Parabens	Domínguez et al. [90]
		Salicylic acid	Rabaaoui and Allagui [91]
Electrolyte	NaCl can potentially favor the removal of PhACs, due to the formation of active chlorine	Berberines	Tu et al. [92]
		Ibuprofen	Ambuludi et al. [93]
		Caffeine	Indermuhle et al. [94]
	Na_2SO_4 performs better than phosphate buffer (0.1 M) for the removal of PhACs	Sulfamethoxazole and diclofenac	Sifuna et al. [95]
	The formation of chlorinated organic compounds is a drawback of using NaCl resulting in a low mineralization rate of PhACs	Ketoprofen	Murugananthan et al. [96]
	The formation of hypochlorite ions in presence of chloride species can have a positive effect on the degradation of PhACs	Acetaminophen, diclofenac, and sulfamethoxazole	Liu et al. [97]
Current density (j value)	Increasing the current density can directly enhance the performance of the system by promoting the generation of oxidative species	Naproxen	González et al. [98]
		Isothiazolin-3-ones	Velappan et al. [99]
		Carbamazepine	Domínguez et al. [100]
	There is a limit for the positive effect of the increase in the current density on the degradation of PhACs	Carboplatin	Barışçı et al. [101]

(continued)

Table 10.2 (continued)

Parameter	Remarks	Examples	References
pH	Although pH is less important than the j value, it can influence the degradation of some PhACs	Isothiazolin-3-ones (no effect)	Velappan et al. [99]
		chloramphenicol (optimum pH = 2)	Sun et al. [102]
Temperature	Elevation of the temperature can enhance the removal of PhACs through a gradual increase in the diffusion coefficient. However, the effect of this parameter is less important than that of the current density	Norfloxacin	Coledam et al. [103]
		N,N-diethyl-m-toluamide	Chen et al. [104]
Stirring rate	A higher mass transfer rate can be achieved by increasing the stirring speed. Such positive effects have been observed for the degradation of some PhACs	Naproxen	Díaz et al. [105]

In addition to the effectiveness of the EO-AOPs for the removal of a wide range of PhACs in a relatively short reaction time, they can also result in high TOC removal due to the generation of sufficient amounts of oxidative species in the medium [106]. This can potentially reduce the toxicity of the treated effluents, satisfying sustainability considerations.

Despite the effectiveness of these processes for the removal of various types of PhACs, there are factors that can limit their application for real (waste)water treatment. Oxidative reactions in such processes normally occur near or at the surface of the electrodes. Hence, the process is naturally mass transfer limited. On the other hand, low Fickian diffusion rates can be expected for PhACs because such compounds are generally present at relatively low concentrations of μg/L or ng/L in most water bodies [107, 108]. Electrode materials and their properties can also cause limitations. Boron-doped diamond (BDD) is currently considered among the most efficient electrodes for applications with properties such as high electrical conductivity and stability. However, high manufacturing costs have seriously limited such types of electrodes for real applications. Providing the energy required for EO-AOPs can also be considered a main source of treatment costs. Recent literature strongly recommends exploring cheap and renewable sources of energy, such as sunlight, to reduce the overall costs of the treatment process [109], making the technology more appealing for real applications.

The effectiveness of EO-AOPs can be further improved through their integration with other AOPs, such as Fenton reaction processes. This can also potentially aid in achieving high mineralization efficiencies using Fenton processes by removing the byproducts formed as a result of advanced oxidation of PhACs [110]. In situ generation of hydrogen peroxide using electrochemical systems through the cathodic reduction of oxygen molecules (Eqs. 10.40, and 10.41) can also aid in reducing the amount of hydrogen peroxide needed for Fenton processes for the removal of PhACs [111].

$$O_2 + H^+ + 2e^- \rightarrow H_2O_2 \tag{10.40}$$

$$Fe^{2+} + H_2O_2 + H^+ \rightarrow Fe^{3+} + {}^{\cdot}OH + H_2O \tag{10.41}$$

Effective removal of organic pollutants can also be achieved by the anodic in situ generation of hydrogen peroxide, according to Eqs. 10.42–10.44 [112].

$$O_2 + e^- \rightarrow {}^{\cdot}O_2 - \tag{10.42}$$

$$\cdot O_2^- + e^- + 2H_2O \rightarrow H_2O_2 + 2OH^- \tag{10.43}$$

$$\equiv M(n)/M(n) + H_2O_2 \rightarrow \equiv M(n+1)/M(n+1) + {}^{\cdot}OH + OH^- \tag{10.44}$$

Although these simple methods normally represent high efficiencies for the removal of a wide range of organic and inorganic pollutants [113], they currently suffer from drawbacks such as the dependence on a narrow range of pH, as well as difficulties in the recovery of the spent catalysts from the treated streams.

It is also evident that effluents containing recalcitrant and nonbiodegradable compounds represent a high chemical oxygen demand (COD) and a low biological oxygen demand (BOD_5) [114]. Hence, the biodegradability index (defined as BOD_5/COD) can be considered an important indirect measure indicating the degree of complexity or toxicity of the degradation products formed in the medium. In this regard, few studies have aimed to measure the biodegradability of pharmaceutical-containing effluents before and after treatment with AOPs. For instance, sonolysis at 520 kHz has been indicated as an efficient method for increasing the biodegradability of fluoroquinolone antibiotic (such as ciprofloxacin)-containing effluents from 0.06 to 0.60, 0.17, and 0.18 depending on the pH (3, 7, and 10, respectively) after 120 min of reaction time [115].

Plasma discharge is another E-AOP technology that has been recently implemented for the removal of PhACs. There are normally three ways of plasma discharge into the water media, including (a) discharge above the water surface, (b) direct discharge into the water medium, and (c) discharge in bubbles or vapour into the water medium [116]. Such a discharge results in the generation of oxidative species that can attack and decompose organic pollutants in the water medium. This happens through the reactions between the electron and neutral water, including momentum transfer, rotational excitation, vibrational excitation, dissociation, ionization, and attachment [117, 118]. Dissociation normally results in the formation of hydroxyl radicals, according to Eq. 10.45. As seen in this equation, such a reaction can also result in the generation of $H^{\cdot}$, which can be involved in the decomposition of organic compounds [118, 119].

$$H_2O + e \rightarrow OH^{\cdot} + H^{\cdot} + e \tag{10.45}$$

Hydroxyl radicals can also be generated under ionization mechanisms (Eqs. 10.46, and 10.47).

$$H_2O + e \rightarrow 2e + H_2O^+ \tag{10.46}$$

$$H_2O^+ + H_2O \rightarrow OH^{\cdot} + H_3O^+ \tag{10.47}$$

Furthermore, vibrational and rotational excitation can result in the formation of $H^{\cdot}$, $OH^{\cdot}$, and $O^{\cdot}$, according to Eqs. 10.48–10.51.

$$H_2O + e \rightarrow H_2O^* + e \tag{10.48}$$

$$H_2O^* + H_2O \rightarrow H_2O + H^{\cdot} + OH^{\cdot} \tag{10.49}$$

$$H_2O^* + H_2O \rightarrow H_2 + O^{\cdot} + H_2O \tag{10.50}$$

$$H_2O^* + H_2O \rightarrow 2H^{\cdot} + O^{\cdot} + H_2O \tag{10.51}$$

In addition to the mentioned mechanisms, $O^{\cdot}$ can directly react with H_2O molecules to produce hydroxyl radicals (Eq. 10.52) [120].

$$O^{\cdot} + H_2O \rightarrow 2OH^{\cdot} \tag{10.52}$$

Ozone can also be produced in these processes through the reaction of $O^{\cdot}$ and O_2 (Eq. 10.53).

$$O^{\cdot} + O_2 \rightarrow O_3 \tag{10.53}$$

The formation of hydrogen peroxide (H_2O_2) is another possibility under plasma discharge, which occurs through the direct reaction of hydroxyl radicals, according to Eq. 10.54 [121, 122].

$$OH^{\cdot} + OH^{\cdot} \rightarrow H_2O_2 \tag{10.54}$$

Other oxidative species, such as $HO_2^{\cdot}$, can also be generated through a chain of reactions involved in the plasma discharge process [123–125]. In addition, reductive species such as aqueous electrons and hydrogen radicals are produced with the irradiation of water with high-energy electrons. These agents can also contribute to the decomposition of PhACs. For instance, H radicals are involved in the reduction reactions of organic compounds under mechanisms such as hydrogen addition into unsaturated bonds and hydrogen abstraction.

Other mechanisms can also be involved in the plasma discharge process. The "exited species relaxation" process normally leads to the emission of UV light, which results in the photolysis of organic compounds [126, 127]. Hydrogen peroxide and ozone molecules generated in the medium can also be activated by the emitted UV light, which is a well-known process for the degradation of PhACs [128]. Shockwaves can also be generated in the medium by plasma discharging, leading to pyrolytic and chemical reactions under cavitation of the generated bubbles [129, 130]. Pyrolysis is also a thermal process in the absence of oxygen with a role in the degradation of organic compounds during the plasma discharge process. This can happen in the nonthermal plasma process through the formation of local hot spots as a result of the irradiation of water using plasma discharge technology [131].

Various types of plasma discharge reactors have been developed and employed for the treatment of polluted (waste)waters, including dielectric barrier discharge (DBD), corona discharge (CD), glow discharge plasma (GDP), microwave-UV plasma (MUVP), and gliding arc discharge (GADP). DBD plasma was first developed in 1857 by Ernst Werner von Siemens [132]. This type of plasma is normally generated under atmospheric pressure by the electrical discharge between the electrodes that have been separated by an insulator barrier. The CD process, as a nonequilibrium discharge, was first studied by Faraday in 1838 [133] and occurs in a highly nonuniform field. Such a pulsed discharge is generated using electrodes with a small curvature radius, such as a thin wire, which supports the ionization processes limited to a local region in the vicinity of the high voltage corona electrode [134]. Under the GDP process, the applied voltage exceeds the amount required for the breakdown of low-pressure gas, resulting in the ionization process. This process can lead to the dissociation of water molecules into hydroxyl radicals ($HO^{\cdot}$) and hydrogen atoms ($^{\cdot}H$) [135]. Such species can actively contribute to the decomposition of organic pollutants. In MUVP, high-intensity UV light is generated in the system. The basis of this method is the injection of the microwave into a resonant cavity and then into the plasma via a quartz tube wall. Finally, GADP is a nonequilibrium plasma that is generated under atmospheric pressure when diverging electrodes are placed in a fast gas flow [136].

Plasma discharge technologies have been used thus far for the degradation of various types of PhACs. Nonthermal plasma technologies such as DBD and CD (which produce oxidative species in the gas phase) have been widely used for the removal of antibiotics from (waste)waters, as reviewed by Magureanu et al. [137]. They concluded that high efficiencies (over 90% for most cases) can be achieved using such technologies for the removal of various types of antibiotics, such as amoxicillin, ampicillin, ceftriaxone, sulfadiazine, sulfathiazole, sulfamethoxazole, sulfamethazine, ciprofloxacin, norfloxacin, enrofloxacin, ofloxacin, flumequine, levofloxacin, tetracycline, doxycycline, oxytetracycline, chlortetracycline, lincomycin, chloramphenicol, and thiamphenicol. Although almost complete removals of the PhACs have been reported in the relevant studies in relatively short reaction times (> 30 min in most cases), relatively low mineralization performances have been observed for the employed technologies, and there are only a few reports for efficient TOC removal from the system. For instance, over 80% mineralization was reported by Singh et al. [138] for ciprofloxacin (10 mg/L) using a multineedle corona above liquid within 10 min. Hence, further studies are required to reach efficient mineralization of PhACs because in some cases, the decomposition of such compounds results in the formation of more toxic and recalcitrant products.

In addition, the operating conditions of plasma discharge technologies can highly influence their performance for the removal of PhACs. The applied voltage amplitude and its frequency are the most important variables determining the efficiency of these technologies. Recent studies have revealed that even a small change in the voltage can highly influence the performance of the system in terms of the removal of PhACs. An example is an increase in the enrofloxacin removal kinetics from 0.019 to 0.114 min^{-1} by increasing the voltage from 18 to 22 kV using a pulsed discharge plasma reactor [139]. The initial pollutant concentration also has a determinant effect on the removal efficiency of PhACs using plasma discharge technologies. This is mainly because, under a higher initial concentration of the pollutant, higher degrees of oxidative species are required to remove the pollutants [140]. Furthermore, the nature of the gaseous atmosphere and gas flow rate have critical impacts on the degradation of PhACs. Air, oxygen, nitrogen, and their combinations are the most widely used atmospheres in plasma discharge reactors. Discharge in an oxygen atmosphere has been identified as the most efficient way to optimize the performance of plasma systems due to the generation of oxidative species such as ozone in the medium [141]. Finally, the solution pH can greatly affect the performance of plasma discharge technologies. Acidic and basic pH values each can favor specific types of oxidative species in the medium, as discussed in various chapters of this book. It is also worth mentioning that the pH is readily acidified when most plasma technologies are applied (such as air-generated plasma in contact with water) [142]. The formation of carboxylic acids or inorganic acids is considered the main reason for such a drop in the pH of the medium. As discussed before, PhACs can be found in cationic, neutral (zwitterionic), or anionic forms, depending on the pH of the medium. The deprotonation of such molecules at alkaline pH values can be explained as the most important reason for their lower removal efficiencies under such pH conditions. On the other hand, the pH can influence the amount and type of oxidative species generated in the reaction medium. For instance, maximum hydroxyl radical generation has been reported at acidic pH values (e.g., pH = 4) [138].

10.3.2.2 Ultrasound-Assisted HO-AOPs

Ultrasound is another useful and sustainable technique for the removal of PhACs from polluted waters. Irradiation of a liquid with ultrasound (20 kHz–10 MHz) results in the growth and instantaneous collapse of bubbles (cavitation), which results in the creation of hot spots in media with extreme temperatures (up to 5000 °C) and pressures (exceeding 500 atm) within a few microseconds [143]. Electron spin resonance (ESR) and spin-trapping studies have revealed that both $H^{\cdot}$ and $^{\cdot}OH$ are formed during the sonolysis of water, according to Eqs. (10.55–10.61) [144].

$$H_2O \rightarrow H^{\cdot} + {}^{\cdot}OH \quad (10.55)$$

$$2H^{\cdot} \rightarrow H_2 \quad (10.56)$$

$$H^{\cdot} + O_2 \rightarrow HO_2^{\cdot} \quad (10.57)$$

$$HO_2^{\cdot} + HO_2^{\cdot} \rightarrow H_2O_2 + O_2 \quad (10.58)$$

$$2{}^{\cdot}OH \rightarrow H_2O_2 \quad (10.59)$$

$$2{}^{\cdot}OH \rightarrow H_2O + O^{\cdot} \quad (10.60)$$

$$H^{\cdot} + H_2O_2 \rightarrow {}^{\cdot}OH + H_2O \quad (10.61)$$

Depending on the operating conditions, various removal efficiencies have been reported using ultrasonic-based HO-AOPs for different types of PhACs. In this regard, the aqueous matrix, ultrasonic frequency, pH, temperature, and reaction time have been considered the most important influencing parameters. Under a higher initial concentration of PhACs, lower reaction rates can be expected because of the lower abundance of generated hydroxyl radicals relative to the pollutant molecules. Hence, a higher energy input may be required to degrade the pollutants. However, under higher ultrasonic frequencies, a decrease occurs in the size of the bubbles formed in the cavitation process with a short acoustic period. This results in the presence of fewer water molecules close to the surface of the bubbles and hence less probability of contact between the hydroxyl radicals and the pollutants [145]. Therefore, there is a need to investigate the optimum conditions under which the ideal amount of hydroxyl radicals are in contact with the PhACs. In addition to the mentioned parameters, pH can greatly affect the effectiveness of such processes. Under $pKa > pH$, the PhACs are present in molecular form. In this situation, mass transfer through the cavitation process is facilitated, resulting in a higher degradation rate of the pollutants [146, 147]. In contrast, when $pKa < pH$, the compound is present in its anionic form by losing a proton. The oxidation of anionic compounds by the

Table 10.3 Further reading suggestions for more detailed coverage of the literature on ozone-based technologies for the removal of pharmaceuticals in (waste)waters

References	Item	Subject
Issaka et al. [150]	Table 6	Detailed values from the reports on the application of ozone-based oxidation and catalytic ozone oxidation processes for the removal of pharmaceutically active compounds
	Table 7	Detailed values from the reports regarding the catalytic ozonation and mineralization of pharmaceutically active compounds
Paucar et al. [40]	Figure 9	Degradability of various pharmaceutically active compounds against ozonation under various initial ozone dosages
Gomes et al. [151]	Table 2	Various catalysts which have been used in the catalytic ozonation of contaminants of emerging concern (CECs)
	Table 3	The effects of operating conditions regarding the application of ozonation-based processes for the degradation of pharmaceutically active compounds
Sousa et al. [152]	Table 1	Boron-doped electrodes for the electrochemical oxidation of PhACs

generated hydroxyl radicals is also a pathway for the removal of PhACs [115]. Higher temperatures can also enhance the mass transfer in the reaction medium, leading to higher degradation rates of the PhACs. Furthermore, the presence of chlorides and inorganic carbon in the composition of effluents can affect the performance of ultrasonic irradiation on the removal of PhACs because they can eliminate the generated hydroxyl radicals and hence affect the removal of PhACs [148].

Studies have confirmed that sonolysis coupled with ozonation can result in a higher degradation rate for PhACs. For instance, higher degradation of ibuprofen was observed when such a combined system was employed compared to ozonation alone or sonolysis alone [149].

10.4 Further Reading

Table 10.3 contains items from the recent literature that the reader can consult for more detailed information regarding the efficiency of ozone-based technologies for the degradation of pharmaceutical micropollutants.

10.5 Summary

Homogenous advanced oxidation processes (HO-AOPs) have been developed rapidly in recent years for the removal of pharmaceutically active compounds (PhACs) from polluted (waste)waters. Various types of HO-AOPs include ozone-based technologies, activation of oxidation agents, light-assisted techniques, electricity-based methods, and those mediated by ultrasonic irradiation in the medium. Each of these technologies has its pros and cons to be considered for real- and large-scale (waste)water applications. Ozonation is an attractive technique to deal with PhACs, especially when coupled with homogeneous catalysts, resulting in the generation of sufficient oxidative species in the medium. However, there are efforts required to minimize the associated treatment costs. Additionally, recovery of the spent catalysts after being used is of high importance to prevent their release to the environment with the treated effluents. Activation of oxidation agents such as hydrogen peroxide, persulfate, iodine, etc., has been considered very efficient for the removal of PhACs, especially under lower initial concentrations. However, the toxic nature of the effluents treated with these techniques (considering the probability of the presence of active species) and the associated treatment costs are points requiring more study. Light-assisted techniques, especially those that utilize visible light and solar irradiation, have also received attention. Their combinations with other AOPs (such as ozonation) have also been considered attractive for further mineralization of PhACs. However, among HO-AOPs, electrical-based technologies, especially plasma discharge processes, are outstanding not only for the removal of parent molecules but also for the mineralization of the containing effluents. Reducing the operating costs using sustainable sources of energy and optimization in the reactor configuration and the operating parameters can further push such technologies for real (waste)water treatment applications.

References

1. Kamali M et al (2022) ZnO/γ-Fe_2O_3/Bentonite: an efficient solar-light active magnetic photocatalyst for the degradation of pharmaceutical active compounds. Molecules 27:3050
2. Peng J et al (2019) Characterizing the removal routes of seven pharmaceuticals in the activated sludge process. Sci Total Environ 650:2437–2445. https://doi.org/10.1016/j.scitotenv.2018.10.004
3. Li J et al (2016) Occurrence and removal of antibiotics and the corresponding resistance genes in wastewater treatment plants: effluents' influence to downstream water environment. Environ Sci Pollut Res 23:6826–6835. https://doi.org/10.1007/s11356-015-5916-2
4. Pallares-Vega R et al (2019) Determinants of presence and removal of antibiotic resistance genes during WWTP treatment: a cross-sectional study. Water Res 161:319–328. https://doi.org/10.1016/j.watres.2019.05.100
5. Diniz MS et al (2015) Ecotoxicity of ketoprofen, diclofenac, atenolol and their photolysis byproducts in zebrafish (Danio rerio). Sci Total Environ 505:282–289. https://doi.org/10.1016/j.scitotenv.2014.09.103
6. Prata JC et al (2018) Influence of microplastics on the toxicity of the pharmaceuticals procainamide and doxycycline on the marine microalgae Tetraselmis chuii. Aquat Toxicol 197:143–152. https://doi.org/10.1016/j.aquatox.2018.02.015
7. Santos A, Veiga F, Figueiras A (2020) Dendrimers as pharmaceutical excipients: synthesis, properties, toxicity and biomedical applications. Materials. https://doi.org/10.3390/ma13010065

8. Seydi E, Tabbati Y, Pourahmad J (2020) Toxicity of atenolol and propranolol on rat heart mitochondria. Drug Research 70:151–157. https://doi.org/10.1055/a-1112-7032
9. Awad AM et al (2019) Adsorption of organic pollutants by natural and modified clays: a comprehensive review. Sep Purif Technol 228:115719. https://doi.org/10.1016/j.seppur.2019.115719
10. Tabana L et al (2020) Adsorption of phenol from wastewater using calcined magnesium-zinc-aluminium layered double hydroxide clay. Sustainability 12:4273. https://doi.org/10.3390/su12104273
11. Ain QU et al (2020) Superior dye degradation and adsorption capability of polydopamine modified Fe_3O_4-pillared bentonite composite. J Hazard Mater 397:122758. https://doi.org/10.1016/j.jhazmat.2020.122758
12. Dolar D et al (2017) Adsorption of hydrophilic and hydrophobic pharmaceuticals on RO/NF membranes: identification of interactions using FTIR. J Appl Polym Sci 134:17–21. https://doi.org/10.1002/app.44426
13. Qu F et al (2019) Tertiary treatment of secondary effluent using ultrafiltration for wastewater reuse: correlating membrane fouling with rejection of effluent organic matter and hydrophobic pharmaceuticals. Environ Sci Water Res Technol Royal Soc Chem 5:672–683. https://doi.org/10.1039/c9ew00022d
14. Garcia-Segura S, Brillas E (2017) Applied photoelectrocatalysis on the degradation of organic pollutants in wastewaters. J Photochem Photobiol, C 31:1–35. https://doi.org/10.1016/j.jphotochemrev.2017.01.005
15. Kawrani S et al (2020) Enhancement of calcium copper titanium oxide photoelectrochemical performance using boron nitride nanosheets. Chem Eng J 389:124326. https://doi.org/10.1016/j.cej.2020.124326
16. da Silva SW et al (2021) Advanced electrochemical oxidation processes in the treatment of pharmaceutical containing water and wastewater: a review. Curr Pollut Rep 7(2):146–159. https://doi.org/10.1007/s40726-021-00176-6
17. Von Sonntag C (2008) Advanced oxidation processes: mechanistic aspects. Water Sci Technol 58(5):1015–1021. https://doi.org/10.2166/wst.2008.467
18. Loeb BL et al (2012) Worldwide ozone capacity for treatment of drinking water and wastewater: a review. Ozone: Sci Eng J Int Ozone Assoc 9512. https://doi.org/10.1080/01919512.2012.640251
19. Rakness KL et al (2005) Cryptosporidium log-inactivation with ozone using effluent CT10, geometric mean CT10, extended integrated CT10 and extended CSTR calculations. Ozone: Sci Eng 27(5):335–350. https://doi.org/10.1080/01919510500250267
20. Thompson CM, Drago JA (2015) North American installed water treatment ozone systems. J Am Water Works Assoc 107(10):45–55. https://doi.org/10.5942/jawwa.2015.107.0157
21. Brillas E (2020) A review on the photoelectro-Fenton process as efficient electrochemical advanced oxidation for wastewater remediation. treatment with UV light, sunlight, and coupling with conventional and other photo-assisted advanced technologies. Chemosphere 250:126198. https://doi.org/10.1016/j.chemosphere.2020.126198
22. Mei Q et al (2019) Sulfate and hydroxyl radicals-initiated degradation reaction on phenolic contaminants in the aqueous phase: mechanisms, kinetics and toxicity assessment. Chem Eng J 373(May):668–676. https://doi.org/10.1016/j.cej.2019.05.095
23. Zhang P et al (2018) Mechanisms of hydroxyl radicals production from pyrite oxidation by hydrogen peroxide: surface versus aqueous reactions. Geochim Cosmochim Acta 238:394–410. https://doi.org/10.1016/j.gca.2018.07.018
24. Battimelli A et al (2010) Combined ozone pretreatment and biological processes for removal of colored and biorefractory compounds in wastewater from molasses fermentation industries. J Chem Technol Biotechnol 85(7):968–975. https://doi.org/10.1002/jctb.2388
25. Carini D et al (2001) Ozonation as pre-treatment step for the biological batch degradation of industrial wastewater containing 3-methyl-pyridine. Ozone: Sci Eng 23(3):189–198. https://doi.org/10.1080/01919510108962002

26. Chávez AM et al (2019) Treatment of highly polluted industrial wastewater by means of sequential aerobic biological oxidation-ozone based AOPs. Chem Eng J 361:89–98. https://doi.org/10.1016/j.cej.2018.12.064
27. Kovalova L et al (2013) Elimination of micropollutants during post-treatment of hospital wastewater with powdered activated carbon, ozone, and UV. Environ Sci Technol 47(14):7899–7908. https://doi.org/10.1021/es400708w
28. Taoufik N et al (2021) Comparative overview of advanced oxidation processes and biological approaches for the removal pharmaceuticals. J Environ Manage 288:112404. https://doi.org/10.1016/j.jenvman.2021.112404
29. Umar M et al (2013) Application of ozone for the removal of bisphenol A from water and wastewater—a review. Chemosphere 90:2197–2207. https://doi.org/10.1016/j.chemosphere.2012.09.090
30. Ikehata, K et al (2006) Degradation of aqueous pharmaceuticals by ozonation and advanced oxidation processes: a review degradation of aqueous pharmaceuticals by ozonation and advanced oxidation processes: a review. Ozone: Sci Eng 28:353–414. https://doi.org/10.1080/01919510600985937
31. Yacouba ZA et al (2021) Removal of organic micropollutants from domestic wastewater: the effect of ozone-based advanced oxidation process on nanofiltration. J Water Process Eng 39:101869. https://doi.org/10.1016/j.jwpe.2020.101869
32. Aghaeinejad-Meybodi A et al (2021) Comparative investigation on catalytic ozonation of Fluoxetine antidepressant drug in the presence of boehmite and γ-alumina nanocatalysts: operational parameters, kinetics and degradation mechanism studies. Chem Papers 75:421–430. https://doi.org/10.1007/s11696-020-01312-0
33. Mcdowell DC et al (2005) Ozonation of carbamazepine in drinking water: identification and kinetic study of major oxidation products. Environ Sci Technol 39:8014–8022. https://doi.org/10.1021/es050043l
34. Zhao Y et al (2017) Ozonation of indomethacin: kinetics, mechanisms and toxicity. J Hazard Mater 323:460–470. https://doi.org/10.1016/j.jhazmat.2016.05.023
35. Alharbi SK et al (2016) Ozonation of carbamazepine, diclofenac, sulfamethoxazole and trimethoprim and formation of major oxidation products. Desalin Water Treat 57:29340–29351. https://doi.org/10.1080/19443994.2016.1172986
36. Edefell E et al (2021) MBBRs as post-treatment to ozonation: degradation of transformation products and ozone-resistant micropollutants. Sci Total Environ 754:142103. https://doi.org/10.1016/j.scitotenv.2020.142103
37. Mojiri A et al (2019) Combined ozone oxidation process and adsorption methods for the removal of acetaminophen and amoxicillin from aqueous solution; kinetic and optimisation. Environ Technol Innov 15:100404. https://doi.org/10.1016/j.eti.2019.100404
38. Márquez G et al (2014) Integration of ozone and solar TiO_2-photocatalytic oxidation for the degradation of selected pharmaceutical compounds in water and wastewater. Sep Purif Technol 136:18–26. https://doi.org/10.1016/j.seppur.2014.08.024
39. Abellán MN, Gebhardt W, Schröder HF (2008) Detection and identification of degradation products of sulfamethoxazole by means of LC/MS and—MSn after ozone treatment. Water Sci Technol 58:1803–1812. https://doi.org/10.2166/wst.2008.539
40. Paucar NE et al (2019) Ozone treatment process for the removal of pharmaceuticals and personal care products in wastewater. Ozone: Sci Eng 41:3–16. https://doi.org/10.1080/01919512.2018.1482456
41. Guo Y, Yang L, Wang X (2012) The application and reaction mechanism of catalytic ozonation in water treatment. J Environ Anal Toxicol 02(06). https://doi.org/10.4172/2161-0525.1000150
42. Qin H et al (2014) Efficient degradation of fulvic acids in water by catalytic ozonation with CeO_2/AC. J Chem Technol Biotechnol 89(9):1402–1409. https://doi.org/10.1002/jctb.4222
43. Khan MH, Jung JY (2008) Ozonation catalyzed by homogeneous and heterogeneous catalysts for degradation of DEHP in aqueous phase. Chemosphere 72:690–696. https://doi.org/10.1016/j.chemosphere.2008.02.037

44. Zeng Z et al (2012) Ozonation of acidic phenol wastewater with O3/Fe(II) in a rotating packed bed reactor: optimization by response surface methodology. Chem Eng Process 60(Ii):1–8. https://doi.org/10.1016/j.cep.2012.06.006
45. Nawrocki J, Kasprzyk-Hordern B (2010) The efficiency and mechanisms of catalytic ozonation. Appl Catal B 99(1–2):27–42. https://doi.org/10.1016/j.apcatb.2010.06.033
46. Koricic K et al (2016) Mineralization of salicylic acid in water by catalytic ozonation. Environ Eng Manag J 15:4597
47. Clarizia L et al (2017) Homogeneous photo-Fenton processes at near neutral pH: a review. Appl Catal B 209:358–371. https://doi.org/10.1016/j.apcatb.2017.03.011
48. Wei Y, Li G, Wang B (2011) Research on harbor oily wastewater treatment by Fenton oxidation. Adv Mater Res 322:164–168. https://doi.org/10.4028/www.scientific.net/AMR.322.164
49. Remucal CK, Lee C, Sedlak DL (2011) Comment on "oxidation of sulfoxides and arsenic(III) in corrosion of nanoscale zero valent iron by oxygen: evidence against ferryl ions (Fe(IV)) as active intermediates in fenton reaction. Environ Sci Technol 45:3177–3178. https://doi.org/10.1021/es104399p
50. Tang J, Wang J (2018) Metal organic framework with coordinatively unsaturated sites as efficient fenton-like catalyst for enhanced degradation of sulfamethazine. Environ Sci Technol 52:5367–5377. https://doi.org/10.1021/acs.est.8b00092
51. Ribeiro JP, Marques CC, Nunes MI (2020) AOX removal from pulp and paper wastewater by Fenton and photo-Fenton processes: a real case-study. Energy Rep 6:770–775. https://doi.org/10.1016/j.egyr.2019.09.068
52. Lee J, Von Gunten U, Kim JH (2020) Persulfate-based advanced oxidation: critical assessment of opportunities and roadblocks. Environ Sci Technol 54:3064–3081. https://doi.org/10.1021/acs.est.9b07082
53. Deng J et al (2013) Thermally activated persulfate (TAP) oxidation of antiepileptic drug carbamazepine in water. Chem Eng J 228:765–771. https://doi.org/10.1016/j.cej.2013.05.044
54. Milh H et al (2020) Degradation of sulfamethoxazole by heat-activated persulfate oxidation: elucidation of the degradation mechanism and influence of process parameters. Chem Eng J 379:122234. https://doi.org/10.1016/j.cej.2019.122234
55. Waldemer RH et al (2007) Oxidation of chlorinated ethenes by heat-activated persulfate: kinetics and products. Environ Sci Technol 41:1010–1015. https://doi.org/10.1021/es062237m
56. Pirsaheb M, Hossaini H, Janjani H (2019) An overview on ultraviolet persulfate based advances oxidation process for removal of antibiotics from aqueous solutions: a systematic review. Desalin Water Treat 165:382–395. https://doi.org/10.5004/dwt.2019.24559
57. Tsitonaki A et al (2010) In situ chemical oxidation of contaminated soil and groundwater using persulfate: a review. Crit Rev Environ Sci Technol 40:55–91. https://doi.org/10.1080/10643380802039303
58. Xiao S et al (2020) Iron-mediated activation of persulfate and peroxymonosulfate in both homogeneous and heterogeneous ways: a review. Chem Eng J 384:123265. https://doi.org/10.1016/j.cej.2019.123265
59. Wu J et al (2020) Nanoscale zero valent iron-activated persulfate coupled with Fenton oxidation process for typical pharmaceuticals and personal care products degradation. Sep Purif Technol 239:116534. https://doi.org/10.1016/j.seppur.2020.116534
60. Nachiappan S, Gopinath KP (2015) Treatment of pharmaceutical effluent using novel heterogeneous fly ash activated persulfate system. J Environ Chem Eng 3:2229–2235. https://doi.org/10.1016/j.jece.2015.07.019
61. Yang Y et al (2015) Production of sulfate radical and hydroxyl radical by reaction of ozone with peroxymonosulfate: a novel advanced oxidation process. Environ Sci Technol 49:73307339. https://doi.org/10.1021/es506362e
62. Yang Y et al (2016) Degradation of bisphenol a using ozone/persulfate process: kinetics and mechanism. Water Air Soil Pollut 227:53. https://doi.org/10.1007/s11270-016-2746-x

63. Deniere E et al (2018) Advanced oxidation of pharmaceuticals by the ozone-activated peroxymonosulfate process: the role of different oxidative species. J Hazard Mater 360:204–213. https://doi.org/10.1016/j.jhazmat.2018.07.071
64. Khashij M, Mehralian M, Goodarzvand Chegini Z (2020) Degradation of acetaminophen (ACT) by ozone/persulfate oxidation process: experimental and degradation pathways. Pigm Resin Technol 49:363–368. https://doi.org/10.1108/PRT-11-2019-0107
65. Haddad T, Kümmerer K (2014) Characterization of photo-transformation products of the antibiotic drug Ciprofloxacin with liquid chromatography-tandem mass spectrometry in combination with accurate mass determination using an LTQ-Orbitrap. Chemosphere 115:40–46. https://doi.org/10.1016/j.chemosphere.2014.02.013
66. Tegze A et al (2019) Radiation induced degradation of ciprofloxacin and norfloxacin: kinetics and product analysis. Radiat Phys Chem 158:68–75. https://doi.org/10.1016/j.radphyschem.2019.01.025
67. Alharbi SK et al (2017) 'Photolysis and UV/H_2O_2 of diclofenac, sulfamethoxazole, carbamazepine, and trimethoprim: Identification of their major degradation products by ESI–LC–MS and assessment of the toxicity of reaction mixtures. Process Safety Environ Protect Instit Chem Eng 112:222–234. https://doi.org/10.1016/j.psep.2017.07.015
68. Abrile MG et al (2021) Degradation and mineralization of the emerging pharmaceutical pollutant sildenafil by ozone and UV radiation using response surface methodology. Environ Sci Pollut Res 28:23868–23886. https://doi.org/10.1007/s11356-020-11717-9
69. Topkaya E, Arslan A, Yatmaz HC (2021) Diclofenac degradation by ozone-based oxidation processes: PROMETHEE method kinetic and cost-effectiveness study. Ozone: Sci Eng 43:136–146. https://doi.org/10.1080/01919512.2020.1765737
70. Farzaneh H et al (2020) Ozone and ozone/hydrogen peroxide treatment to remove gemfibrozil and ibuprofen from treated sewage effluent: factors influencing bromate formation. Emerg Contam 6:225–234. https://doi.org/10.1016/j.emcon.2020.06.002
71. Liu Z et al (2019) Combining ozone with UV and H_2O_2 for the degradation of micropollutants from different origins: lab-scale analysis and optimization. Environ Technol (United Kingdom) 40:3773–3782. https://doi.org/10.1080/09593330.2018.1491630
72. Macías-Quiroga IF et al (2021) Bibliometric analysis of advanced oxidation processes (AOPs) in wastewater treatment: global and Ibero-American research trends. Environ Sci Pollut Res 28:23791–23811. https://doi.org/10.1007/s11356-020-11333-7
73. Panizza M, Cerisola G (2009) Direct and mediated anodic oxidation of organic pollutants. Chem Rev 109:6541–6569. https://doi.org/10.1021/cr9001319
74. Martínez-Huitle CA et al (2015) Single and coupled electrochemical processes and reactors for the abatement of organic water pollutants: a critical review. Chem Rev 115:13362–13407. https://doi.org/10.1021/acs.chemrev.5b00361
75. Martínez-Huitle CA, Ferro S (2006) Electrochemical oxidation of organic pollutants for the wastewater treatment: direct and indirect processes. Chem Soc Rev 35:1324–1340. https://doi.org/10.1039/b517632h
76. Dewil R et al (2017) New perspectives for advanced oxidation processes. J Environ Manage 195:93–99. https://doi.org/10.1016/j.jenvman.2017.04.010
77. Marselli B et al (2003) Electrogeneration of hydroxyl radicals on boron-doped diamond electrodes. J Electrochem Soc 150:D79. https://doi.org/10.1149/1.1553790
78. Simond O, Schaller V, Comninellis C (1997) Theoretical model for the anodic oxidation of organics on metal oxide electrodes. Electrochim Acta 42:2009–2012. https://doi.org/10.1016/S0013-4686(97)85475-8
79. Lan Y et al (2017) On the role of salts for the treatment of wastewaters containing pharmaceuticals by electrochemical oxidation using a boron doped diamond anode. Electrochim Acta 231:309–318. https://doi.org/10.1016/j.electacta.2017.01.160
80. Medeiros De Araújo D et al (2014) Electrochemical conversion/combustion of a model organic pollutant on BDD anode: Role of sp3/sp2 ratio. Electrochem Commun 47:37–40. https://doi.org/10.1016/j.elecom.2014.07.017

81. da Silva SW et al (2019) Using p-Si/BDD anode for the electrochemical oxidation of norfloxacin. J Electroanal Chem 832:112–120. https://doi.org/10.1016/j.jelechem.2018.10.049
82. Souza FL et al (2016) The effect of the sp3/sp2 carbon ratio on the electrochemical oxidation of 2,4-D with p-Si BDD anodes. Electrochim Acta 187:119–124. https://doi.org/10.1016/j.electacta.2015.11.031
83. Hosseini M et al (2020) Degradation of ciprofloxacin antibiotic using photo-electrocatalyst process of Ni-doped ZnO deposited by RF sputtering on FTO as an anode electrode from aquatic environments: synthesis, kinetics, and ecotoxicity study. Microchem J 154:104663. https://doi.org/10.1016/j.microc.2020.104663
84. Zhang G et al (2021) Efficient photoelectrocatalytic degradation of tylosin on TiO_2 nanotube arrays with tunable phosphorus dopants. J Environ Chem Eng 9:104742. https://doi.org/10.1016/j.jece.2020.104742
85. Feng L et al (2019) Evaluation of process influencing factors, degradation products, toxicity evolution and matrix-related effects during electro-Fenton removal of piroxicam from waters. J Environ Chem Eng 7:103400. https://doi.org/10.1016/j.jece.2019.103400
86. Haidar M et al (2013) Electrochemical degradation of the antibiotic sulfachloropyridazine by hydroxyl radicals generated at a BDD anode. Chemosphere 91:1304–1309. https://doi.org/10.1016/j.chemosphere.2013.02.058
87. Orimolade BO et al (2020) Coupling cathodic electro-fenton with anodic photo-electrochemical oxidation: a feasibility study on the mineralization of paracetamol. J Environ Chem Eng 8:104394. https://doi.org/10.1016/j.jece.2020.104394
88. López-Guzmán M, Flores-Hidalgo MA, Reynoso-Cuevas L (2021) Electrocoagulation process: an approach to continuous processes, reactors design, pharmaceuticals removal, and hybrid systems—a review. Processes 9:1831. https://doi.org/10.3390/pr9101831
89. Wang C et al (2016) Insights of ibuprofen electro-oxidation on metal-oxide-coated Ti anodes: kinetics, energy consumption and reaction mechanisms. Chemosphere 163:584–591. https://doi.org/10.1016/j.chemosphere.2016.08.057
90. Domínguez JR et al (2016) Parabens abatement from surface waters by electrochemical advanced oxidation with boron doped diamond anodes. Environ Sci Pollut Res 23:20315–20330. https://doi.org/10.1007/s11356-016-7175-2
91. Rabaaoui N, Allagui MS (2012) Anodic oxidation of salicylic acid on BDD electrode: variable effects and mechanisms of degradation. J Hazard Mater 243:187–192. https://doi.org/10.1016/j.jhazmat.2012.10.016
92. Tu X et al (2015) Treatment of simulated berberine wastewater by electrochemical process with Pt/Ti anode. Environ Earth Sci 73:4957–4966. https://doi.org/10.1007/s12665-015-4323-9
93. Ambuludi SL et al (2013) Kinetic behavior of anti-inflammatory drug ibuprofen in aqueous medium during its degradation by electrochemical advanced oxidation. Environ Sci Pollut Res 20:2381–2389. https://doi.org/10.1007/s11356-012-1123-6
94. Indermuhle C et al (2013) Degradation of caffeine by conductive diamond electrochemical oxidation. Chemosphere 93:1720–1725. https://doi.org/10.1016/j.chemosphere.2013.05.047
95. Sifuna FW et al (2016) Comparative studies in electrochemical degradation of sulfamethoxazole and diclofenac in water by using various electrodes and phosphate and sulfate supporting electrolytes. J Environ Sci Health Part A Toxic/Hazard Subst Environ Eng 51:954–961. https://doi.org/10.1080/10934529.2016.1191814
96. Murugananthan M et al (2010) Anodic oxidation of ketoprofen-an anti-inflammatory drug using boron doped diamond and platinum electrodes. J Hazard Mater 180:753–758. https://doi.org/10.1016/j.jhazmat.2010.05.007
97. Liu YJ, Hu CY, Lo SL (2019) Direct and indirect electrochemical oxidation of amine-containing pharmaceuticals using graphite electrodes. J Hazard Mater 366:592–605. https://doi.org/10.1016/j.jhazmat.2018.12.037
98. González T et al (2010) Conductive-diamond electrochemical advanced oxidation of naproxen in aqueous solution: optimizing the process. Foundations 86:121–127

99. Velappan K et al (2016) Anodic oxidation of isothiazolin-3-ones in aqueous medium by using boron-doped diamond electrode. Diam Relat Mater 69:152–159. https://doi.org/10.1016/j.diamond.2016.08.008
100. Domínguez JR et al (2010) Electrochemical advanced oxidation of carbamazepine on boron-doped diamond anodes. Influence of operating variables. Ind Eng Chem Res 49:8353–8359. https://doi.org/10.1021/ie101023u
101. Barışçı S et al (2018) Electrochemical treatment of anti-cancer drug carboplatin on mixed-metal oxides and boron doped diamond electrodes: density functional theory modelling and toxicity evaluation. J Hazard Mater 344:316–321. https://doi.org/10.1016/j.jhazmat.2017.10.029
102. Sun Y et al (2017) Electrochemical treatment of chloramphenicol using Ti-Sn/γ-Al_2O_3 particle electrodes with a three-dimensional reactor. Chem Eng J 308:1233–1242. https://doi.org/10.1016/j.cej.2016.10.072
103. Coledam DAC et al (2016) Electrochemical mineralization of norfloxacin using distinct boron-doped diamond anodes in a filter-press reactor, with investigations of toxicity and oxidation by-products. Electrochim Acta 213:856–864. https://doi.org/10.1016/j.electacta.2016.08.003
104. Chen TS, Chen PH, Huang KL (2014) 'Electrochemical degradation of N, N-diethyl-m-toluamide on a boron-doped diamond electrode. J Taiwan Instit Chem Eng 45:2615–2621. https://doi.org/10.1016/j.jtice.2014.06.020
105. Díaz E et al (2019) Electrochemical degradation of naproxen from water by anodic oxidation with multiwall carbon nanotubes glassy carbon electrode. Water Sci Technol 79:480–488. https://doi.org/10.2166/wst.2019.070
106. Sirés I, Brillas E (2012) Remediation of water pollution caused by pharmaceutical residues based on electrochemical separation and degradation technologies: a review. Environ Int 40:212–229. https://doi.org/10.1016/j.envint.2011.07.012
107. Montes-Grajales D, Fennix-Agudelo M, Miranda-Castro W (2017) Occurrence of personal care products as emerging chemicals of concern in water resources: a review. Sci Total Environ 595:601–614. https://doi.org/10.1016/j.scitotenv.2017.03.286
108. Rivera-Utrilla J et al (2013) Pharmaceuticals as emerging contaminants and their removal from water. a review. Chemosphere 93:1268–1287. https://doi.org/10.1016/j.chemosphere.2013.07.059
109. Feng L et al (2013) Removal of residual anti-inflammatory and analgesic pharmaceuticals from aqueous systems by electrochemical advanced oxidation processes. a review. Chem Eng J 228:944–964. https://doi.org/10.1016/j.cej.2013.05.061
110. Joshi M et al (2009) Chlorophyll-based photocatalysts and their evaluations for methyl orange photoreduction. J Photochem Photobiol, A 204:83–89. https://doi.org/10.1016/j.jphotochem.2009.01.016
111. Ghasemi M et al (2020) In-situ electro-generation and activation of hydrogen peroxide using a CuFeNLDH-CNTs modified graphite cathode for degradation of cefazolin. J Environ Manage 267:110629. https://doi.org/10.1016/j.jenvman.2020.110629
112. Tian L et al (2022) Mineralization of cyanides via a novel Electro-Fenton system generating ·OH and ·O2−. Water Res 209:117890
113. Zhang D et al (2020) Selective H_2O_2 production on N-doped porous carbon from direct carbonization of metal organic frameworks for electro-Fenton mineralization of antibiotics. Chem Eng J 383:123184. https://doi.org/10.1016/j.cej.2019.123184
114. Kamali M, Khodaparast Z (2015) Review on recent developments on pulp and paper mill wastewater treatment. Ecotoxicol Environ Saf 114:326–342. https://doi.org/10.1016/j.ecoenv.2014.05.005
115. De Bel E et al (2009) Influence of pH on the sonolysis of ciprofloxacin: biodegradability, ecotoxicity and antibiotic activity of its degradation products. Chemosphere 77:291–295. https://doi.org/10.1016/j.chemosphere.2009.07.033
116. Locke BR et al (2006) Electrohydraulic discharge and nonthermal plasma for water treatment. Ind Eng Chem Res 45:882–905. https://doi.org/10.1021/ie050981u

117. Cui Y et al (2018) The types of plasma reactors in wastewater treatment. IOP Conf Series: Earth Environ Sci 208:012002. https://doi.org/10.1088/1755-1315/208/1/012002
118. Joshi RP, Thagard SM (2013) Streamer-like electrical discharges in water: Part II. Environmental applications. Plasma Chem Plasma Process 33:17–49. https://doi.org/10.1007/s11090-013-9436-x
119. Jiang B et al (2014) Review on electrical discharge plasma technology for wastewater remediation. Chem Eng J 236:348–368. https://doi.org/10.1016/j.cej.2013.09.090
120. Malik MA, Ghaffar A, Malik SA (2001) Water purification by electrical discharges. Plasma Sour Sci Technol 10:82–91. https://doi.org/10.1088/0963-0252/10/1/311
121. Reddy PMK, Subrahmanyam C (2012) Green approach for wastewater treatment-degradation and mineralization of aqueous organic pollutants by discharge plasma. Ind Eng Chem Res 51:11097–11103. https://doi.org/10.1021/ie301122p
122. Wang Y et al (2014) Addition of hydrogen peroxide for the simultaneous control of bromate and odor during advanced drinking water treatment using ozone. J Environ Sci (China) 26:550–554. https://doi.org/10.1016/S1001-0742(13)60409-X
123. Ajo P, Kornev I, Preis S (2015) Pulsed corona discharge in water treatment: the effect of hydrodynamic conditions on oxidation energy efficiency. Ind Eng Chem Res 54:7452–7458. https://doi.org/10.1021/acs.iecr.5b01915
124. Wang M et al (2017) 'In situ degradation of antibiotic residues in medical intravenous infusion bottles using high energy electron beam irradiation. Sci Rep 7:1–8. https://doi.org/10.1038/srep39928
125. Xu X (2001) Dielectric barrier discharge—properties and applications. Thin Solid Films 390:237–242. https://doi.org/10.1016/S0040-6090(01)00956-7
126. Lukes P, Locke BR, Brisset JL (2012) Aqueous-phase chemistry of electrical discharge plasma in water and in gas-liquid environments. Plasma Chem Catal Gases Liquids (Chapter 6), pp 243–308. https://doi.org/10.1002/9783527649525.ch7
127. Neta P (1972) Reactions of hydrogen atoms in aqueous solutions. Chem Rev 72:533–543. https://doi.org/10.1021/cr60279a005
128. Anpilov AM et al (2001) Electric discharge in water as a source of UV radiation, ozone and hydrogen peroxide. J Phys D Appl Phys 34:993–999. https://doi.org/10.1088/0022-3727/34/6/322
129. Bruggeman P et al (2008) DC electrical breakdown in a metal pin-water electrode system. IEEE Trans Plasma Sci 36:1138–1139. https://doi.org/10.1109/TPS.2008.917294
130. Šunka P et al (2004) Potential applications of pulse electrical discharges in water. Acta Physica Slovaca 54:135–145
131. Bruggeman P et al (2007) Plasma characteristics in air and vapor bubbles in water. In: Proceedings International Conference on Phenomena in Ionized Gases (Prague, Czech Republic), pp 899–962. Available at: http://icpig2007.ipp.cas.cz/files/download/cd-cko/ICPIG2007/pdf/3P10-10.pdf
132. Kogelschatz U, Eliasson B, Egli W (1999) From ozone generators to flat television screens: history and future potential of dielectric-barrier discharges. Pure Appl Chem 71:1819–1828. https://doi.org/10.1351/pac199971101819
133. Stracqualursi E, Araneo R, Celozzi S (2021) The corona phenomenon in overhead lines: critical overview of most common and reliable available models. Energies 14:6612. https://doi.org/10.3390/en14206612
134. Parvulescu VI et al (2022) recent progress and prospects in catalytic water treatment. Chem Rev 122:2981–3121. https://doi.org/10.1021/acs.chemrev.1c00527
135. Wang X, Zhou M, Jin X (2012) Application of glow discharge plasma for wastewater treatment. Electrochim Acta 83:501–512. https://doi.org/10.1016/j.electacta.2012.06.131
136. Fridman A et al (1998) Gliding arc gas discharge. Prog Energy Combust Sci 25:211–231
137. Magureanu M et al (2021) A review on non-thermal plasma treatment of water contaminated with antibiotics. J Hazard Mater 417:125481. https://doi.org/10.1016/j.jhazmat.2021.125481
138. Singh RK, Philip L, Ramanujam S (2017) Rapid degradation, mineralization and detoxification of pharmaceutically active compounds in aqueous solution during pulsed corona discharge treatment. Water Res 121:20–36. https://doi.org/10.1016/j.watres.2017.05.006

139. Guo H et al (2019) Pulsed discharge plasma assisted with graphene-WO_3 nanocomposites for synergistic degradation of antibiotic enrofloxacin in water. Chem Eng J 372:226–240. https://doi.org/10.1016/j.cej.2019.04.119
140. He D et al (2015) Decomposition of tetracycline in aqueoussolution by corona discharge plasma combined with a Bi_2MoO_6 nanocatalyst. J Chem Technol Biotechnol 015:2249–2256
141. Aggelopoulos CA et al (2020) Degradation of antibiotic enrofloxacin in water by gas-liquid nsp-DBD plasma: parametric analysis, effect of H_2O_2 and CaO_2 additives and exploration of degradation mechanisms. Chem Eng J 398:125622. https://doi.org/10.1016/j.cej.2020.125622
142. Xu Z et al (2020) Degradation effect and mechanism of gas-liquid phase dielectric barrier discharge on norfloxacin combined with H_2O_2 or Fe^{2+}. Sep Purif Technol 230:115862. https://doi.org/10.1016/j.seppur.2019.115862
143. Kamali M et al (2021) Nanostructured materials via green sonochemical routes e sustainability aspects. Chemosphere 276:130146. https://doi.org/10.1016/j.chemosphere.2021.130146
144. Miljevic B et al (2014) To sonicate or not to sonicate PM filters: reactive oxygen species generation upon ultrasonic irradiation. Aerosol Sci Technol 48:1276–1284. https://doi.org/10.1080/02786826.2014.981330
145. Capocelli M et al (2012) Sonochemical degradation of estradiols: Incidence of ultrasonic frequency. Chem Eng J 210:9–17. https://doi.org/10.1016/j.cej.2012.08.084
146. Huang T et al (2017) Effects and mechanism of diclofenac degradation in aqueous solution by US/Zn0. Ultrason Sonochem 37:676–685. https://doi.org/10.1016/j.ultsonch.2017.02.032
147. Lonappan L et al (2016) Diclofenac and its transformation products: environmental occurrence and toxicity—a review. Environ Int 96:127–138. https://doi.org/10.1016/j.envint.2016.09.014
148. Reggiane de Carvalho Costa L, Guerra Pacheco Nunes K, Amaral Féris L (2021) Ultrasound as an advanced oxidative process: a review on treating pharmaceutical compounds. Chem Eng Technol 44:1744–1758. https://doi.org/10.1002/ceat.202100090
149. Yargeau V, Danylo F (2015) Removal and transformation products of ibuprofen obtained during ozone- and ultrasound-based oxidative treatment. Water Sci Technol 72:491–500. https://doi.org/10.2166/wst.2015.234
150. Issaka E et al (2022) Advanced catalytic ozonation for degradation of pharmaceutical pollutants—a review. Chemosphere 289:133208. https://doi.org/10.1016/j.chemosphere.2021.133208
151. Gomes J et al (2017) Application of ozonation for pharmaceuticals and personal care products removal from water. Sci Total Environ 586:265–283. https://doi.org/10.1016/j.scitotenv.2017.01.216
152. Sousa CP et al (2019) Electroanalysis of pharmaceuticals on boron-doped diamond electrodes: a review. ChemElectroChem 6:2350–2378. https://doi.org/10.1002/celc.201801742

Chapter 11
Heterogeneous Advanced Oxidation Processes (HE-AOPs) for the Removal of Pharmaceutically Active Compounds—Pros and Cons

11.1 Introduction

Advanced oxidation processes (AOPs) have been widely studied and implemented in recent years for the removal of various organic pollutants present in the content of (waste)waters [1, 2]. The basis of these methods is the generation of highly reactive species (e.g., hydroxyl radicals, singlet oxygen, H^+, and e^-) in the medium with the ability to oxidize pollutants.

AOPs can be divided into heterogeneous and homogenous techniques, as illustrated in Fig. 11.1. Heterogeneous AOPs (HE-AOPs) mainly include the application of catalytic materials for the generation of active species in the medium. Catalytic ozonation [3], photocatalytic systems [4], photoelectrochemical processes [5], and activation of oxidation agents (e.g., persulfate (PS), peroxymonosulfate (PMS), iodine, and chlorine [6]) are the most widely studied heterogeneous AOPs. It is also evident from Fig. 11.1 that H-AOPs can be classified based on the use of an external source of energy (e.g., light, heat, ultrasound, or microwave).

The efficiency of these methods is directly dependent on the applied method and the active species generated in the medium, the type and molecular structure of the PhACs, and the operating conditions involved in the removal of PhACs [7–10]. Such parameters can also determine the degree of mineralization (removal of total organic carbon, TOC), the degradation products, and the toxic effects that can be expected from the treated effluents using such techniques. This chapter has aimed to discuss the fundamentals and the mechanisms involved in the implementation of HE-AOPs for the elimination of PhACs and to explore the latest findings in the literature and the recommendations for future studies on the efficient and cost-effective elimination of PhACs.

M. Kamali et al., *Advanced Wastewater Treatment Technologies for the Removal of Pharmaceutically Active Compounds*, Green Energy and Technology,
https://doi.org/10.1007/978-3-031-20806-5_11

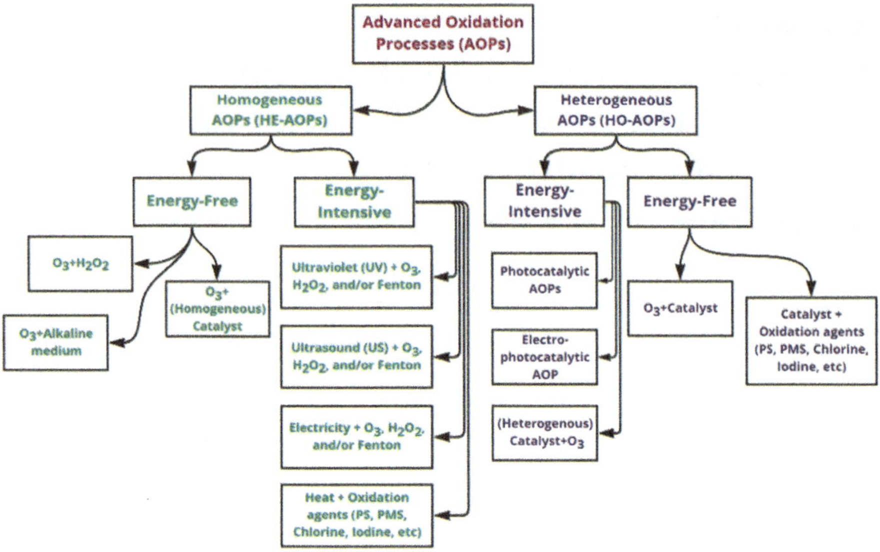

Fig. 11.1 Most widely studied and implemented AOPs for the removal of organic pollutants from (waste)waters

11.2 Energy-Free HE-AOPs

As indicated in Fig. 11.1, energy-free HE-AOPs include methods such as catalytic ozonation and activation of various oxidation agents in the medium for the generation of oxidative species, as discussed in Chap. 2, for the degradation of PhACs.

11.2.1 Catalytic Ozonation

Both homogeneous and heterogeneous catalytic processes have been recently studied for the degradation of pharmaceutical micropollutants. Transition metal ions such as Cu^{2+}, Mn^{2+}, Fe^{3+}, Zn^{2+} are normally employed in homogenous ozonation processes for the generation of oxidative species in the treatment system [11, 12]. Ferrous or manganese catalytic ozonation processes are among the most widely used homogenous catalytic ozonation processes for the removal of PhACs [13].

Heterogeneous catalysis has been more studied in the literature for the degradation of PhACs compared to homogeneous catalysis processes. In such methods, solid catalysts are used to enhance the performance of the ozonations system [8]. In this regard, a number of heterogeneous catalysts, such as magnesium oxide, aluminum oxide, iron-based materials, and polymetallic/bimetallic oxides, have been studied and examined (see, e.g., [14]). Stabilization of solid catalysts on low-cost and environmentally friendly materials such as waste-derived carbonaceous compounds (e.g.,

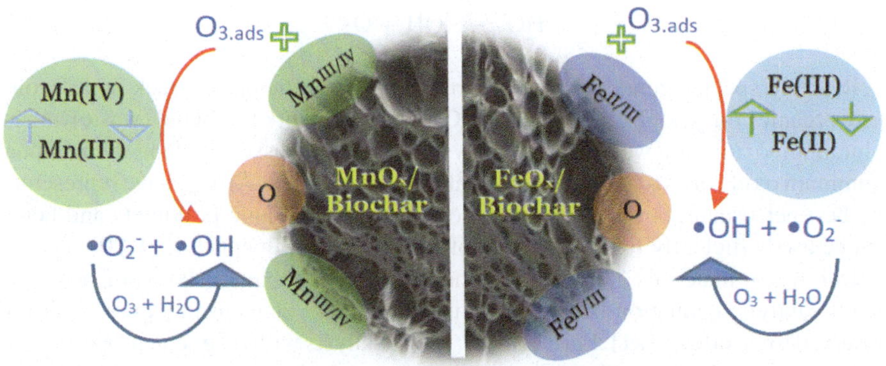

Fig. 11.2 Biochar-supported MnO_x or FeO_x are efficient heterogeneous catalysts to enhance the ozonation degradation of PhACs such as atrazine. Only 48% degradation of this compound was achieved with 2.5 mh/L O_3 (at pH 7 in 30 min). However, an increase in atrazine removal to 83% and 100% was observed when Mn-loaded biochar and Fe-loaded biochar, respectively, were used as catalysts under identical treatment conditions, as reported by Tian et al. [17]

biochar) has also been indicated as an interesting method to enhance the available specific surface area of the catalysts by preventing their rapid agglomeration when introduced into the water medium. As an example, Fe-loaded or Mn-loaded biochars have excellent activities for the degradation of PhACs in heterogeneous catalytic ozonation systems (Fig. 11.2). In such composite materials, the carbon-based compartment can also efficiently contribute to the adsorption of PhACs and facilitate their decomposition mediated by the catalytic compartment [15]. Here, the pH of the solution can play an important role by affecting the surface charge of the catalyst, which can determine the extent of the PhACs adsorbed on the surface of the materials for further decomposition reactions. Other carbonaceous materials, such as graphene oxide (GO) and reduced graphene oxide (rGO), have also been used as appropriate materials to host metallic compounds for the efficient catalytic ozonation of pollutants [16].

In heterogeneous catalytic ozonation, metal oxides (M) catalyze the ozonation process through the reactions formulated in Eqs. 11.1–11.5 [18].

$$O_3 + M - OH_2^+ \rightarrow M - OH^{\cdot +} + HO_3 \quad (11.1)$$

$$2O_3 + M - OH \rightarrow M - O2^{\cdot -} + HO_3^{\cdot} + O_2 \quad (11.2)$$

$$M - OH^{+\cdot} + H_2O \rightarrow M - OH_2^+ + {}^{\cdot}OH \quad (11.3)$$

$$M - O_2^{-\cdot} + O_3 + H_2O \rightarrow M - OH + O_2 + HO_3^{\cdot} \quad (11.4)$$

$$HO_3^{\cdot} \rightarrow {}^{\cdot}OH + O_2 \quad (11.5)$$

Reactive species generated in these reactions can efficiently attack and decompose organic compounds, including PhACs, in the content of polluted (waste)waters. Aluminum oxides are well-known catalysts in such processes. Among them, alpha-aluminum oxide (α-Al_2O_3) and gamma-aluminum oxide (γ-Al_2O_3) have represented the best performances for the degradation of various organic pollutants and hence can be used efficiently for the removal of PhACs from effluents [3, 19, 20].

Iron-based materials can also be considered sustainable catalysts for heterogeneous catalytic ozonation because iron is an abundant element in nature, with low observed toxic effects [21]. Furthermore, such catalysts with magnetic properties can be recovered and reused after the treatment process using a magnetic field. This can be beneficial in terms of economic and environmental considerations. Among iron-based catalysts, FeOOH has the highest efficiency in catalytic ozonation processes compared to other iron oxides, such as Fe_3O_4 and Fe_2O_3, due to having more catalytic constituents [3].

Manganese oxides, especially MnO_2, are other catalytic materials that have been widely employed for the catalytic ozonation of organic pollutants [22, 23]. Composite solid catalysts have also presented excellent efficiencies for the catalytic ozonation of organic compounds. For instance, it has been indicated that the synergistic effects of surface $\equiv Cu^{2+}$ and $\equiv Al^{3+}$ active sites in $CuAl_2O_4$ can result in its superior catalytic activities for the generation of powerful oxidation radicals in combination with ozone (Fig. 11.3). In the involved mechanisms, $\equiv Al^{3+}$ hosts surface active sites (e.g., hydroxyl groups) and Lewis acid sites. Additionally, $\equiv Cu^{2+}$ facilitates the electron transfer process to promote the generation of oxidative species.

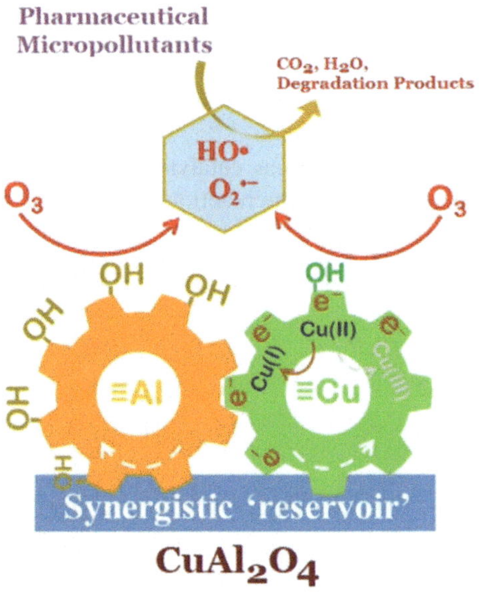

Fig. 11.3 Typical mechanisms involved in catalytic ozonation using $CuAl_2O_4$ for the degradation of PhACs, adapted from Xu et al. [16]

11.2.2 Activation of Oxidation Agents

Activation of hydrogen peroxide using iron-based materials is a well-known process, called Fenton, for the generation of hydroxyl radicals with a high oxidation potential (2.8 eV). This process has been widely used in recent decades for the removal of various refractory organic compounds [24, 25]. Ferrous ions have been conventionally used as hydrogen peroxide activators (Eq. 11.6). However, the release of iron ions after the treatment process, as well as the dependence of the respective reactions on the working pH (optimum pH is 2–4), are the most important limiting parameters for homogenous Fenton reactions [26].

$$H_2O_2 + Fe^{2+} \rightarrow \cdot OH + Fe^{3+} + OH^- \qquad (11.6)$$

The development of heterogeneous catalytic Fenton-like processes has been an effort to address such issues. Iron oxides and zero-valent iron (ZVI) have been frequently used in such processes to conduct degradation processes [27, 28]. Studies are also available for the stabilization of iron-based materials on appropriate substrates, such as carbonaceous (e.g., activated carbon and biochar) and clay-based materials (e.g., bentonite or zeolite), to limit the rapid agglomeration of such heterogeneous catalysts when introduced to water media [29–31].

Two main mechanisms can be involved in the removal of pollutants when heterogeneous catalytic Fenton reactions are employed. Organic (or inorganic) molecules can be adsorbed on the surface of the catalyst under mechanisms such as ion exchange, hydrogen bonds, π-π interactions, hydrophobic interactions, and electrostatic interactions [32]. Fe compartments on the surface of the catalyst (or those leached into the medium) can induce Fenton reactions in the presence of hydrogen peroxide to generate hydroxyl radicals [33, 34].

Although there are several studies on the application of iron oxides (e.g., α-Fe_2O_3 [35] and Fe_3O_4 [36]), the loading of iron-based materials on low-cost materials, such as carbonaceous structures or clay-based materials, has attracted attention due to inherent advantages, such as preventing rapid agglomeration in water media, which can ensure the maintenance of the available surface area of the materials for oxidative reactions. Such a strategy has been adopted in recent years to deal with various PhACs. For instance, diatomite-supported iron was pelletized for the degradation of a mixture of PhACs (i.e., carbamazepine, sulfamethazine, ketoprofen, clindamycin, and gemfibrozil) and was reported by Ulloa-Ovares et al. [37] in fixed-bed and fluidized-bed reactors. The system represented the optimum performance for the removal of gemfibrozil (up to 100%) and clindamycin (up to 90%). However, the efficiency of the system for carbamazepine, sulfamethazine, and florfenicol was reported to be below 20%.

There are innovative studies developed to enhance the applicability of heterogeneous catalytic Fenton reactions. Recovery and reuse of the catalysts spent in such a process are of high importance to minimize the operating costs and to prevent the release of the materials together with the treated effluents. Coupling this type of HE-AOP with membrane separation has been recommended as a way to satisfy such a need. For instance, Lan et al. [38] indicated that immersing a hollow-fiber membrane in a heterogeneous catalytic Fenton using an Fe-loaded zeolite catalyst can retain the materials inside the system for the efficient treatment of ibuprofen-containing effluents. However, the stability of the membrane in such a system is a critical point to be considered due to the generation of powerful oxidative species in the system that can potentially damage the structure of the membrane [39].

Activation of oxidation agents, such as persulfate (PS), peroxymonosulfate (PMS), chlorine and iodine, using heterogeneous catalysts has also been considered recently as an efficient method for the degradation of PhACs [40]. Active species generated either in the medium or on the active sites of the catalysts can be involved in the degradation of organic pollutants. This section introduces the recently explored oxidation agents used for such purposes and the mechanisms involved in the decomposition of PhACs using the generated oxidative species.

Sulfate radicals with a high redox potential (2.5–3.1 V) and a relatively long half-life (30–40 μs) have also been considered a suitable oxidant to deal with various organic pollutants. Such radicals can also represent high efficiencies over a wide range of pH values from 2 to 8 [41]. Hence, activation of persulfate can be considered a viable treatment for effluents from various origins.

Persulfate can be self-activated through its direct reaction with organic molecules to generate sulfate radicals (Eq. 11.7). The creation of organic radicals can also be expected by the reaction of PS with organic molecules, which can then act as oxidation agents in the medium (Eq. 11.8) [42–44].

$$S_2O_8^{2-} + R \rightarrow SO_4^{2-} + SO_4^{-\cdot} + R^* \tag{11.7}$$

$$S_2O_8^{2-} + R \rightarrow 2SO_4^{-\cdot} + R^{\cdot} \tag{11.8}$$

However, self-activation of PS is a slow-kinetic reaction, and hence, there is a need for activators to accelerate the generation of active species in the medium.

Heterogeneous catalysis has also been employed in recent years for the efficient activation of PS. As stated by Du et al. [45], various mechanisms are involved in the activation of PS using copper oxide nanomaterials (Eq. 11.9 to Eq. 11.15) in which activated persulfate (Eq. 10.15) has the dominant role in the decomposition of PhACs (Eq. 10.16).

$$\equiv Cu^{II}_{sites} + S_2O_8^{2-} \rightarrow \equiv Cu^{II}{}_{sites} \ldots S_2O_8^{2-} \tag{11.9}$$

$$\begin{aligned} &\equiv Cu^{II}_{sites} \ldots S_2O_8^{2-} + PhACs \rightarrow \equiv Cu^{II}_{sites} \\ &+ S_2O_8^{2-} + \text{Degradation products} \end{aligned} \tag{11.10}$$

$$\equiv Cu^{II}_{sites} + S_2O_8^{2-}(\text{trace}) \rightarrow \equiv Cu^{III}_{sites} + SO_4^{2-} + SO_4^{\cdot -} \tag{11.11}$$

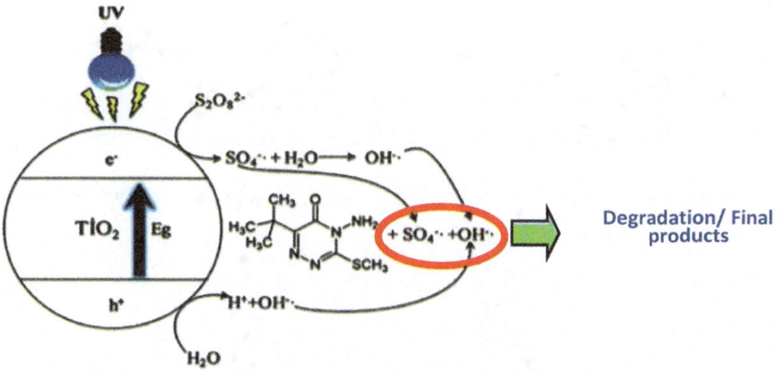

Fig. 11.4 Mechanism involved in the activation of PS for the degradation of metribuzin, adapted from Sabri et al. [46]. Photogenerated electrons and holes contribute to the formation of active species, including hydroxyl radicals, H^+, and sulfate radicals

$$\equiv Cu^{III}_{sites} + H_2O \rightarrow \equiv Cu^{II}_{sites} + H^+ + HO^{\cdot} \quad (11.12)$$

$$Cu^{2+} + S_2O_8^{2-} \rightarrow Cu^{3+} + SO_4^{2-} + SO_4^{\cdot-} \quad (11.13)$$

$$Cu^{3+} + H_2O \rightarrow Cu^{2+} + H^+ + HO^{\cdot} \quad (11.14)$$

$$SO_4^{\cdot-} + H_2O \rightarrow SO_4^{2-} + H^+ + HO^{\cdot} \quad (11.15)$$

Metal oxide photocatalysts, such as titania, have also presented acceptable performance for the activation of PS through a chain of photocatalytic reactions that result in the formation of various oxidative species, such as hydroxyl and sulfate radicals (Fig. 11.4).

However, there are currently limitations to the wider application of such efficient oxidation systems, including the relatively high fabrication costs and the probable environmental and health concerns related to the release of the spent nanomaterials into the environment [47]. A hotspot in this field is hence to explore sustainable and low-cost photocatalysts (e.g., carbonaceous visible-light active materials such as carbon nitride [48, 49]) to replace the conventional semiconductor materials (e.g., TiO_2 and ZnO) used for the treatment of various organic compounds from polluted (waste)waters. There has also been a trend for the application of waste-derived carbonaceous materials such as biochar, which can considerably reduce the overall treatment costs. It has been indicated that biochar components such as defects of graphitic structures and π electrons play important roles in the activation of PS (Fig. 11.5). These properties of biochar can be well controlled, for instance, through a thermal treatment process. Complete degradation of acetaminophen (50 mg/L)

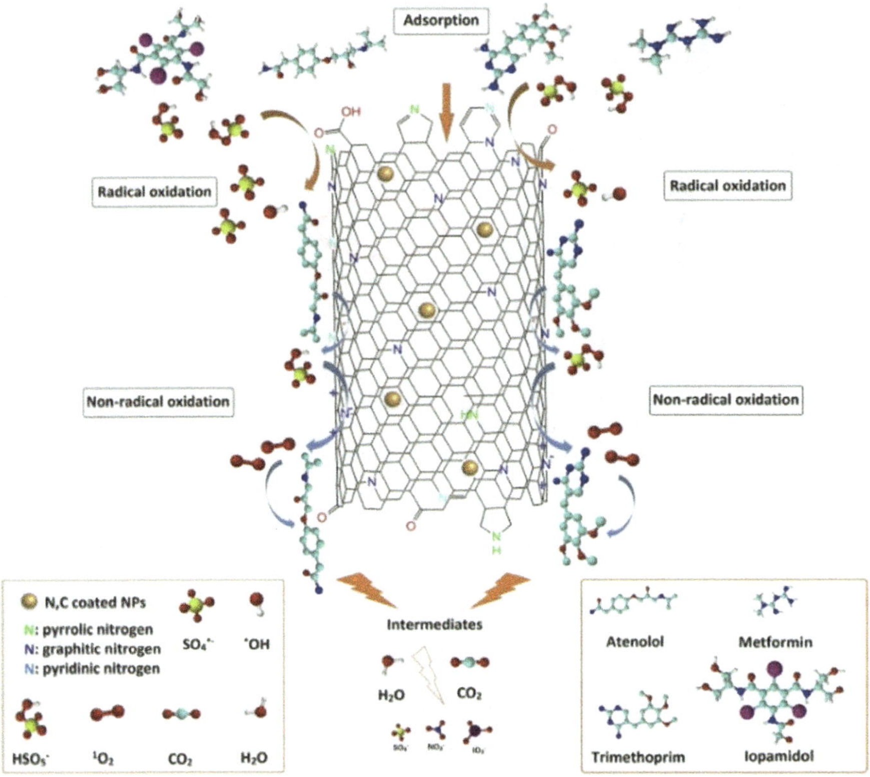

Fig. 11.5 Typical mechanisms involved in the activation of PS using carbonaceous materials for the decomposition of PhACs. Both nonradical (activated persulfate) and radical pathways play roles in this oxidation system, resulting in the transformation of the mother pollutants to the final products (CO_2, H_2O) or the intermediate products, reprinted with permission from Minh et al. [51]

over a thermally modified biochar/PS system [50] is an example of the successful application of such materials.

From a mechanistic point of view, N species such as pyridinic N, graphitic N, and pyrrolic N are considered the main catalytic sites for persulfate activation when carbonaceous materials[1] are used for such applications (Fig. 11.6) [52–54].

There have been other innovative treatment processes for the activation of PS using other oxidation agents. For instance, the combination of PS with ozone has been reported to be an efficient way to simultaneously generate various types of oxidative species (including sulfate and hydroxyl radicals) (Fig. 11.7) to promote the degradation of pharmaceutically active compounds.

From the recent literature, it can be concluded that the activation of persulfate using metal-based catalysts, especially Cu-based nanomaterials, is an efficient way to degrade organic compounds, including PhACs [57, 58]. Recent studies

[1] Such as N-doped or N-rich carbonaceous materials.

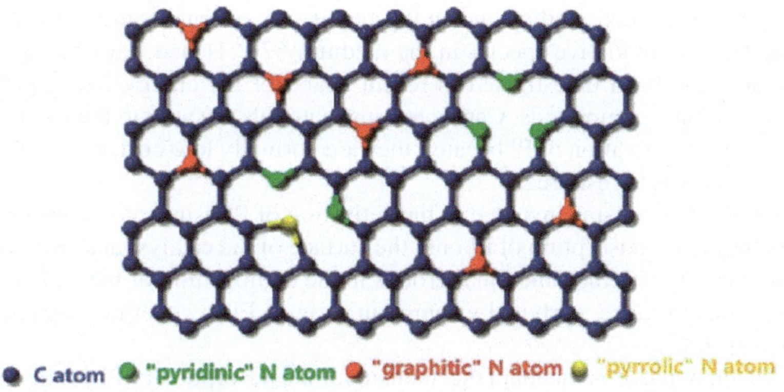

Fig. 11.6 Pyridinic N, graphitic N, and pyrrolic N sites in carbonaceous materials for the activation of persulfate, adapted from Tang et al. [55]

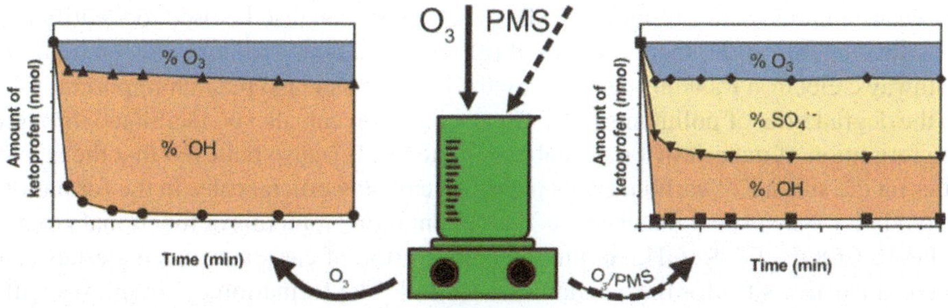

Fig. 11.7 A practical approach for the simultaneous generation of sulfate and hydroxyl radicals for the decomposition of PhACs, including atrazine, metronidazole, ketoprofen, and venlafaxine, adapted from Deniere et al. [56]

have confirmed that such methods can result in a considerable decrease in the toxicity of treated effluents. For instance, it has been recently indicated that the $Cu0.84Bi_{2.08}O_4$/persulfate system under visible-light irradiation is not only able to remove over 90% of ciprofloxacin but also results in a set of degradation products that represent a lower degree of toxicity to *Staphylococcus aureus* [59]. However, there is a need for more studies on the toxicity of real effluents treated by such a system to accelerate the transfer of the technology for real wastewater treatment applications.

Periodate (PI, $IO4^-$) is another oxidation agent with a reduction potential of + 1.60 V, which has been recently used for the generation of various oxidation agents, such as $IO_3^{\cdot}$, $IO_4^{\cdot}$, $^{\cdot}OH$, and $^1O_2$, to decompose organic compounds, including PhACs [60].

PI is a stable compound; hence, a very low degree of self-activation is normally expected when introduced into aqueous media [61, 62]. Hence, there is a need to adopt efficient methods to activate this oxidation agent. Various techniques have been examined thus far to this end, such as ultrasound, microwave, UV irradiation, and freezing [63–66]. Such methods generally require an external source of energy for the generation of oxidative species in the medium[67].[2] Hence, catalytic activation of periodate has been investigated in recent years for the energy-free degradation of various organic compounds. Carbonaceous materials are among the most studied catalysts for the activation of PI because they are normally low-cost, especially when prepared from waste resources.

There are mechanisms involved in the activation of PI using carbonaceous materials, including (a) adsorption of PI onto the surface of the catalyst and (b) activation of PI in reaction with the functional groups in the composition of the carbonaceous materials. For instance, carbonyl groups can activate PI to generate active species, mainly $^1O_2$ [68].

Hydrogen bonding is the main mechanism for the adsorption of PI by the catalyst, while the pollutant can be adsorbed onto the surface of carbonaceous materials under mechanisms such as π-π conjugation, hydrophobic forces, or electrostatic interactions.[3] Quenching experiments, as discussed earlier in Chap. 2, can be performed to identify the share of the mechanisms involved in the decomposition of the pollutants. Electron spin resonance trapping (ESR) experiments can also be used to identify the oxidative agents in the medium. In addition to the radical and nonradical oxidation pathways, electron transfer mediated by the catalyst can also play an important role in the degradation of pollutants [54]. This mechanism can also be facilitated through the formation of material-PI metastable complexes. It is also believed that the active sites on the surface of carbonaceous materials can play crucial roles in the formation of oxidative species. For instance, Fe- and S-containing minerals or functional groups (–Fe–R–COOH, Fe–R–OH, etc.) in the composition of carbonaceous materials can act as the active sites for the activation of PI [69]. The formation of complexes with PI (as an initial step of the activation process) can be considered the main role of such elements [70]. Doping can also be considered an efficient way to optimize the PI activation process. For instance, doping granular activated carbon with I_3^- and I_5^- can introduce a positive charge density on the surface of the material, which promotes the adsorption of PI [71].

It has also been indicated that carbonaceous materials containing N species can actively be involved in the adsorption and activation of PI. Similar to what was mentioned for sulfate radicals, graphitic-N and pyrrolic-N with a positive surface charge can easily adsorb PI and react with it to form oxidative radicals [72]. This fact is schematically demonstrated in Fig. 11.8. Defects in the structure of carbonaceous materials can also enhance their potential for the activation of oxidation agents. π-electrons can be donated from such positions to the oxidation agent for the generation of oxidative species [73].

There are also reports on the role of other types of oxidative radicals in the degradation of PhACs. For instance, HCO_3^- and CO_3^{2-} can react with hydroxyl, chlorine, or sulfate radicals to generate secondary carbonate radicals ($CO_3^{\cdot-}$), according to

---

[2] Homogenous AOPs have been discussed in Chap. 10.

[3] See Chap. 9 for various mechanisms involved in the adsorption of the pollutants by carbonaceous materials.

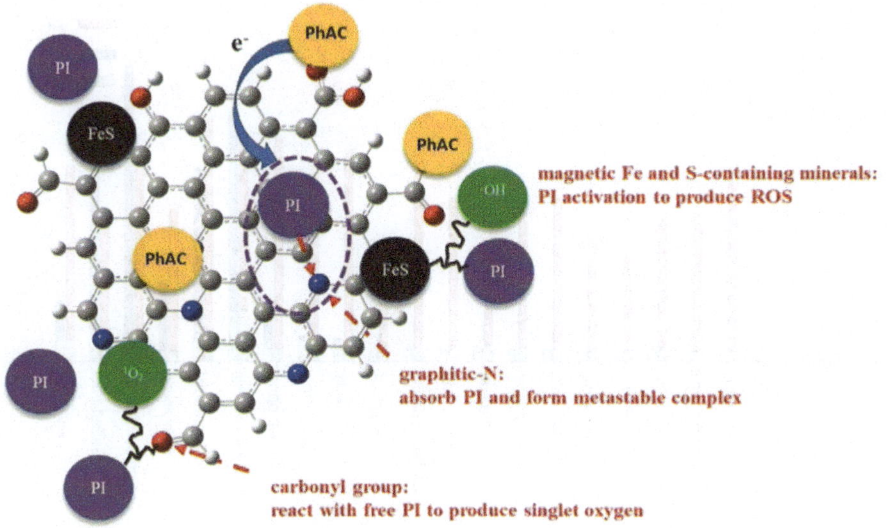

Fig. 11.8 Mechanisms involved in the activation of PI using carbonaceous materials containing N species, adapted from Xiao et al. [72]

Eqs. 11.16 and 11.17 [74, 75].

$$HCO_3^-/CO_3^{2-} + HO^{\cdot} \rightarrow CO_3^{\cdot -} + H_2O/OH^- \quad (11.16)$$

$$HCO_3^-/CO_3^{2-} + Cl^{\cdot} \rightarrow CO_3^{\cdot -} + HCl/Cl^- \quad (11.17)$$

However, this type of radical can selectively inhibit or promote the degradation of different types of PhACs. For instance, Zhou et al. [76] indicated that the combined UV/H_2O_2/HCO_3 process is less efficient for PhACs such as caffeine, atrazine, and atenolol, while the degradation of sulfamethazine and sulfamethoxazole was improved in the presence of carbonate ions (Fig. 11.9). Here, it can be assumed that carbonate radicals selectively react with PhACs that contain phenolic hydroxyl groups, aniline groups, or naphthalene rings.

Other routes have also been reported for the generation of carbonate radicals. For instance, Guo et al. [77] discussed that the reaction that occurs between HCO_3^- and the excited triplet-state propranolol can lead to the generation of carbonate radicals ($CO_3^{\cdot -}$), resulting in the decomposition of this pharmaceutical. However, they indicated that the toxicity of the water after this process increases to *Vibrio fischeri* due to the formation of more toxic degradation products.

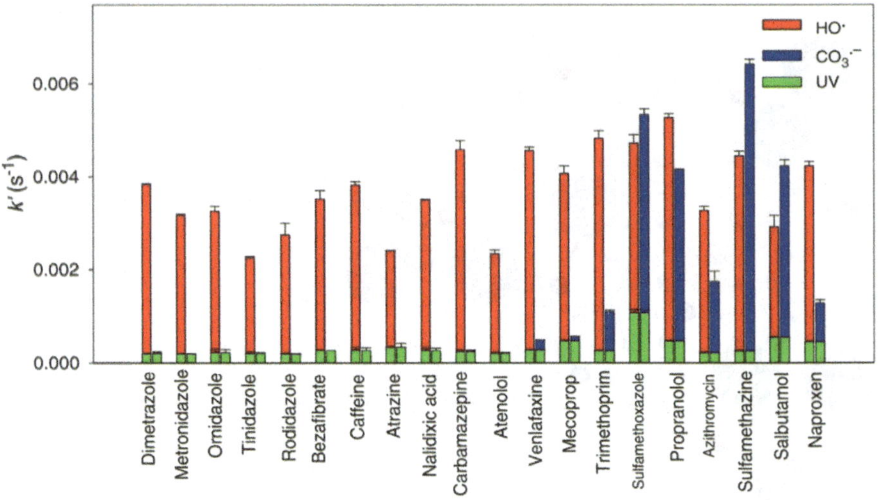

Fig. 11.9 Kinetics of the degradation of various PhACs using different oxidation systems, including photolysis (UV alone), UV + H_2O_2 ($^{\cdot}OH$ radicals), and UV/H_2O_2/HCO_3 ($CO_3^{\cdot-}$), adapted from Zhou et al. [76]

11.3 Energy-Intensive HE-AOPs

11.3.1 Photocatalysis

In photocatalytic processes, the electrons from the valence band (VB) of photocatalysts are excited to their conduction band (CB) by an external source of light (UV or visible), leading to the generation of charge carriers for the decomposition of organic compounds, including PhACs (Fig. 11.10) [78–80]. Various steps are involved in the photocatalytic degradation of pollutants. Initially, the pollutants are adsorbed by the catalyst (mass transfer), resulting in the filling of the active sites on the surface (or within the pores) of the catalyst. Then, the reactions will occur on the surface of the photocatalysts, mediated by the generated active species. Mass transfer of the PhAC decomposition products into the liquid phase is the final step of such processes [81, 82]. Normally, adsorption of the PhACs on the catalyst surface is a fast kinetic process, while the breakdown of the pollutants is considered a rate-limiting step. Hence, two main parameters can define the efficiency of the photocatalytic processes for the removal of PhACs, including the specific surface area of the materials and the extent of the active species generated on the surface of the photocatalysts. According to Eqs. 11.18–11.25, various active species can be generated in the medium under photocatalytic processes that can directly or indirectly contribute to the decomposition of organic compounds.

$$\text{Photocatalyst} \rightarrow \text{Photocatalyst}\left(e^{-(CB)} + h^{+(VB)}\right) \quad (11.18)$$

$$\text{Photocatalyst}\left(e^{-(CB)}\right) + O_2 \rightarrow \cdot O_2^- + \text{photocatalyst (Regeneration)} \quad (11.19)$$

$$\cdot O_2^- + H^+ \rightarrow HO_2\cdot \quad (11.20)$$

$$HO_2 \cdot + HO_2\cdot \rightarrow H_2O_2 + O_2 \quad (11.21)$$

$$\text{Photocatalyst}\left(e^{-(CB)}\right) + H_2O_2 \rightarrow OH\cdot + OH^- \quad (11.22)$$

$$\text{Photocatalyst}\left(h^{+(VB)}\right) + H_2O \rightarrow OH\cdot + H^+ + \text{Photocatalyst (Regeneration)} \quad (11.23)$$

$$\text{Photocatalyst}\left(h^{+(VB)}\right) + OH^- \rightarrow \text{Photocatalyst (Regeneration)} + OH\cdot \quad (11.24)$$

$$\cdot O_2^- + H_2O_2 \rightarrow \cdot OH + OH^- + O_2 \quad (11.25)$$

As can be observed in the abovementioned equations, various oxidative species can be generated, including $e^{-(CB)}$, $h^{+(VB)}$, $OH^{\cdot}$, and $^{\cdot}O^{-}{}_{2,}$ which can be involved in the decomposition of organic compounds [83, 84].

Photocatalysts such as TiO_2, CeO, WO_3, Fe_2O_3, SnO_2, and ZnO have been extensively studied in recent years for the decomposition of various organic compounds [85, 86]. Among them, TiO_2 is a popular n-type semiconductor with acceptable photocatalytic performance [87, 88]. The valence and conduction bands of this photocatalyst are composed of oxygen 2p and titanium 3d orbitals, respectively [89].

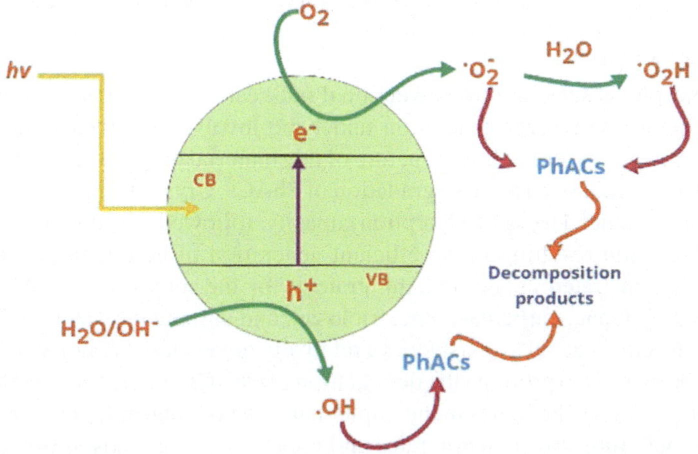

Fig. 11.10 A schematic of the mechanisms involved in the generation of reactive species for the decomposition of PhACs under photocatalytic processes

There are three different crystalline structures for TiO_2, including anatase, rutile, and brucite, and among them, the {101} facets of anatase TiO_2 can represent better photocatalytic activity than the {100} and {001} facets. This is due to the accumulation of photogenerated electrons and holes on this facet, which can facilitate the degradation of organic pollutants [90].

Carbonaceous materials such as graphene, biochar, and carbon nanotubes have also been investigated extensively for the photocatalytic degradation of PhACs [91]. Such a strategy can also enhance the photocatalytic performance of the system by providing the possibility of an extended photoabsorption range. Recent studies have also indicated the efficient charge separation capability of such structures. Loading of well-known photocatalysts such as TiO_2 on carbonaceous structures has also been shown to have the potential for simultaneous adsorption and degradation of PhACs. The application of titanium dioxide-loaded reduced graphene oxide (TiO_2-rGO) for the efficient photocatalytic degradation of carbamazepine, ibuprofen, and sulfamethoxazole (54%, 81%, and 92%, respectively, after 180 min UV irradiation) is an example of such approaches, as reported by Lin et al. [92]. The addition of magnetic compartments has also been considered a possible way to grant collectability and enhance the overall performance of photocatalytic materials (e.g., 90% azithromycin under UV irradiation [93]).

Novel carbonaceous materials, such as graphitic carbon nitride (g-C_3N_4), with a stable visible-light active structure have also been used for the removal of PhACs. This heavy metal-free structure is easy to prepare and low cost, with excellent electronic and optical properties. The current trend in the scientific community is to decrease the particle size of g-C_3N_4 and enhance its charge separation efficiency [94] to promote the photocatalytic decomposition of PhACs. Various strategies have been adopted recently to this end, such as the incorporation of doping agents in the structure of this material or the creation of nitrogen or carbon defects. Efficient degradation of lidocaine using carbon-defective graphitic carbon nitride is an example of the effectiveness of such an approach with 2.5-fold higher performance compared to pristine g-C_3N_4 [94].

The amorphous structure of conventional bulk carbon nitride is another limiting factor for hydroxyl radical generation under the involved photocatalytic reaction. Hence, improving the crystalline structure of this material has also been very recently explored for the enhanced photodegradation of PhACs. Crystalline carbon nitride can represent an enhanced oxygen adsorption capacity, followed by a direct two-electron reduction reaction resulting in the efficient generation of hydrogen peroxide. This has been demonstrated as an efficient strategy for the removal of PhACs such as naproxen, diclofenac, carbamazepine, triclosan, and sulfamethoxazole [95].

Heterostructures are also considered a relatively novel class of photocatalysts with advanced properties to promote the degradation of PhACs. Such structures have been developed to address the issues in the implementation of conventional photocatalytic materials, including slow reaction rates and poor visible-light adsorption due to the wide energy band gap of conventional semiconductors (e.g., 3.2 eV for TiO_2) [96]. Depending on the differences in the energy band gap of the semiconductors in the

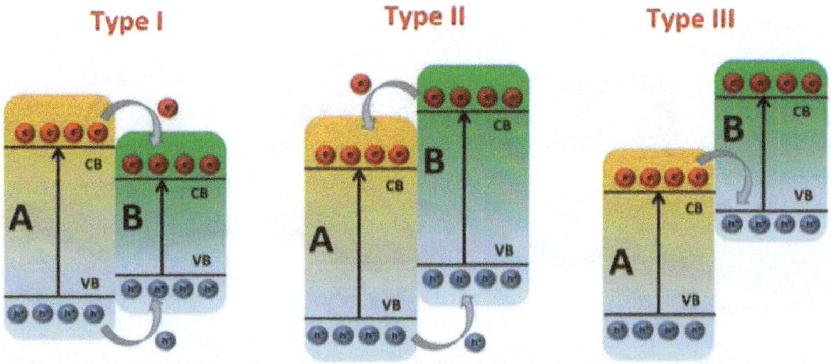

Fig. 11.11 Various types of heterojunctions for the efficient separation of photogenerated electrons and holes, adapted from Kumar et al. [97]. Novel heterojunction structures have also been developed very fast in recent years, such as Z-scheme and S-scheme structures with efficient charge separation potential (see [98])

structure of heterojunctions, there are three types of such structures (i.e., type I, type II, and type III (Fig. 11.11)).

There are reports for the removal of various PhACs using such structures (e.g., 97% tetracycline using $WO_3/g\text{-}C_3N_4$ [99]). High efficiencies for the removal of recalcitrant pharmaceuticals have also been reported in recent literature. For instance, almost complete degradation of carbamazepine (10 mg/L) was reported by $g\text{-}C_3N_4/TiO_2$ [100].

Although photocatalytic processes are considered efficient for dealing with a wide range of PhACs, there are still doubts about the long-term fate and ecotoxicological effects of photocatalysts used in the environment. In addition to the type of materials, their properties, such as size, shape, specific surface area and crystalline structure, need to be considered in related studies to have a realistic evaluation of the sustainability of such processes [101].

Reactor design is also of high importance for the efficient and cost-effective application of photocatalytic systems. There are generally two main types of reactor configurations: (a) mixing the suspended photocatalytic particles to create contact with the pollutants and (b) immobilization of the photocatalysts on the solid pieces. Each of these approaches has pros and cons. While the addition of suspended particles can maximize the contact between the particles and the pollutant molecules by providing the whole available surface area of the photocatalyst, collection of the spent particles is required to prevent their release into the environment after the treatment process with possible economic and ecotoxicological issues. To address this need, the development of fixed-bed photoreactors has been considered a way to push the commercialization of such technologies. Such strategies require precise engineering tools to optimize the efficiency of the system.

3D printing technology (3DPT) is a novel and powerful structure manufacturing tool that has been very recently introduced for various applications, especially in the construction industry [102], medical engineering [103, 104], and food processing

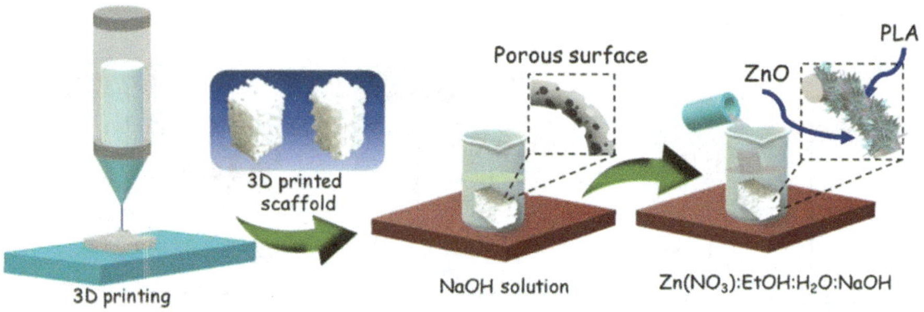

Fig. 11.12 A schematic of the ZnO 3D-printed scaffolds for the treatment of polluted waters, adapted from Kumbhakar et al. [112]

[105]. Recent developments in 3DPT have allowed the fabrication of concepts and devices made of various materials, including ceramic, polymer, or metallic compounds [106]. Such technologies have also been spreading rapidly in various environmental applications, such as fabricating membrane structures with antifouling properties [107] and designing components of microbial fuel cells, including anodes [108], cathodes [109], proton-exchange membranes [110], and chassis [111], to optimize bioelectricity generation from effluents from various origins. Figure 11.12 represents ZnO 3D-printed scaffolds used for the photocatalytic processes.

11.3.2 Photoelectrocatalysis

In photoelectrocatalytic (PEC) processes, a semiconductor is generally used for the photogeneration of electrons and holes under sufficient irradiation [113]. As discussed before, the reaction between holes and water molecules results in the generation of hydroxyl radicals for the decomposition of organic compounds [114]. However, the rapid recombination of electron–hole pairs can considerably limit the efficiency of photocatalytic systems, which can be overcome through the implementation of bias potential. This can aid in separating the electrons from the anode made of a semiconductor [115, 116]. Figure 11.13 illustrates the mechanisms involved in the PEC process using semiconductor materials.

Such strategies can also help address another issue that exists in the application of semiconductor powders, which have been traditionally used in photocatalytic processes. In PEC processes, the photocatalytic materials are stabilized in the electrode structures; hence, they can be reused without the need for additional steps, such as filtration or magnetic separation, which is required for photocatalytic processes [47]. This can also considerably reduce the total treatment costs, which is essential for sustainable (waste)water treatment methods. Similar to photocatalysis, semiconductors such as WO_3, ZnO, and TiO_2 are the most widely used materials for PEC processes [117–119]. Most of these materials can bring advantages to these processes

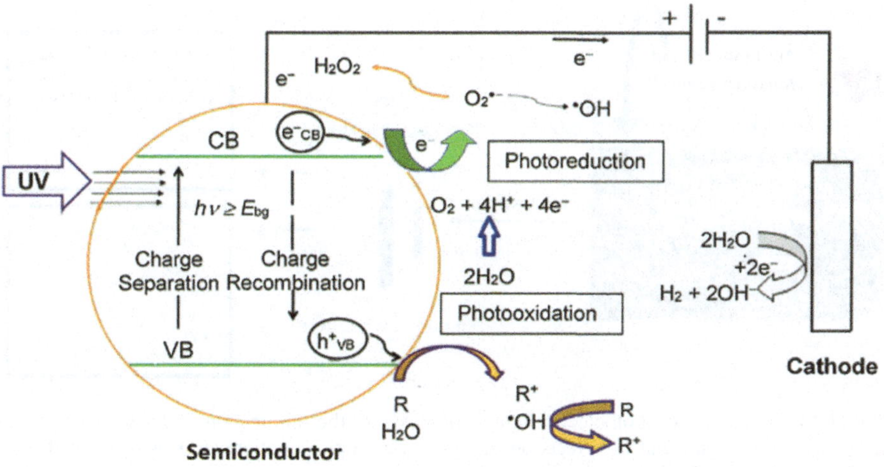

Fig. 11.13 Typical mechanisms involved in the PEC process utilizing semiconductors for the degradation of organic compounds, adapted from Garcia-Segura and Brillas [115]

because they are easy to fabricate and have acceptable electrical conductivity and large specific surface areas to facilitate surface reactions [120]. However, pure semiconductors usually suffer from a large energy band gap (e.g., 3.5 eV for ZnO), which results in better efficiency of these materials under high irradiation intensities (e.g., UV irradiation). There are studies in recent years to address this issue by adopting strategies such as doing with metals and fabrication of heterojunction structures[4] with the ability to adsorb visible light, which can lead to the reduction of the (waste)water treatment costs.

There are successful reports for the elimination of PhACs using PEC processes, especially using modified semiconductor nanomaterials such as Ni-doped ZnO (ciprofloxacin, 100%, 90 min [122]) and FTO/$BiVO_4$/BiOI (68% and 62%, for acetaminophen and ciprofloxacin, respectively, with a bias potential of 1.5 V after 2 h) [123]. The application of carbonaceous materials has also received considerable attention in recent years for the treatment of polluted (waste)waters. This is mainly due to their low cost and green nature, especially when they are prepared from low-cost materials such as biomass wastes. They can also provide interesting properties, such as high electrical conductivity and large specific surface area, which are critical for efficient PECs [124, 125]. They present superior performance when loaded with efficient semiconductors. For instance, carbon fibers (CFs) doped with Pd-ZnO/N have been used for the complete removal of paracetamol within 3 h (current density = 10 $mAcm^{-2}$). The high TOC removal of this process (over 70%) is also an indication of the safe nature of the treated effluents to the receiving environment [126].

[4] Such as FTO/$BiVO_4$/Ag_2S heterojunction anode which has been used efficiently for the ciprofloxacin and sulfamethoxazole (80% and 86%, respectively) [121].

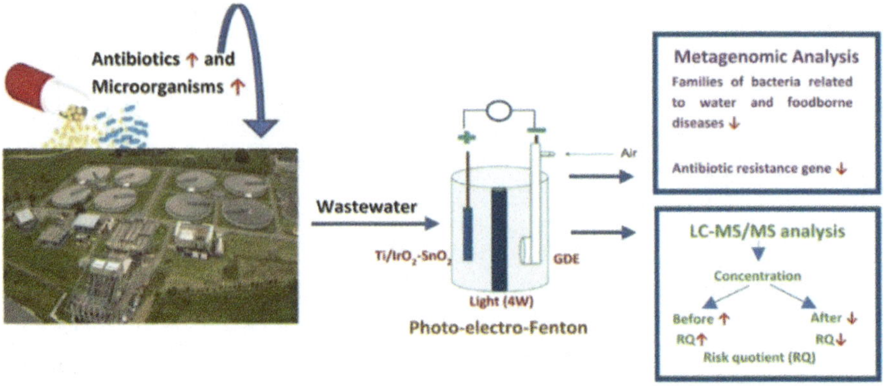

Fig. 11.14 A combined photoelectro-Fenton process for the elimination of bacteria and pharmaceutical compounds and its effects on the reduction of risk quotient (RQ) was adopted from Martínez-Pachón et al. [128]

There are also studies indicating the ability of these methods for the removal of biological agents as well as PhACs to minimize the risk of the discharged effluents to the environment (Fig. 11.14). Other studies have also demonstrated the efficiency of PEC processes for the removal of toxicity from PhAC-containing effluents. For instance, thin films (64 cm^2) made of nanostructured TiO_2 exhibited a high potential for the removal of cefotaxime under a PEC process [127]. The authors suggested a mechanism for the degradation of the pharmaceutical compound, including the cleavage of the cephem nucleus. The toxicity tests indicated the safe nature of the treated effluents, indicating the formation of less toxic degradation products under the applied PEC method.

The superior performance of the PEC process compared to other photocatalytic systems can make them appealing for real (waste)water treatment applications. For instance, Collivignarelli et al. [19] indicated that PEC can represent higher efficiencies for the removal of COD and color compared to photolysis (PL) and photocatalysis (PC) for effluents containing PhACs. However, some issues still need to be addressed for the wider application of such methods, including the reduction of the total treatment costs, mainly associated with electrode material fabrication, and the electrical energy consumed for performing such processes. This can be, for instance, satisfied by developing low-cost and sustainable fabrication processes or by supplying the electricity required for providing the required current density from natural and renewable sources of energy such as solar irradiation.

Durability is among the most important advantages of electrodes made of nanostructure materials for the PEC process. This can satisfy the economic considerations that have been conventionally involved in the application of powdered nanomaterials with difficulties in recovery and reuse. It has also been discussed in the literature that the released nanomaterials (e.g., TiO_2 or ZnO) can cause toxic effects on microorganisms in the receiving environment by, for instance, the generation of reactive oxygen species (ROS) [129–131]. The PEC strategy hence helps environmental considerations by preventing the release of nanomaterials into the environment.

11.3.3 Photocatalytic Ozonation

Photocatalytic treatment processes using semiconductors in the presence of ozone are another class of AOPs called photocatalytic ozonation (PCOz) [132]. This results in synergistic effects for the generation of various types of oxidative species, leading to the efficient removal of complex organic compounds from polluted (waste)waters. The possible reactions in the presence of photocatalyst and ozone are presented in Eqs. 11.26–11.29 [133, 134].

$$\mathrm{PCt} + h\upsilon \rightarrow \mathrm{PCt} + \mathrm{e}^- + \mathrm{h}^+ \tag{11.26}$$

$$\mathrm{O_3} + \mathrm{PCt} \rightarrow {}^{\cdot}\mathrm{O} + \mathrm{O_2} \tag{11.27}$$

$$\mathrm{O_3} + h\upsilon \rightarrow {}^{\cdot}\mathrm{O} + \mathrm{O_2} \tag{11.28}$$

$$\mathrm{H_2O_2} + h\upsilon \rightarrow 2\mathrm{OH}^{\cdot} \tag{11.29}$$

As discussed in the previous sections, the electrons and holes generated on the surface of PCt can also lead to the formation of active species such as super oxides and hydroxyl radicals, which can be involved in the decomposition of organic compounds. The electrons generated on the photocatalyst surface can also be more efficiently involved in the generation of hydroxyl radicals in the presence of ozone because only one electron is required to generate a molecule of hydroxyl radical from each O_3 molecule, while 3 electrons should be trapped for the same output when O_2 is involved in the chain reactions [135, 136].

The large amount of radical and nonradical species generated in the medium can also result in efficient mineralization of organic compounds, which is of high importance in terms of environmental and ecotoxicological considerations. There are studies indicating the superior efficiency of PCOz compared to ozonation or photocatalytic processes alone. For instance, the following order has been reported to compare the efficiencies of various AOPs for the removal/mineralization of refractory organic compounds [137]: UV/air < O_3 < TiO_2/O_3 < UV/O_3 < $TiO_2/UV/O_2$ < $TiO_2/UV/O_3$.

In addition to conventional photocatalytic materials, novel visible-light active materials such as graphitic carbon nitride (g-C_3N_4) and their composites with metal oxides have also received attention for photocatalytic ozonation processes. An example is the application of MgO/g-C_3N_4 in photocatalysis ozonation in which MgO plays a dual role of (a) separation of the photogenerated electron–hole pairs and

(b) MgO surface reactions resulting in the formation of hydroxyl radicals for further improvement of the efficiency of the system for the degradation of the pollutants.

Despite the superior efficiency of photocatalytic ozonation processes over conventional AOP processes, they are still too expensive to be implemented for real (waste)water treatment applications. The high operating costs are mainly associated with the fabrication of engineering catalytic materials with very well-defined properties and the generation of ozone. However, to reduce the operating costs of these processes, strategies can be adopted. Similar to other catalytic systems, the separation and reuse of spent materials is an issue involved in the application of such processes. Stabilization of the catalytic materials using advanced techniques such as 3D printing can also be considered to overcome this drawback. Furthermore, there is a need to develop low-cost ozone generation technologies to make the technology more viable for real applications.

11.4 Further Reading

Table 11.1 contains items from the recent literature that the reader can consult for more detailed information regarding the efficiency of ozone-based technologies for the degradation of pharmaceutically active compounds.

11.5 Summary

Heterogeneous advanced oxidation processes are among the fast-growing (waste)water treatment technologies for the degradation of recalcitrant organic compounds, including PhACs. Various types of these techniques, such as the activation of oxidation agents (e.g., persulfate, proxymonosulfate, and iodine), photocatalytic processes, photoelectrocatalytic processes, catalytic ozonation, and photocatalytic ozonation, have been developed and used in recent years to address the need to remove the contaminants of emerging concerns and to supply clean water resources. Each of these techniques has its pros and cons regarding their efficiencies and the respective sustainability considerations. Activation of oxidation agents using heterogeneous catalysis is an efficient way to generate various oxidative species based on the oxidation agent used. An example is the efficient activation of persulfate using copper oxide nanomaterials for the simultaneous generation of sulfate and hydroxyl radicals. Despite the applicability of these methods for the removal of a wide range of PhACs, there are concerns regarding the toxic nature of the treated effluents due to the presence of compounds such as activated persulfate. Catalytic ozonation is also categorized among the popular energy-free heterogenous AOPs with high efficiency to deal with a wide range of PhACs but still suffers from the relatively high operating costs for real applications. Heterogeneous catalytic materials have also been widely employed in recent years for the photocatalytic decomposition of PhACs. The latest

Table 11.1 Further reading suggestions for more detailed coverage of the literature on ozone-based technologies for the removal of pharmaceuticals in (waste)waters

References	Item	Subject
Issaka et al. [3]	Table 6	Detailed values from the reports on the application of ozone-based oxidation and catalytic ozone oxidation processes for the removal of pharmaceutical micropollutants
	Table 7	Detailed values from the reports regarding the catalytic ozonation and mineralization of pharmaceutical micropollutants
Paucar et al. [138]	Figure 9	Degradability of various pharmaceutical compounds against ozonation under various initial ozone dosages
Gomes et al. [139]	Table 2	Various catalysts used in the catalytic ozonation of contaminants of emerging concern (CECs)
	Table 3	The effects of operating conditions regarding the application of ozonation-based processes for the degradation of pharmaceutical micropollutants
Li et al. [101]	Table 6	Degradation kinetics of pharmaceutically active compounds using graphene-based photocatalytic nanocomposites
Mirzaei et al. [140]	Table 1	ZnO nanomaterials for the photocatalytic removal of various types of PhACs
Matzek and Carter [141]	Table 3	PS activation for the degradation of various organic compounds
Gao et al. [142]	Table 2	The effects of the operating conditions on the efficiency of Fe^{2+}/PS system

trend in this area is to fabricate materials with enhanced properties, such as high stability, well-adjusted energy band gap, and the capability of efficient separation of photogenerated electrons and holes. Stabilization of the well-known photocatalysts using sustainable carbonaceous materials, doping with secondary elements, and fabrication of heterostructures are the latest efforts in the literature to promote photocatalytic processes with interesting achievements for the degradation of recalcitrant pharmaceutically active compounds such as carbamazepine. Coupling ozonation or electrooxidation with photocatalytic processes can also be considered among the most efficient AOPs for the removal of PhACs. Another trend in the implementation of heterogeneous catalytic AOPs is to implement advanced techniques such as 3D printing for the fabrication of catalytic structures. Such an approach will aim at preventing the release of the spent materials as well as creating the possibility of reusing the catalysts, which can be highly beneficial in terms of economic considerations.

Future studies need to be directed on the minimization of the treatment costs involved in the application of heterogeneous catalytic AOPs to deal with PhACs. Coupling advanced analytical techniques (such as LC–MS) with ecotoxicological studies is also required to better understand the toxic nature of effluents before and after treatment with such advanced technologies.

References

1. da Silva SW et al (2021) Advanced electrochemical oxidation processes in the treatment of pharmaceutical containing water and wastewater: a review. Curr Pollut Rep 7(2):146–159. https://doi.org/10.1007/s40726-021-00176-6
2. Von Sonntag C (2008) Advanced oxidation processes: mechanistic aspects. Water Sci Technol 58(5):1015–1021. https://doi.org/10.2166/wst.2008.467
3. Issaka E et al (2022) Advanced catalytic ozonation for degradation of pharmaceutical pollutants—a review. Chemosphere 289:133208. https://doi.org/10.1016/j.chemosphere.2021.133208
4. Bagheri S, Termehyousefi A, Do TO (2017) Photocatalytic pathway toward degradation of environmental pharmaceutical pollutants: structure, kinetics and mechanism approach. Catal Sci Technol Royal Soc Chem 7:4548–4569. https://doi.org/10.1039/c7cy00468k
5. Chair K et al (2017) Combining bioadsorption and photoelectrochemical oxidation for the treatment of soil-washing effluents polluted with herbicide 2,4-D. J Chem Technol Biotechnol 92:83–89. https://doi.org/10.1002/jctb.5001
6. Zhou Z et al (2019) Persulfate-based advanced oxidation processes (AOPs) for organic-contaminated soil remediation: a review. Chem Eng J 372:836–851. https://doi.org/10.1016/j.cej.2019.04.213
7. Clarizia L et al (2017) Homogeneous photo-Fenton processes at near neutral pH: a review. Appl Catal B 209:358–371. https://doi.org/10.1016/j.apcatb.2017.03.011
8. Guo Y et al (2018) Prediction of micropollutant abatement during homogeneous catalytic ozonation by a chemical kinetic model. Water Res 142:383–395. https://doi.org/10.1016/j.watres.2018.06.019
9. Khan MH, Jung JY (2008) Ozonation catalyzed by homogeneous and heterogeneous catalysts for degradation of DEHP in aqueous phase. Chemosphere 72:690–696. https://doi.org/10.1016/j.chemosphere.2008.02.037
10. Luz I, Llabrés i Xamena FX, Corma A (2012) Bridging homogeneous and heterogeneous catalysis with MOFs: Cu-MOFs as solid catalysts for three-component coupling and cyclization reactions for the synthesis of propargylamines, indoles and imidazopyridines. J Catal 285:285–291
11. Guo Y, Yang L, Wang X (2012) The application and reaction mechanism of catalytic ozonation in water treatment. J Environ Anal Toxicol 02(06). https://doi.org/10.4172/2161-0525.1000150
12. Qin H et al (2014) Efficient degradation of fulvic acids in water by catalytic ozonation with CeO_2/AC. J Chem Technol Biotechnol 89(9):1402–1409. https://doi.org/10.1002/jctb.4222
13. Koricic K et al (2016) Mineralization of salicylic acid in water by catalytic ozonation. Environ Eng Manag J 15:4597
14. Mao L et al (2018) Plasma-catalyst hybrid reactor with $CeO_2/\Gamma\text{-}Al_2O_3$ for benzene decomposition with synergetic effect and nano particle by-product reduction. J Hazard Mater 347:150–159. https://doi.org/10.1016/j.jhazmat.2017.12.064
15. Li X et al (2018) Relationship between the structure of Fe-MCM-48 and its activity in catalytic ozonation for diclofenac mineralization. Chemosphere 206:615–621. https://doi.org/10.1016/j.chemosphere.2018.05.066
16. Xu Y et al (2019) Mechanism and kinetics of catalytic ozonation for elimination of organic compounds with spinel-type $CuAl_2O_4$ and its precursor. Sci Total Environ 651:2585–2596. https://doi.org/10.1016/j.scitotenv.2018.10.005

17. Tian SQ et al (2021) Heterogeneous catalytic ozonation of atrazine with Mn-loaded and Fe-loaded biochar. Water Res 193:116860. https://doi.org/10.1016/j.watres.2021.116860
18. Wang Z et al (2020) ZIF-8-modified $MnFe_2O_4$ with high crystallinity and superior photo-Fenton catalytic activity by Zn-O-Fe structure for TC degradation. Chem Eng J 392:124851. https://doi.org/10.1016/j.cej.2020.124851
19. Collivignarelli MC et al (2021) Wastewater treatment plants effluents: photoelectrocatalysis vs. Water 13:821
20. Rashid R et al (2021) A state-of-the-art review on wastewater treatment techniques: the effectiveness of adsorption method. Environ Sci Pollut Res 28:9050–9066. https://doi.org/10.1007/s11356-021-12395-x
21. Harshiny M et al (2017) Biosynthesized FeO nanoparticles coated carbon anode for improving the performance of microbial fuel cell. Int J Hydrogen Energy 42:26488–26495. https://doi.org/10.1016/j.ijhydene.2017.07.084
22. Wang J et al (2016) Magnetic lanthanide oxide catalysts: an application and comparison in the heterogeneous catalytic ozonation of diethyl phthalate in aqueous solution. Sep Purif Technol 159:57–67. https://doi.org/10.1016/j.seppur.2015.12.031
23. Wang J, Chen H (2020) Catalytic ozonation for water and wastewater treatment: recent advances and perspective. Sci Total Environ 704:135249. https://doi.org/10.1016/j.scitotenv.2019.135249
24. Li R et al (2015) Heterogeneous Fenton oxidation of 2,4-dichlorophenol using iron-based nanoparticles and persulfate system. Chem Eng J 264:587–594. https://doi.org/10.1016/j.cej.2014.11.128
25. Wang W et al (2020) Electro-Fenton and photoelectro-Fenton degradation of sulfamethazine using an active gas diffusion electrode without aeration. Chemosphere 250:126177. https://doi.org/10.1016/j.chemosphere.2020.126177
26. Koba O, Biro L (2015) Fenton-like reaction: a possible way to efficiently remove illicit drugs and pharmaceuticals from wastewater. Environ Toxicol Pharmacol 9:483–488. https://doi.org/10.1016/j.etap.2014.12.016
27. Babuponnusami A, Muthukumar K (2014) A review on Fenton and improvements to the Fenton process for wastewater treatment. J Environ Chem Eng 2:557–572. https://doi.org/10.1016/j.jece.2013.10.011
28. Wu J et al (2020) Nanoscale zero valent iron-activated persulfate coupled with Fenton oxidation process for typical pharmaceuticals and personal care products degradation. Sep Purif Technol 239:116534. https://doi.org/10.1016/j.seppur.2020.116534
29. Diao Z-H et al (2016) Bentonite-supported nanoscale zero-valent iron/persulfate system for the simultaneous removal of Cr(VI) and phenol from aqueous solutions. Chem Eng J 302:213–222. https://doi.org/10.1016/j.cej.2016.05.062
30. Lin Y et al (2014) Decoloration of acid violet red B by bentonite-supported nanoscale zero-valent iron: reactivity, characterization, kinetics and reaction pathway. Appl Clay Sci 93–94:56–61. https://doi.org/10.1016/j.clay.2014.02.020
31. Wang S et al (2019) Biochar-supported nZVI (nZVI/BC) for contaminant removal from soil and water: a critical review. J Hazard Mater 373:820–834. https://doi.org/10.1016/j.jhazmat.2019.03.080
32. De Andrade JR et al (2018) Adsorption of pharmaceuticals from water and wastewater using nonconventional low-cost materials: a review. Ind Eng Chem Res 57:3103–3127. https://doi.org/10.1021/acs.iecr.7b05137
33. Hartmann M, Keller H (2010) Wastewater treatment with heterogeneous Fenton-type catalysts based on porous materials. J Mater Chem 20:9002–9017. https://doi.org/10.1039/c0jm00577k
34. Lin SS, Gurol MD (1998) Catalytic decomposition of hydrogen peroxide on iron oxide: kinetics, mechanism, and implications. Environ Sci Technol 32(10):1417–1423. https://doi.org/10.1021/es970648k
35. Cong Y et al (2012) Synthesis of α-Fe_2O_3/TiO_2 nanotube arrays for photoelectro-Fenton degradation of phenol. Chem Eng J 191:356–363

36. Sun SP et al (2014) Enhanced heterogeneous and homogeneous Fenton-like degradation of carbamazepine by nano-Fe_3O_4/H_2O_2 with nitrilotriacetic acid. Chem Eng J 244:44–49. https://doi.org/10.1016/j.cej.2014.01.039
37. Ulloa-Ovares D et al (2021) Simultaneous degradation of pharmaceuticals in fixed and fluidized bed reactors using iron-modified diatomite as heterogeneous Fenton catalyst. Process Saf Environ Prot 152:97–107. https://doi.org/10.1016/j.psep.2021.05.032
38. Lan Y et al (2020) Feasibility of a heterogeneous Fenton membrane reactor containing a Fe-ZSM5 catalyst for pharmaceuticals degradation: membrane fouling control and long-term stability. Sep Purif Technol 231:115920. https://doi.org/10.1016/j.seppur.2019.115920
39. Chang H et al (2017) Hydraulic backwashing for low-pressure membranes in drinking water treatment: a review. J Membr Sci 540:362–380. https://doi.org/10.1016/j.memsci.2017.06.077
40. Lee J, Von Gunten U, Kim JH (2020) Persulfate-based advanced oxidation: critical assessment of opportunities and roadblocks. Environ Sci Technol 54:3064–3081. https://doi.org/10.1021/acs.est.9b07082
41. Nidheesh PV, Rajan R (2016) 'Removal of rhodamine B from a water medium using hydroxyl and sulphate radicals generated by iron loaded activated carbon. RSC Adv Royal Society of Chemistry 6:5330–5340. https://doi.org/10.1039/c5ra19987e
42. Evseev AK et al (2008) Electrochemical synthesis of peroxodisulfates from dilute sulfate solutions for detoxification of biological media. Russ J Electrochem 44:901–909. https://doi.org/10.1134/S1023193508080041
43. He X et al (2014) Degradation kinetics and mechanism of β-lactam antibiotics by the activation of H_2O_2 and $Na_2S_2O_8$ under UV-254nm irradiation. J Hazard Mater 279:375–383. https://doi.org/10.1016/j.jhazmat.2014.07.008
44. Wang X et al (2014) Degradation of acid orange 7 by persulfate activated with zero valent iron in the presence of ultrasonic irradiation. Sep Purif Technol 122:41–46. https://doi.org/10.1016/j.seppur.2013.10.037
45. Du X et al (2017) Insight into reactive oxygen species in persulfate activation with copper oxide: activated persulfate and trace radicals. Chem Eng J 313:1023–1032. https://doi.org/10.1016/j.cej.2016.10.138
46. Sabri M et al (2021) Titania-activated persulfate for environmental remediation: the-state-of-the-art. Catal Rev Sci Eng 00:1–56. https://doi.org/10.1080/01614940.2021.1996776
47. Kamali M, Persson KM et al (2019) Sustainability criteria for assessing nanotechnology applicability in industrial wastewater treatment: current status and future outlook. Environ Int 125:261–276. https://doi.org/10.1016/j.envint.2019.01.055
48. Kumar A et al (2021) Construction of dual Z-scheme g-$C_3N_4/Bi_4Ti_3O_{12}/Bi_4O_5I_2$ heterojunction for visible and solar powered coupled photocatalytic antibiotic degradation and hydrogen production: Boosting via I−/I3− and Bi^{3+}/Bi^{5+} redox mediators. Appl Catal B 284:119808. https://doi.org/10.1016/j.apcatb.2020.119808
49. Wei T et al (2022) Au tailored on g-C3N4/TiO_2 heterostructure for enhanced photocatalytic performance. J Alloy Compd 894:162338. https://doi.org/10.1016/j.jallcom.2021.162338
50. Kim DG, Ko SO (2020) Effects of thermal modification of a biochar on persulfate activation and mechanisms of catalytic degradation of a pharmaceutical. Chem Eng J 399:125377. https://doi.org/10.1016/j.cej.2020.125377
51. Minh TD et al (2019) Gingerbread ingredient-derived carbons-assembled CNT foam for the efficient peroxymonosulfate-mediated degradation of emerging pharmaceutical contaminants. Appl Catal B 244:367–384. https://doi.org/10.1016/j.apcatb.2018.11.064
52. Cai S et al (2021) Pyrrolic N-rich biochar without exogenous nitrogen doping as a functional material for bisphenol A removal: performance and mechanism. Appl Catal B 291:1–10. https://doi.org/10.1016/j.apcatb.2021.120093
53. Ren W et al (2020) The intrinsic nature of persulfate activation and N-doping in carbocatalysis. Environ Sci Technol 54:6438–6447. https://doi.org/10.1021/acs.est.0c01161
54. Wang H et al (2019) Edge-nitrogenated biochar for efficient peroxydisulfate activation: an electron transfer mechanism. Water Res 160:405–414. https://doi.org/10.1016/j.watres.2019.05.059

55. Tang Q, Zhou Z, Chen Z (2013) Graphene-related nanomaterials: tuning properties by functionalization. Nanoscale 5:4541–4583. https://doi.org/10.1039/c3nr33218g
56. Deniere E et al (2018) Advanced oxidation of pharmaceuticals by the ozone-activated peroxymonosulfate process: the role of different oxidative species. J Hazard Mater 360:204–213. https://doi.org/10.1016/j.jhazmat.2018.07.071
57. Du X et al (2019) Persulfate non-radical activation by nano-CuO for efficient removal of chlorinated organic compounds: reduced graphene oxide-assisted and CuO. Chem Eng J 356(August 2018):178–189. https://doi.org/10.1016/j.cej.2018.08.216
58. Xing S et al (2020) Removal of ciprofloxacin by persulfate activation with CuO: a pH-dependent mechanism. Chem Eng J 382:122837. https://doi.org/10.1016/j.cej.2019.122837
59. Tang H et al (2019) Promotion of peroxydisulfate activation over Cu0.84Bi2.08O4 for visible light induced photodegradation of ciprofloxacin in water matrix. Chem Eng J 356:472–482. https://doi.org/10.1016/j.cej.2018.09.066
60. Lee Y et al (2016) Efficient sonochemical degradation of perfluorooctanoic acid using periodate. Ultrasonics Sonochem 31:499–505. https://doi.org/10.1016/j.ultsonch.2016.01.030
61. Djaballah ML et al (2021) Development of a free radical-based kinetics model for the oxidative degradation of chlorazol black in aqueous solution using periodate photoactivated process. J Photochem Photobiol, A 408:113102. https://doi.org/10.1016/j.jphotochem.2020.113102
62. Zong Y et al (2021) Enhanced oxidation of organic contaminants by Iron(II)-activated periodate: the significance of high-valent iron-oxo species. Environ Sci Technol 55:7634–7642. https://doi.org/10.1021/acs.est.1c00375
63. Hamdaoui O, Merouani S (2017) Improvement of sonochemical degradation of Brilliant blue R in water using periodate ions: implication of iodine radicals in the oxidation process. Ultrasonics Sonochem 37:344–350. https://doi.org/10.1016/j.ultsonch.2017.01.025
64. Heger D, Kim K, Kim J (2018) Activation of periodate by freezing for the degradation of aqueous organic pollutants. Environ Sci Technol 52:5378–5385. https://doi.org/10.1021/acs.est.8b00281
65. Mohammadi AMS et al (2016) Oxidation of phenol from synthetic wastewater by a novel advance oxidation process: microwave-assisted periodate. J Sci Ind Res 75:267–272
66. Zhang X et al (2021) Efficiency and mechanism of 2,4-dichlorophenol degradation by the UV/IO_4-process. Sci Total Environ 782:146781. https://doi.org/10.1016/j.scitotenv.2021.146781
67. Li X et al (2016) Activation of periodate by granular activated carbon for acid orange 7 decolorization. J Taiwan Inst Chem Eng 68:211–217. https://doi.org/10.1016/j.jtice.2016.08.039
68. Shao P et al (2018) Identification and regulation of active sites on nanodiamonds: establishing a highly efficient catalytic system for oxidation of organic contaminants. Foundations 28:1705295
69. He L et al (2022) Fe, N-doped carbonaceous catalyst activating periodate for micropollutant removal: significant role of electron transfer. Appl Catal B: Environ. 303:120880. https://doi.org/10.1016/j.apcatb.2021.120880
70. Long Y et al (2021) Atomically dispersed cobalt sites on graphene as efficient periodate activators for selective organic pollutant degradation. Environ Sci Technol 55(8):5357–5370. https://doi.org/10.1021/acs.est.0c07794
71. Li X et al (2017) Enhanced activation of periodate by iodine-doped granular activated carbon for organic contaminant degradation. Chemosphere 181:609–618. https://doi.org/10.1016/j.chemosphere.2017.04.134
72. Xiao P et al (2022) Catalytic performance and periodate activation mechanism of anaerobic sewage sludge-derived biochar. J Hazard Mater 424:127692. https://doi.org/10.1016/j.jhazmat.2021.127692
73. Ouyang D et al (2019) Activation mechanism of peroxymonosulfate by biochar for catalytic degradation of 1,4-dioxane: important role of biochar defect structures. Chem Eng J 370:614–624. https://doi.org/10.1016/j.cej.2019.03.235

74. Buxton GV et al (1988) Critical review of rate constants for reactions of hydrated electrons, hydrogen atoms and hydroxyl radicals (·OH/·O− in Aqueous Solution. J Phys Chem Ref Data 17:513–886. https://doi.org/10.1063/1.555805
75. Mertens R, von Sonntag C (1995) Photolysis (λ = 354 nm of tetrachloroethene in aqueous solutions. J Photochem Photobiol, A 85:1–9. https://doi.org/10.1016/1010-6030(94)03903-8
76. Zhou Y et al (2020) Kinetics and pathways of the degradation of PPCPs by carbonate radicals in advanced oxidation processes. Water Res 185:116231. https://doi.org/10.1016/j.watres.2020.116231
77. Guo Y et al (2022) Photodegradation of propranolol in surface waters: an important role of carbonate radical and enhancing toxicity phenomenon. Chemosphere 297:134106. https://doi.org/10.1016/j.chemosphere.2022.134106
78. Karthik KV et al (2022) Green synthesis of Cu-doped ZnO nanoparticles and its application for the photocatalytic degradation of hazardous organic pollutants. Chemosphere 287:132081. https://doi.org/10.1016/j.chemosphere.2021.132081
79. Zheng H et al (2021) In situ phase evolution of TiO_2/Ti_3C2Tx heterojunction for enhancing adsorption and photocatalytic degradation. Appl Surf Sci 545:149031. https://doi.org/10.1016/j.apsusc.2021.149031
80. Zyoud AH et al (2020) Raw clay supported ZnO nanoparticles in photodegradation of 2-chlorophenol under direct solar radiations. J Environ Chem Eng 8(5):104227. https://doi.org/10.1016/j.jece.2020.104227
81. Al-Mamun MR et al (2019) Photocatalytic activity improvement and application of UV-TiO_2 photocatalysis in textile wastewater treatment: a review. J Environ Chem Eng 7:103248. https://doi.org/10.1016/j.jece.2019.103248
82. Sharma M et al (2022) TiO_2 based photocatalysis: a valuable approach for the removal of pharmaceuticals from aquatic environment. Int J Environ Sci Technol, https://doi.org/10.1007/s13762-021-03894-y, https://doi.org/10.1007/s13762-021-03894-y
83. Gomez-Ruiz B et al (2018) Photocatalytic degradation and mineralization of perfluorooctanoic acid (PFOA) using a composite TiO_2–rGO catalyst. J Hazard Mater 344:950–957. https://doi.org/10.1016/j.jhazmat.2017.11.048
84. Oves M et al (2020) Modern age waste water problems. Modern Age Waste Water Problems. https://doi.org/10.1007/978-3-030-08283-3
85. Asif AH, Wang S, Sun H (2021) Hematite-based nanomaterials for photocatalytic degradation of pharmaceuticals and personal care products (PPCPs): a short review. Curr Opin Green Sustain Chem 28:100447. https://doi.org/10.1016/j.cogsc.2021.100447
86. Javaid A et al (2022) Nanohybrids-assisted photocatalytic removal of pharmaceutical pollutants to abate their toxicological effects—a review. Chemosphere 291:133056. https://doi.org/10.1016/j.chemosphere.2021.133056
87. Awfa D et al (2018) Photodegradation of pharmaceuticals and personal care products in water treatment using carbonaceous-TiO_2 composites: a critical review of recent literature. Water Res 142:26–45. https://doi.org/10.1016/j.watres.2018.05.036
88. Isari AA et al (2020) N, Cu co-doped TiO_2@functionalized SWCNT photocatalyst coupled with ultrasound and visible-light: an effective sono-photocatalysis process for pharmaceutical wastewaters treatment. Chem Eng J 392:123685. https://doi.org/10.1016/j.cej.2019.123685
89. Chen X, Mao SS (2007) Titanium dioxide nanomaterials: synthesis, properties, modifications and applications. Chem Rev 107:2891–2959
90. Wang X et al (2013) Roles of (001) and (101) facets of anatase TiO_2 in photocatalytic reactions. Wuli Huaxue Xuebao/ Acta Physico - Chimica Sinica 29:1566–1571. https://doi.org/10.3866/PKU.WHXB201304284
91. Mestre AS, Carvalho AP (2019) Photocatalytic degradation of pharmaceuticals wastewater. Molecules 24:3702. Available at: https://www.mdpi.com/1420-3049/24/20/3702/pdf
92. Lin L, Wang H, Xu P (2017) Immobilized TiO_2-reduced graphene oxide nanocomposites on optical fibers as high performance photocatalysts for degradation of pharmaceuticals. Chem Eng J 310:389–398. https://doi.org/10.1016/j.cej.2016.04.024

93. Sayadi MH, Sobhani S, Shekari H (2019) Photocatalytic degradation of azithromycin using GO@Fe_3O_4/ZnO/SnO_2 nanocomposites. J Clean Prod 232:127–136. https://doi.org/10.1016/j.jclepro.2019.05.338
94. Tao L et al (2022) Photocatalytic degradation of pharmaceuticals by pore-structured graphitic carbon nitride with carbon vacancy in water: identification of intermediate degradants and effects of active species. Sci Total Environ 824:153845. https://doi.org/10.1016/j.scitotenv.2022.153845
95. Wang Y et al (2020) Mechanism insight into enhanced photodegradation of pharmaceuticals and personal care products in natural water matrix over crystalline graphitic carbon nitrides. Water Res 180:115925. https://doi.org/10.1016/j.watres.2020.115925
96. Wang Y et al (2013) Visible light driven type II heterostructures and their enhanced photocatalysis properties: a review. Nanoscale 5:8326–8339. https://doi.org/10.1039/c3nr01577g
97. Kumar S et al (2020) Nanoscale zinc oxide based heterojunctions as visible light active photocatalysts for hydrogen energy and environmental remediation. Catal Rev Sci Eng 62:346–405. https://doi.org/10.1080/01614940.2019.1684649
98. Xu Q et al (2020) S-scheme heterojunction photocatalyst. Chemistry 6:1543–1559. https://doi.org/10.1016/j.chempr.2020.06.010
99. Yan H et al (2019) Single-source-precursor-assisted synthesis of porous WO_3/g-C3N4 with enhanced photocatalytic property. Colloids Surf, A 582:123857. https://doi.org/10.1016/j.colsurfa.2019.123857
100. Hu Z et al (2019) Construction of carbon-doped supramolecule-based g-C_3N_4/TiO_2 composites for removal of diclofenac and carbamazepine: a comparative study of operating parameters, mechanisms, degradation pathways. J Hazard Mater 380:120812. https://doi.org/10.1016/j.jhazmat.2019.120812
101. Li C et al (2022) Graphene-based photocatalytic nanocomposites used to treat pharmaceutical and personal care product wastewater: a review. Environ Sci Pollut Res https://doi.org/10.1007/s11356-022-19469-4, https://doi.org/10.1007/s11356-022-19469-4
102. Hager I, Golonka A, Putanowicz R (2016) 3D printing of buildings and building components as the future of sustainable construction? Procedia Eng 151:292–299. https://doi.org/10.1016/j.proeng.2016.07.357
103. Distler T, Boccaccini AR (2020) 3D printing of electrically conductive hydrogels for tissue engineering and biosensors—a review. Acta Biomater 101:1–13. https://doi.org/10.1016/j.actbio.2019.08.044
104. Surmen HK, Ortes F, Arslan YZ (2020) Fundamentals of 3D printing and its applications in biomedical engineering, pp 23–41. https://doi.org/10.1007/978-981-15-5424-7_2
105. Chen Y et al (2022) Improving 3D/4D printing characteristics of natural food gels by novel additives: a review. Food Hydrocolloids 123(July 2021):107160. https://doi.org/10.1016/j.foodhyd.2021.107160
106. Arbaoui Y et al (2016) Full 3-D printed microwave termination: a simple and low-cost solution. IEEE Trans Microw Theory Tech 64:271–278
107. Wang X et al (2017) 3D printing of polymer matrix composites: a review and prospective. Compos B Eng 110:442–458. https://doi.org/10.1016/j.compositesb.2016.11.034
108. Bian B et al (2018) 3D printed porous carbon anode for enhanced power generation in microbial fuel cell. Nano Energy 44:174–180. https://doi.org/10.1016/j.nanoen.2017.11.070
109. Jayapiriya US, Goel S (2021) Influence of cellulose separators in coin-sized 3D printed paper-based microbial fuel cells. Sustain Energy Technol Assess 47:101535. https://doi.org/10.1016/j.seta.2021.101535
110. Philamore H et al (2015) Cast and 3D printed ion exchange membranes for monolithic microbial fuel cell fabrication. J Power Sources 289:91–99. https://doi.org/10.1016/j.jpowsour.2015.04.113
111. Theodosiou P, Greenman J, Ieropoulos IA (2020) Developing 3D-printable cathode electrode for monolithically printed microbial fuel cells (MFCs). Molecules 25:25163635. https://doi.org/10.3390/molecules25163635

112. Kumbhakar P et al (2021) Quantifying instant water cleaning efficiency using zinc oxide decorated complex 3D printed porous architectures. J Hazard Mater 418:126383. https://doi.org/10.1016/j.jhazmat.2021.126383
113. Murgolo S et al (2021) Novel TiO_2-based catalysts employed in photocatalysis and photoelectrocatalysis for effective degradation of pharmaceuticals (PhACs) in water: a short review. Curr Opin Green Sustain Chem 30:100473. https://doi.org/10.1016/j.cogsc.2021.100473
114. Saratale RG et al (2018) Photocatalytic activity of $CuO/Cu(OH)_2$ nanostructures in the degradation of reactive green 19A and textile effluent, phytotoxicity studies and their biogenic properties (antibacterial and anticancer). J Environ Manage 223:1086–1097. https://doi.org/10.1016/j.jenvman.2018.04.072
115. Garcia-Segura S, Brillas E (2017) Applied photoelectrocatalysis on the degradation of organic pollutants in wastewaters. J Photochem Photobiol, C 31:1–35. https://doi.org/10.1016/j.jphotochemrev.2017.01.005
116. Kawrani S et al (2020) Enhancement of calcium copper titanium oxide photoelectrochemical performance using boron nitride nanosheets. Chem Eng J 389:124326. https://doi.org/10.1016/j.cej.2020.124326
117. Daghrir R, Drogui P, Robert D (2012) Photoelectrocatalytic technologies for environmental applications. J Photochem Photobiol, A 238:41–52. https://doi.org/10.1016/j.jphotochem.2012.04.009
118. Umukoro EH et al (2017) Towards wastewater treatment: photo-assisted electrochemical degradation of 2-nitrophenol and orange II dye at a tungsten trioxide-exfoliated graphite composite electrode. Chem Eng J 317:290–301. https://doi.org/10.1016/j.cej.2017.02.084
119. Zhang M et al (2015) Photoelectrocatalytic activity of liquid phase deposited α-Fe_2O_3 films under visible light illumination. J Alloys Compd. 648:719–725. https://doi.org/10.1016/j.jallcom.2015.07.026
120. Ma QL et al (2015) Ultrasonic synthesis of fern-like ZnO nanoleaves and their enhanced photocatalytic activity. Appl Surf Sci 324:842–848. https://doi.org/10.1016/j.apsusc.2014.11.054
121. Orimolade BO, Arotiba OA (2020) Towards visible light driven photoelectrocatalysis for water treatment: application of a $FTO/BiVO_4/Ag_2S$ heterojunction anode for the removal of emerging pharmaceutical pollutants. Sci Rep 10:1–13. https://doi.org/10.1038/s41598-020-62425-w
122. Feng J et al (2019) Liquid phase deposition of nickel-doped ZnO film with enhanced visible light photoelectrocatalytic activity. J Electrochem Soc 166:H685–H690. https://doi.org/10.1149/2.0361914jes
123. Orimolade BO et al (2019) Visible light driven photoelectrocatalysis on a $FTO/BiVO_4/BiOI$ anode for water treatment involving emerging pharmaceutical pollutants. Electrochim Acta 307:285–292. https://doi.org/10.1016/j.electacta.2019.03.217
124. Najem M et al (2020) Palladium/carbon nanofibers by combining atomic layer deposition and electrospinning for organic pollutant degradation. Materials 13:1–18. https://doi.org/10.3390/MA13081947
125. Wang J et al (2016) Small and well-dispersed Cu nanoparticles on carbon nanofibers: self-supported electrode materials for efficient hydrogen evolution reaction. Int J Hydrogen Energy 41:18044–18049. https://doi.org/10.1016/j.ijhydene.2016.08.058
126. Nada AA et al (2021) Photoelectrocatalysis of paracetamol on Pd–ZnO/N-doped carbon nanofibers electrode. Appl Mater Today 24:101129. https://doi.org/10.1016/j.apmt.2021.101129
127. Kondalkar VV et al (2014) Photoelectrocatalysis of cefotaxime using nanostructured TiO_2 photoanode: identification of the degradation products and determination of the toxicity level. Ind Eng Chem Res 53:18152–18162. https://doi.org/10.1021/ie501821a
128. Martínez-Pachón D, Echeverry-Gallego RA, Serna-Galvis EA, Villarreal JM, Botero-Coy AM, Hernández F, Torres-Palma RA, Moncayo-Lasso A (2021) Treatment of wastewater effluents from Bogotá–Colombia by the photo-electro-Fenton process: elimination of bacteria and pharmaceutical. Sci Total Environ 772: 144890. https://doi.org/10.1016/j.scitotenv.2020.144890

129. Matteis VD et al (2016) Toxicology in vitro toxicity assessment of anatase and rutile titanium dioxide nanoparticles: the role of degradation in different pH conditions and light exposure. Toxicol In Vitro 37:201–210. https://doi.org/10.1016/j.tiv.2016.09.010
130. Nogueira V et al (2020) Evaluation of the toxicity of nickel nanowires to freshwater organisms at concentrations and short-term exposures compatible with their application in water treatment. Aquat Toxicol 227:105595. https://doi.org/10.1016/j.aquatox.2020.105595
131. Ribeiro F et al (2014) Silver nanoparticles and silver nitrate induce high toxicity to Pseudokirchneriella subcapitata, Daphnia magna and Danio rerio. Sci Total Environ 466–467:232–241. https://doi.org/10.1016/j.scitotenv.2013.06.101
132. Agustina TE, Ang HM, Vareek VK (2005) A review of synergistic effect of photocatalysis and ozonation on wastewater treatment. J Photochem Photobiol, C 6:264–273. https://doi.org/10.1016/j.jphotochemrev.2005.12.003
133. Beltrán FJ et al (2012) Kinetic studies on black light photocatalytic ozonation of diclofenac and sulfamethoxazole in water. Ind Eng Chem Res 51:4533–4544. https://doi.org/10.1021/ie202525f
134. Sein MM et al (2008) Oxidation of diclofenac with ozone in aqueous solution. Environ Sci Technol 42:6656–6662. https://doi.org/10.1021/es8008612
135. Of D et al (1999) degradation of nitrogen containing organic compounds by combined photocatalysis and ozonation. Chemosphere 38:2013–2027
136. Sánchez L, Peral J, Domènech X (1998) Aniline degradation by combined photocatalysis and ozonation. Appl Catal B 19:59–65. https://doi.org/10.1016/S0926-3373(98)00058-7
137. Ye M et al (2009) Ozone enhanced activity of aqueous titanium dioxide suspensions for photodegradation of 4-chloronitrobenzene. J Hazard Mater 167:1021–1027. https://doi.org/10.1016/j.jhazmat.2009.01.091
138. Paucar NE et al (2019) Ozone treatment process for the removal of pharmaceuticals and personal care products in wastewater. Ozone Sci Eng 41:3–16. https://doi.org/10.1080/01919512.2018.1482456
139. Gomes J et al (2017) Application of ozonation for pharmaceuticals and personal care products removal from water. Sci Total Environ 586:265–283. https://doi.org/10.1016/j.scitotenv.2017.01.216
140. Mirzaei A et al (2016) Removal of pharmaceuticals and endocrine disrupting compounds from water by zinc oxide-based photocatalytic degradation: a review. Sustain Cities Soc 27:407–418. https://doi.org/10.1016/j.scs.2016.08.004
141. Matzek LW, Carter KE (2016) Activated persulfate for organic chemical degradation: a review. Chemosphere 151:178–188. https://doi.org/10.1016/j.chemosphere.2016.02.055
142. Gao Y et al (2020) Activated persulfate by iron-based materials used for refractory organics degradation: a review. Water Sci Technol 81:853–875. https://doi.org/10.2166/wst.2020.190

The manufacturer's authorised representative in the EU is Springer Nature Customer Service Centre GmbH, Europaplatz 3, 69115 Heidelberg, Germany. If you have any concerns regarding our products, please contact ProductSafety@springernature.com

Printed and bound by CPI Group (UK) Ltd, Croydon, CR0 4YY
30/09/2026
02225244-0004